Functional Inorganic Biomaterials for Molecular Sensing and Biomedical Applications

Functional Inorganic Biomaterials for Molecular Sensing and Biomedical Applications

Guest Editor

Nabanita Saikia

Basel • Beijing • Wuhan • Barcelona • Belgrade • Novi Sad • Cluj • Manchester

Guest Editor
Nabanita Saikia
Department of Chemistry
New Mexico Highlands University
Las Vegas, NV
USA

Editorial Office
MDPI AG
Grosspeteranlage 5
4052 Basel, Switzerland

This is a reprint of the Special Issue, published open access by the journal *Inorganics* (ISSN 2304-6740), freely accessible at: https://www.mdpi.com/journal/inorganics/special_issues/F8PVDE01FN.

For citation purposes, cite each article independently as indicated on the article page online and as indicated below:

Lastname, A.A.; Lastname, B.B. Article Title. *Journal Name* **Year**, *Volume Number*, Page Range.

ISBN 978-3-7258-6658-8 (Hbk)
ISBN 978-3-7258-6659-5 (PDF)
https://doi.org/10.3390/books978-3-7258-6659-5

Contents

About the Editor

Nabanita Saikia

Nabanita Saikia is an Assistant Professor of the Chemistry Department, School of Science, Ivan Hilton Science Center, New Mexico Highlands University. Her research interests include computational biophysics, RNA-targeted therapeutics, and functional nanomaterials for molecular sensing and drug delivery. She has published over 45 peer-reviewed research articles and book chapters. She currently serves as Vice President (Presidential Line) of the New Mexico Academy of Science. Dr. Saikia serves as an Editorial Board Member for *Scientific Reports* (Nature) and *Discover Materials* (Springer Nature), Academic Editor for *PLoS ONE*, and Guest Editor for MDPI journals.

Preface

This Reprint presents advances in functional inorganic biomaterials, highlighting synthesis, structural characterization, biophysical and biochemical properties, and applications in imaging, sensing, drug delivery, and regenerative medicine for researchers in chemistry, materials science, and biomedicine.

Nabanita Saikia
Guest Editor

inorganics

MDPI

Editorial

Functional Inorganic Biomaterials for Molecular Sensing and Biomedical Applications

Nabanita Saikia

Department of Chemistry, New Mexico Highlands University, Las Vegas, NM 87701, USA; nsaikia@nmhu.edu

Inorganic biomaterials comprise a broad array of materials that include metals, polymers, ceramics, and composites [1–6], and play a pivotal role across disciplines such as chemistry, materials science, biology, medicine [7], and biomedical engineering [8–14]. The development of inorganic biomaterials with tunable intrinsic properties [15,16] and functionality has positioned them at the forefront of applications from biosensing and imaging [17,18] to therapy [19,20] regenerative medicine, and wound healing [21–28]. The field of tissue engineering and regenerative medicine can vastly benefit from advancements in nanoscience and technology [29–32], particularly in the fabrication and application of inorganic-based nanoparticles and bioactive materials [1,33–35]. Techniques like small-angle X-rays and neutron scattering and light scattering have enabled the structural characterization (size, shape and morphological transitions) of nanostructured biomaterials in the sub-millisecond time scale [36]. Coupled with traditional characterization techniques, scattering methods can facilitate their design and application in advanced bio-nanotechnological fields and disease therapy [36,37]. Mitić et al. [38] reviewed recent developments in the application of instrumental techniques for structural and physicochemical characterization of biomaterials and bone tissue. The same study discussed the application of instrumental techniques, namely, X-ray, IR and Raman vibrational spectroscopy, NMR, ESR, mass spectrometry, atomic absorption spectrometry, inductively coupled plasma-atomic emission spectrometry, thermogravimetric analysis, differential thermal analysis, SEM, and high-resolution TEM, for structural and morphological characterization, and predicting the reactivity, thermodynamic stability, crystalline behavior, and phase transformations of biomaterials.

The Special Issue combines original research and critical reviews that showcase the versatility of inorganic biomaterials and nanoparticles in molecular sensing, targeted therapy [39,40], drug delivery [24], and regenerative medicine [41] utilizing current state-of-the-art techniques [41,42]. The synergy between experiment and computation that is evident throughout the collection highlights the integrative approaches essential in developing next-generation biomaterials.

The study by Cuperová et al. [43] investigated the effect of severe plastic deformation on the microstructure and mechanical properties of an as-cast Zn–0.1 wt.% Mg alloy using the Equal Channel Angular Pressing (ECAP) technique. Zn-based alloys showed strong potential as bioactive materials for intracorporeal implants due to their lower cytotoxicity, reduced inflammatory responses compared to pure Zn, and improved biocompatibility and optimal degradation rate in physiological environments. This study underscored ECAP's effectiveness in simultaneously enhancing strength and ductility in Zn-based alloys, thereby advancing their potential applications as orthopedic implant materials.

The synthesis of a novel silver(I) complex, $[Ag(HL^1)_2]NO_3$ (**AgHL^1**), with a coumarin derivative, (3*E*)-3-(1-{[(pyridin-2-yl)methyl]amino} ethylidene)-3,4-dihydro-2*H*-benzopyran-

2,4-dione (**HL¹**) was reported Kurjan et al. [44] to study their anticancer and antimicrobial activity. The **HL¹** ligand exhibited moderate activity, while **AgHL¹** displayed a pronounced, time- and dose-dependent inhibitory effect on the metabolic activities of both cancerous cell lines. The flexural strength, color change and antimicrobial effect of silver–zeolite NPs in acrylic resin materials was reported by Yaman et al. [45] using three different rates of NPs (0%, 2%, 4%, 5%). Discs with 2% silver–zeolite NPs showed a significant antimicrobial effect and were more effective against *C. albicans* than those with rates of 4% and 5%. While adding silver–zeolite NPs provided antimicrobial properties, it reduced the material's structural integrity and altered its esthetic appearance, highlighting the necessity of balancing antimicrobial efficacy with mechanical and esthetic properties in dental materials.

Marabello et al. [46], employed a computational method to investigate three new solid salicylatoborate compounds, $[Ca(H_2O)_6](C_{14}H_8O_6B)_2$ (CaSB), $[Cu(C_{14}H_8O_6B)]$ (CuSB), and $[Li(C_{14}H_8O_6B)(H_2O)]$ (LiSB), as potential boron and/or lithium delivery agents for the development of ^{10}B-enriched NPs in neutron capture therapy (NCT). LiSB exhibited the highest dose absorbed by the tumor mass and resulted in being the most efficient compound. Thus, to design more efficient compounds for NCT, the density of ^{10}B should be enhanced rather than that of ^{6}Li.

Doxorubicin (DOX)-loaded Ca-Mg-doped mesoporous silica nanoparticles (MSNs) were developed by Zhang et al. [47] as tumor-targeted drug delivery systems for breast cancer. The incorporation of calcium and magnesium demonstrated an increase in the degradation rate under acidic conditions and an accelerated release, which reduced the toxicity of DOX and promoted cellular uptake. Molecular dynamic simulation was used to study the intermolecular interactions and physicochemical properties of novel mesoporous silica. Doping with calcium and magnesium led to an increase in the free energy of the system; the free energy of the Ca_2Mg_2@MSNs group was the largest, which indicated that the stability of the MSN system decreased after being doped with calcium and magnesium. Hemolysis experiments verified that MSNs could be used as a carrier for medical purposes. Liu et al. [48] synthesized multifunctional Ca^{2+} doped mesoporous silica nanoparticles (CMSNs) using the sol–gel method. DOX was loaded into the as-prepared CMSNs (CMSNs@Dox). The doping of Ca^{2+} endowed the MSNs with the ability to carry out excellent specific degradation and pH-responsive drug release, and enabled the synergy of chemotherapy and calcicoptosis. This newly prepared Ca^{2+}-doped nanoplatform provided a practical and feasible strategy for improving the efficacy of chemotherapy.

Barnash et al. [49] studied the effect of $CoFe_2O_4$ NPs on the properties of an electro-optical liquid crystal (LC) cell based on the nematic composition of 4-Cyano-4′-pentylbiphenyl (5CB). At the input signal, with a frequency of 500 kHz, a resonant current increase was obtained in the electrical circuit, followed by a decrease in the current with an increase in the frequency. An increase in the resonant current caused a decrease in LC cell impedance. This indicated the formation of a consistent oscillatory circuit. Giacomet et al. [50] developed a simple and cheap electrochemical method for diagnosing antibodies against SARS-Cov-2. Disposable screen-printed carbon electrodes (SPEs) covered with gold microblobs (AuMBs) were synthesized. The SPE-AuMBs were coated with cysteamine, which allowed the N-hydroxysuccinimide-activated SARS-CoV-2 antigen to be immobilized. The antigen-activated SPE could qualitatively and quantitatively evaluate the presence of SARS-CoV-2 antibodies. The antibody analysis took around 15 min and was cheaper than traditional rapid tests.

Ntoupis et al. [51] measured the luminescence efficiency of a cerium fluoride (CeF_3) inorganic scintillator in crystal form as a possible alternative to high-luminescence but hygroscopic cerium bromide ($CeBr_3$). The obtained luminescence efficiency results showed that CeF_3 was not suitable for medical imaging applications covering the range of 50–140 kVp;

however, an examination of its luminescence output in the nuclear medicine energy range (~70 to 511 keV) could reveal possible applicability in these modalities. Lastly, Saikia [33] reviewed the emerging applications of inorganic-based NPs and biomaterials in stem cell regenerative research, tissue engineering, artificial skin and cartilage regeneration, neural nerve injuries, 3D bioprinting, and the development of novel bio-scaffolds. The review also addressed critical challenges in the clinical application of inorganic biomaterials in regenerative medicine such as their biocompatibility and bioaccumulation, toxicity, assessment tools, optimization of mechanical properties, and their long-term fate.

The contributions in the Special Issue range from advanced molecular sensing to innovative therapeutic strategies for the treatment of diseases, and bridge the interfaces between chemistry, biology, engineering, and medicine. My hope is that this Special Issue collection, *Functional Inorganic Biomaterials for Molecular Sensing and Biomedical Applications*, will not only stimulate ongoing research advancements in the field but also inspire interdisciplinary collaborative efforts aimed at translating fundamental research into real-world applications. I take this opportunity to invite researchers to contribute to the second volume of the Special Issue, *Functional Inorganic Biomaterials for Molecular Sensing and Biomedical Applications, 2nd Edition*, that will continue to focus on recent advances in inorganic biomaterials and nanoparticles. The forthcoming volume includes all stages of the process, from rational design and structural characterization to applications in molecular sensing, clinical diagnosis, drug delivery, and medicine.

Conflicts of Interest: The author declares no conflict of interest.

References

1. Zhang, H.; Zhao, Z.; Wu, C. Bioactive Inorganic Materials for Innervated Multi-Tissue Regeneration. *Adv. Sci.* **2025**, *12*, e2415344. [CrossRef]
2. Sharma, A.; Kokil, G.R.; He, Y.; Lowe, B.; Salam, A.; Altalhi, T.A.; Ye, Q.; Kumeria, T. Inorganic/Organic Combination: Inorganic Particles/Polymer Composites for Tissue Engineering Applications. *Bioact. Mater.* **2023**, *24*, 535. [CrossRef]
3. Brokesh, A.M.; Gaharwar, A.K. Inorganic Biomaterials for Regenerative Medicine. *ACS Appl. Mater. Interfaces* **2020**, *12*, 5319. [CrossRef]
4. dos Santos, V.I.; Chevalier, J.; Fredel, M.C.; Henriques, B.; Gremillard, L. Ceramics and Ceramic Composites for Biomedical Engineering Applications via Direct Ink Writing: Overall Scenario, Advances in the Improvement of Mechanical and Biological Properties and Innovations. *Mater. Sci. Eng. R Rep.* **2024**, *161*, 100841. [CrossRef]
5. Vaiani, L.; Boccaccio, A.; Uva, A.E.; Palumbo, G.; Piccininni, A.; Guglielmi, P.; Cantore, S.; Santacroce, L.; Charitos, I.A.; Ballini, A. Ceramic Materials for Biomedical Applications: An Overview on Properties and Fabrication Processes. *J. Funct. Biomater.* **2023**, *14*, 146. [CrossRef]
6. Khan, T.; Vadivel, G.; Ramasamy, B.; Murugesan, G.; Sebaey, T.A. Biodegradable Conducting Polymer-Based Composites for Biomedical Applications—A Review. *Polymers* **2024**, *16*, 1533. [CrossRef] [PubMed]
7. Lee, H.P.; Gaharwar, A.K. Light-Responsive Inorganic Biomaterials for Biomedical Applications. *Adv. Sci.* **2020**, *7*, 2000863. [CrossRef] [PubMed]
8. Huebsch, N.; Mooney, D.J. Inspiration and Application in the Evolution of Biomaterials. *Nature* **2009**, *462*, 426. [CrossRef] [PubMed]
9. Huang, H.; Feng, W.; Chen, Y. Two-dimensional Biomaterials: Material Science, Biological Effect and Biomedical Engineering Applications. *Chem. Soc. Rev.* **2021**, *50*, 11381. [CrossRef]
10. Bhat, S.; Kumar, A. Biomaterials and Bioengineering Tomorrow's Healthcare. *Biomatter* **2013**, *3*, e24717. [CrossRef] [PubMed]
11. Guo, Z.; Liu, Z.; Tang, R. Applications of Amorphous Inorganics as Novel Functional Materials. *Mater. Chem. Front.* **2024**, *8*, 1703. [CrossRef]
12. Mao, W.; Yoo, H.S. Inorganic Nanoparticle Functionalization Strategies in Immunotherapeutic Applications. *Biomater. Res.* **2024**, *28*, 0086. [CrossRef] [PubMed]
13. Li, J.; Zeng, H.; Zeng, Z.; Zeng, Y.; Xie, T. Promising Graphene-Based Nanomaterials and Their Biomedical Applications and Potential Risks: A Comprehensive Review. *ACS Biomater. Sci. Eng.* **2021**, *7*, 5363. [CrossRef]
14. Huang, Y.; Guo, X.; Wu, Y.; Chen, X.; Feng, L.; Xie, N.; Shen, G. Nanotechnology's Frontier in Combatting Infectious and Inflammatory Diseases: Prevention and Treatment. *Signal Transduct. Target. Ther.* **2024**, *9*, 34. [CrossRef]

15. Liu, F.; Yao, S.; Li, J.; Huang, K.; Zhang, D.; Wong, T.M.; Tang, R.; Yeung, K.W.K.; Wu, J. Inorganic-organic Hybrid Metamaterials with Switchable High Stiffness and Elasticity. *Nat. Commun.* **2025**, *16*, 4423. [CrossRef]
16. Zimina, T.M.; Sitkov, N.O.; Gareev, K.G.; Fedorov, V.; Grouzdev, D.; Koziaeva, V.; Gao, H.; Combs, S.E.; Shevtsov, M. Biosensors and Drug Delivery in Oncotheranostics Using Inorganic Synthetic and Biogenic Magnetic Nanoparticles. *Biosensors* **2022**, *12*, 789. [CrossRef]
17. Trucillo, P. Biomaterials for Drug Delivery and Human Applications. *Materials* **2024**, *17*, 456. [CrossRef] [PubMed]
18. Fu, Y.; Liu, T.; Wang, H.; Wang, Z.; Hou, L.; Jiang, J.; Xu, T. Applications of Nanomaterial Technology in Biosensing. *J. Sci. Adv. Mater. Devices* **2024**, *9*, 100694. [CrossRef]
19. Shi, S.; Vissapragada, R.; Abi Jaoude, J.; Huang, C.; Mittal, A.; Liu, E.; Zhong, J.; Kumar, V. Evolving Role of Biomaterials in Diagnostic and Therapeutic Radiation Oncology. *Bioact. Mater.* **2020**, *5*, 233. [CrossRef]
20. Wang, D.; Peng, Y.; Li, Y.; Kpegah, J.K.S.K.; Chen, S. Multifunctional Inorganic Biomaterials: New Weapons Targeting Osteosarcoma. *Front. Mol. Biosci.* **2023**, *9*, 1105540. [CrossRef]
21. Kharbikar, B.N.; Mohindra, P.; Desai, T.A. Biomaterials to Enhance Stem Cell Transplantation. *Cell Stem Cell* **2022**, *29*, 692. [CrossRef]
22. Nikolova, M.P.; Chavali, M.S. Recent Advances in Biomaterials for 3D Scaffolds: A Review. *Bioact. Mater.* **2019**, *4*, 271. [CrossRef]
23. Richbourg, N.R.; Peppas, N.A.; Sikavitsas, V.I. Tuning the Biomimetic Behavior of Scaffolds for Regenerative Medicine Through Surface Modifications. *J. Tissue Eng. Regen. Med.* **2019**, *13*, 1275. [CrossRef]
24. Sahoo, R.N.; Rout, S.; Parmanik, A.; Satapathy, B.S.; Pattnaik, S.; Maharana, L.; Nayak, A.K. Inorganic Biomaterials for Ocular Drug Delivery: A Comprehensive Review. *J. Drug Deliv. Sci. Technol.* **2025**, *106*, 106667. [CrossRef]
25. Murali, A.; Brokesh, A.M.; Cross, L.M.; Kersey, A.L.; Jaiswal, M.K.; Singh, I.; Gaharwar, A. Inorganic Biomaterials Shape the Transcriptome Profile to Induce Endochondral Differentiation. *Adv. Sci.* **2024**, *11*, e2402468. [CrossRef]
26. Zheng, Y.; Wu, J.; Zhu, Y.; Wu, C. Inorganic-based Biomaterials for Rapid Hemostasis and Wound Healing. *Chem. Sci.* **2022**, *14*, 29–53. [CrossRef] [PubMed]
27. Ferraro, A. Biomaterials and Therapeutic Applications. *IOP Conf. Ser. Mater. Sci. Eng.* **2016**, *108*, 012021. [CrossRef]
28. Dai, T.; He, W.; Yao, C.; Ma, X.; Ren, W.; Mai, Y.; Wu, A. Applications of Inorganic Nanoparticles in the Diagnosis and Therapy of Atherosclerosis. *Biomater. Sci.* **2020**, *8*, 3784–3799. [CrossRef] [PubMed]
29. Chen, F.M.; Liu, X. Advancing Biomaterials of Human Origin for Tissue Engineering. *Prog. Polym. Sci.* **2016**, *53*, 86. [CrossRef]
30. Kim, H.-S.; Baby, T.; Lee, J.-H.; Shin, U.S.; Kim, H.-W. Biomaterials-Enabled Electrical Stimulation for Tissue Healing and Regeneration. *Med-X* **2024**, *2*, 7. [CrossRef]
31. Alshangiti, D.M.; El-Damhougy, T.K.; Zaher, A.; Madani, M.; Mohamady Ghobashy, M. Revolutionizing Biomedicine: Advancements, Applications, and Prospects of Nanocomposite Macromolecular Carbohydrate-based Hydrogel Biomaterials: A Review. *RSC Adv.* **2023**, *13*, 35251. [CrossRef]
32. Shi, J.; Votruba, A.R.; Farokhzad, O.C.; Langer, R.; Gaharwar, A.K. Nanotechnology in Drug Delivery and Tissue Engineering: From Discovery to Applications. *Nano Lett.* **2010**, *10*, 3223–3230. [CrossRef]
33. Saikia, N. Inorganic-Based Nanoparticles and Biomaterials as Biocompatible Scaffolds for Regenerative Medicine and Tissue Engineering: Current Advances and Trends of Development. *Inorganics* **2024**, *12*, 292. [CrossRef]
34. Ma, J.; Wu, C. Bioactive Inorganic Particles-based Biomaterials for Skin Tissue Engineering. *Exploration* **2022**, *2*, 20210083. [CrossRef]
35. Song, P.; Zhou, D.; Wang, F.; Li, G.; Bai, L.; Su, J. Programmable Biomaterials for Bone Regeneration. *Mater. Today Bio.* **2024**, *29*, 101296. [CrossRef]
36. Lombardo, D.; Calandra, P.; Kiselev, M.A. Structural Characterization of Biomaterials by Means of Small Angle X-rays and Neutron Scattering (SAXS and SANS), and Light Scattering Experiments. *Molecules* **2020**, *25*, 5624. [CrossRef]
37. Carvalho, P.M.; Felício, M.R.; Santos, N.C.; Gonçalves, S.; Domingues, M.M. Application of Light Scattering Techniques to Nanoparticle Characterization and Development. *Front. Chem.* **2018**, *6*, 237. [CrossRef] [PubMed]
38. Mitić, Ž.; Stolić, A.; Stojanović, S.; Najman, S.; Ignjatović, N.; Nikolić, G.; Trajanović, M. Instrumental Methods and Techniques for Structural and Physicochemical Characterization of Biomaterials and Bone Tissue: A Review. *Mater. Sci. Eng. C* **2017**, *79*, 930–949. [CrossRef] [PubMed]
39. Han, X.; Alu, A.; Liu, H.; Shi, Y.; Wei, X.; Cai, L.; Wei, Y. Biomaterial-assisted Biotherapy: A Brief Review of Biomaterials Used in Drug Delivery, Vaccine Development, Gene Therapy, and Stem Cell Therapy. *Bioact. Mater.* **2022**, *17*, 29–48. [CrossRef]
40. Aly, R.M. Current State of Stem Cell-based Therapies: An Overview. *Stem Cell Investig.* **2020**, *7*, 8. [CrossRef]
41. Jarrige, M.; Frank, E.; Herardot, E.; Martineau, S.; Darle, A.; Benabides, M.; Domingues, S.; Chose, O.; Habeler, W.; Lorant, J.; et al. The Future of Regenerative Medicine: Cell Therapy Using Pluripotent Stem Cells and Acellular Therapies Based on Extracellular Vesicles. *Cells* **2021**, *10*, 240. [CrossRef]
42. Lele, M.; Kapur, S.; Hargett, S.; Sureshbabu, N.M.; Gaharwar, A.K. Global Trends in Clinical Trials Involving Engineered Biomaterials. *Sci. Adv.* **2024**, *10*, eabq0997. [CrossRef]

43. Cuperová, I.; Fujda, M.; Kočiško, R.; Petroušek, P.; Molčanová, Z.; Matvija, M.; Džunda, R.; Ballóková, B.; Csík, D.; Gáborová, K.; et al. Significant Enhancement of Strength and Ductility in Bioresorbable Zn–0.1Mg Alloy via ECAP Processing. *Inorganics* **2025**, *13*, 193. [CrossRef]
44. Kurjan, J.; Jendželovsk, Z.; Dečmanov., V.; Vilkov, M.; Ćirković, K.; Radojević, I.; Liteck, M.; Jendželovský, R.; Potočň.k, I. The Coumarin-Based Silver(I) Complex Showed Enhanced Antitumor and Antimicrobial Activity than Ligand Itself. *Inorganics* **2025**, *13*, 164. [CrossRef]
45. Yaman, K.; Demir, N.; Arslan, U.; Çiftçi, N. Antimicrobial, Optical, and Mechanical Properties of Saliva-Contaminated Silver–Zeolite Nanoparticle-Incorporated Dental Acrylic Resins. *Inorganics* **2024**, *12*, 258. [CrossRef]
46. Marabello, D.; Benzi, P.; Canepa, C.; Cioci, A. Ca-, Li-, and Cu-Salicylatoborates for Potential Applications in Neutron Capture Therapy: A Computational Method for the Preliminary Discrimination of the More Promising Compounds. *Inorganics* **2025**, *13*, 136. [CrossRef]
47. Zhang, Q.; Huang, J.; Liu, C.; Chen, R.; Jiang, T.; Hailili, Y.; Bahetibieke, T.; Tang, X.; Wang, M. Preparation of Ca-Mg Double-Doped Mesoporous Silica Nanoparticles and Their Drug- Loading and Drug-Releasing Properties. *Inorganics* **2025**, *13*, 12. [CrossRef]
48. Liu, C.; Tang, X.; Huang, G. Biodegradable Ca^{2+} Doped Mesoporous Silica Nanoparticles Promote Chemotherapy Synergism with Calcicoptosis and Activate Anti-Tumor Immunity. *Inorganics* **2024**, *12*, 152. [CrossRef]
49. Barnash, Y.; Jovanović, S.; Jovanović, Z.; Kamanina, N. Electrical Features of Liquid Crystal Composition Doped with Cobalt Ferrite: Possible Sensing Applications. *Inorganics* **2025**, *13*, 107. [CrossRef]
50. Giacomet, M.M.; Buzzetti, P.H.M.; Junior, O.O.S.; Martins, A.F.; Bonafe, E.G.; Monteiro, J.P. Portable Electrochemical Immunosensor Based on a Gold Microblobs-Optimized Screen- Printed Electrode for SARS-CoV-2 Diagnosis. *Inorganics* **2024**, *12*, 252. [CrossRef]
51. Ntoupis, V.; Michail, C.; Kalyvas, N.; Bakas, A.; Kandarakis, I.; Fountos, G.; Valais, I. Luminescence Efficiency and Spectral Compatibility of Cerium Fluoride (CeF3) Inorganic Scintillator with Various Optical Sensors in the Diagnostic Radiology X-ray Energy Range. *Inorganics* **2024**, *12*, 230. [CrossRef]

inorganics

MDPI

Review

Inorganic-Based Nanoparticles and Biomaterials as Biocompatible Scaffolds for Regenerative Medicine and Tissue Engineering: Current Advances and Trends of Development

Nabanita Saikia

Department of Chemistry, New Mexico Highlands University, Las Vegas, NM 87701, USA; nsaikia@nmhu.edu

Abstract: Regenerative medicine amalgamates stem cell technology and tissue engineering strategies to replace tissues and organs damaged by injury, aging, ailment, and/or chronic conditions by leveraging the innate self-healing mechanism of the body. The term 'regenerative medicine' was coined by William A. Haseltine during a 1999 conference on Lake Como. Since its inception in 1968, the field has offered clinical benefits for the regeneration, repair, and restoration of bones, skin, cartilage, neural tissue, and the heart, as well as scaffold fabrication. The field of tissue engineering and regenerative medicine can vastly benefit from advancements in nanoscience and technology, particularly in the fabrication and application of inorganic-based nanoparticles and bionanomaterials. Due to the tunable intrinsic properties, i.e., size, topography, surface charge, and chemical stability, inorganic-based nanoparticles and biomaterials have surpassed traditional synthetic materials. Given the wide gamut of near-future applications of inorganic nanoparticles and biomaterials, this article gives an overview of the emerging roles in stem cell regenerative research, tissue engineering, artificial skin and cartilage regeneration, neural nerve injuries, 3D bioprinting, and development of new inorganic bio-scaffolds. The review also addresses the challenges related to the clinical application and tissue compatibility of inorganic nanoparticles and biomaterials, utilizing current state-of-the-art techniques.

Keywords: regenerative medicine; inorganic biomaterials; nanoparticles; tissue engineering; nanotechnology

1. Introduction

Regenerative medicine is an emerging interdisciplinary area focused on developing methods to replace or repair damaged cells and restore tissue or organ function affected by trauma, chronic conditions, disease, aging, or congenital deformities [1,2]. Regenerative medicine amalgamates the principles of chemistry, biology, and applied engineering to develop processes and therapies that can promote the regeneration of human cells, functional organs, and injured tissues to restore normal body functions [3,4]. By harnessing the human body's innate ability to self-heal, regenerative medicine is poised to revolutionize science, healthcare, and medicine [1]. Regenerative medicine is also used synonymously with 'tissue engineering', which relies on developing scaffolds that can mimic the extracellular matrix (ECM) of tissues [5]. Examples of regenerative medicine include stem cell-based therapy, immunomodulation therapy, and tissue engineering [6]. Tissue engineering is the application of principles and methods of engineering and life sciences toward the fundamental understanding of structure–function relationships in normal and pathological mammalian tissues and the development of biological substitutes to restore, maintain, or improve tissue function [7,8]. The primal goal of tissue engineering is the regeneration of patients' tissues and organs devoid of poor biocompatibility, functionality, and immune rejection [9].

In tissue engineering, cells, scaffolds, and growth factors are crucial for promoting the regeneration of bone and cartilage [10]. Cells synthesize the matrix for the growth of new tissues [9]. Scaffold materials, templates, or artificial extracellular matrices serve as substrates for loading cells and growth factors, which stimulate cell proliferation, differentiation, biosynthesis, and regeneration [10]. Growth factors are signaling molecules that guide cell behavior to stimulate the growth of new tissues and provide chemical cues to stem cells, regulating their biological responses and tissue differentiation [9,11]. Tissue engineering takes place in the presence of a scaffold which serves as a template for regeneration. Bone tissue engineering and enhancement of bone healing for orthopedic, craniofacial, and periodontal applications can be achieved by a group of multi-functional growth factors called bone morphogenetic proteins (BMPs) [12]. BMPs are members of the transforming growth factor beta (TGFβ) superfamily of cell regulatory proteins that control osteoblastogenesis, embryonic development, cellular function, differentiation, proliferation, morphogenesis, and apoptosis [13]. A signal transduction cascade is initiated upon BMP ligand binding to a transmembrane, heterotetrametric receptor complex composed of type I BMP receptors (BMPR) (ACVR1/ALK2, BMPR1A/ALK3, and BMPR1B/ALK6) and type II BMPR (BMPR2, ActR-2A and ActR-2B) [14,15].

Stem cells are undifferentiated, non-specialized cells that can self-renew and differentiate into other cell types such as liver cells, nerve cells, and cardiomyocytes [3,16,17]. Stem cells are categorized into embryonic stem cells (ESCs), adult stem cells (ASCs), and induced pluripotent stem cells (iPSCs) [18,19]. ASCs include mesenchymal stem cells (MSCs), myogenic stem cells, neural stem cells, and umbilical cord stem cells [20]. ESCs and iPSCs have high self-regeneration ability and differentiation potential [21–23]. Stem cells can be classified based on their ability to differentiate into various cell types [24,25]. The classifications are as follows:

- Totipotent: stem cells can develop into any cell type, including embryonic and adult lineages.
- Pluripotent: stem cells can become any cell type in an adult.
- Multipotent: stem cells can differentiate into multiple cell types within a specific lineage.
- Unipotent: stem cells are limited to differentiating into just one specific cell type.

A current thrust in regenerative medicine research is in cord blood stem cells that can offer patients the much-needed means to life-saving stem cell transplants [26]. Umbilical cord blood, like bone marrow and peripheral blood collected from the umbilical cord and placenta of healthy newborns, contains a rich source of blood-forming stem cells, which can renew and differentiate into other cell types. Cord blood stem cells find applications in transplants for patients suffering from leukemia, lymphoma, aplastic anemia, multiple myeloma, and immune deficiency disorders, and for people facing difficulty finding matched adult donors [27]. Human umbilical cord blood stem cells, an alternative source of hematopoietic stem cells, have expanded transplant eligibility for patients with neurodegenerative and cardiovascular disorders and hematologic conditions, especially from minority groups (racial and ethnic) across the U.S. and worldwide [28–30]. Umbilical cord blood has advantages over bone marrow because cord blood does not require human leukocyte antigen tissue matching, has less incidence of graft vs. host disease, and may be used in allogeneic cell therapy [31,32]. MSCs are widely used for wound healing due to their abundance, high proliferation rate, multilineage differentiation capacity, and expression of paracrine factors [21]. MSCs hold promise in the regeneration of cardiac tissues for improving cardiac function [33–35]. Despite limitations in treatment modalities, costs involved, and limited success, stem cell-based therapeutic strategies have emerged as a promising new approach to enhancing regenerative wound healing [36–38].

The skin is the largest organ in the human body that is prone to damage, such as wear and tear, bruises, scars, ulcers, burns, wounds (infectious or chronic), and aging, which can compromise its integrity and function [21,39]. Cytokines, chemokines, growth factors, and cell signaling cascades are crucial in maintaining the health, integrity, and epidermal regenerative ability of the skin [21,39–41]. The Wnt (Wingless and INT-1) signal transduction cascade is involved in embryonic development, stem cell function, tissue homoeostatic self-renewal in adult organisms, and skin development through canonical and noncanonical signaling pathways, and is linked to the regulation of cell proliferation, differentiation, migration, and polarity of stem and cancer stem cells [42–44]. Due to the slow recovery and limited self-healing abilities of human skin, there is limited knowledge about preventing skin degeneration and identifying effective therapies for skin regeneration and rejuvenation [45].

The peripheral nervous system (PNS) consists of a network of 43 pairs of motor and sensory nerves that control the voluntary functions of sensation, movement, and motor coordination. The PNS comprises of three types of cells: neuronal, glial, and stromal. Traumatic and infectious diseases can lead to peripheral nerve damage and long-lasting disability affecting both sensory and motor functions [46]. Recovery from peripheral nerve damage is an intractable challenge for clinical researchers, as it is a complex process typically managed through surgical nerve autografting. However, this method has disadvantages, as it often fails to achieve complete nerve recovery and nerve axon regrowth is inhibited due to the presence of nerve gaps [46]. Stem cells such as adipose-derived stem cells (ADSCs), MSCs—in particular, bone marrow stem cells (BMSCs)—and umbilical cord stem cells (endothelial progenitor cells and hematopoietic stem cells) with high proliferative capabilities are suitable candidates for regenerating neural tissues [47–50]. Stem cell-based therapy in peripheral nerve injury can induce nerve regeneration and axonal remyelination by providing the microenvironment for peripheral nerve regeneration and regulating inflammatory cascade after injury [46,51]. Zeng and Zhang discussed innovative approaches for in vivo gene expression reprogramming in neural regeneration and repair [52]. However, continuous efforts are desirable to develop safe and effective therapies to understand the underlying mechanisms that regulate neuronal regeneration in spinal cord repair.

Inorganic nanoparticles have been at the forefront of research in designing novel biomaterial hybrids for versatile biomedical applications, including, but not limited to, cancer, tumor biomarkers for targeted therapy [53], nervous system repair, nanocarriers for antimicrobial drugs [54], nanoprobes or imaging probes for magnetic resonance imaging (MRI), computed tomography (CT), anti-Stokes shift-based optical imaging [55], and engineering technology [56]. This is attributed to its controllable size and composition, unique physical, mechanical, optical, and electronic properties at the nanoscale level, making them outstanding candidates for biomedical research [55,57,58]. Inorganic-based biomaterials, on the other hand, are of particular interest in rapid hemostasis and wound healing [59], bacterial activity, dental restorations, and skin tissue engineering [60], due to improved biocompatibility, higher bioactivity, promotion of blood coagulation, and ability to release bioactive ions [59,61]. The design and development of inorganic-based biomaterials rely on two components: active chemical components—which include gelatin, chitosan, natural and synthetic polymers—and alginic acid. The active inorganic components primarily include silicates, phosphates, and natural minerals. Inorganic biomaterials are typically hydrophilic or super-hydrophilic, have high surface energy, and can readily absorb water to achieve ultra-rapid hemostasis [59]. Furthermore, tunable magnetic, optical, electronic, biophysical, and biochemical properties, along with biocompatibility [62], endow inorganic-based nanoparticles and biomaterials with prospective applications in regenerative medicine and tissue engineering [63], stem cell differentiation, in-growth tissue integration, in vivo stem cell tracking, and other diverse cellular functions [5,58,64]. Hence, a synergistic approach that combines the best of the two research fields can accelerate scientific advancements in disease treatment and therapy, the development of next-generation cell-based therapeutics, and precision medicine. This review discusses the emergent applications and scalability of inorganic-based biomaterials and nanoparticles as novel bio-scaffolds in stem cell regenerative research, tissue engineering, artificial skin and cartilage regeneration, and neural nerve injury. Although excellent reviews and studies cover developments in tissue engineering and regenerative medicine [7,62,64–66], the primary focus of this review is to highlight current progress in the application of inorganic-based biomaterials and nanoparticles in stem cell regenerative research, tissue engineering, artificial skin and cartilage regeneration, cardiovascular disease, and wound healing. The review also accounts for the challenges in clinical application and tissue compatibility of inorganic nanoparticles and biomaterials utilizing current state-of-the-art techniques.

2. Inorganic-Based Nanoparticles and Biomaterials as Scaffolds in Regenerative Medicine

Contemporary regenerative medicine strategies employ inorganic-based biomaterial scaffolds that mimic the biological and mechanical features of the ECM, thereby enhancing tissue regeneration and healing [1]. Scaffolding materials constitute the structural basis that, at best, can fully, or in part, mimic the native function of the ECM and help cells adhere, proliferate, and differentiate in vitro and in vivo following implantation [67,68]. The fabrication of bio-based inorganic material scaffolds enables the construction of substitute tissues that are biocompatible, safe, and easily biodegradable, without producing toxic byproducts [69]. Polymeric nanoparticles, lipid-based nanoparticles, and inorganic nanoparticles are widely studied in drug delivery, tissue engineering, and biomedical imaging [70–72] to enhance cell viability and achieve optimal cell-directing capabilities [35,72].

Bone regeneration relies on biomaterials that are osteoconductive (supports bone growth and encourages the ingrowth of surrounding bone) as well as capable of osseointegration (osteointegration or integration into surrounding bone). Inorganic biomaterials such as hydroxyapatite (HAP), HAP/poly(lactic-co-glycolic acid) composites, calcium phosphate-based biomaterials, and polymer/ceramic composites have the potential to be osteoinductive, meaning they can induce ectopic bone formation by instructing the surrounding in vivo environment to form bone [19,68,73,74].

Hasan et al. reviewed the diverse applications of inorganic nanoparticles (magnetic, metallic, metal oxide, and metallic) in tissue engineering [75]. Table 1 summarizes the composition of different inorganic-based nanoparticles and biomaterials along with biocomposite materials reviewed in this study. Inorganic-based nanoparticles and biomaterials can be broadly classified based on the material from which they are synthesized or fabricated: carbon nanomaterials, inorganic nanoparticles, inorganic biomaterials, biomimetic natural biomaterials/biopolymers, and biocomposite materials. Carbon nanomaterials include graphene oxide, reduced graphene oxide, fullerene, fullerol, carbon dot, carbon black, and carbon nanotube. The inorganic nanoparticles can be metals, such as gold or silver nanoparticles, or metal oxides (TiO_2, silica, MgO, Fe_3O_4, alumina, and bioactive glass) and are the most widely expended type of nanoparticles in biomedical fields. Biomimetic materials are classified based on their structure, function, manufacturing process, and molecular characteristics [76]. Biomimetic natural biomaterials and biopolymers are typically biocompatible, biodegradable, non-toxic, and have been shown to improve the pharmacokinetic properties and bioavailability for targeted drug delivery [77]. These biomaterials find applications in robotics, superhydrophobic materials design, self-cleaning materials, adhesives, biomedical, and nanotechnology [76].

Table 1. Classification of inorganic-based nanoparticles and biomaterials based on the chemical composition or structure.

Type	Composition/Structure
Carbon nanomaterial	Graphene oxide (GO) Single-wall carbon nanotube (SWNT) Multi-wall carbon nanotube (MWCNT) Carbon dot, Carbon black (Printex 90) Fullerene (C60) Fullerol [$C60(OH)_n$]

Table 1. *Cont.*

Type	Composition/Structure
Inorganic nanoparticle	Silica-iron oxide (SiO_2-Fe_3O_4) Titanium dioxide (TiO_2), Ti-6Al-4V alloy Manganese dioxide (MnO_2), Manganese silicate (MnS) Titanium dioxide-silica (TiO_2/SiO_2) Magnesium oxide (MgO) Bioactive glass (BG) Si−Ca−P−Mo glass-ceramic (BBGN) Gold (AuNP) Zinc oxide (ZnO)
Inorganic biomaterial	Hydroxyapatite (HAP) Calcium phosphate (CaP) Ceramic
Biomimetic natural biomaterial/biopolymer	Poly (L-lactic acid) (PLLA), Polycaprolactone (PCL) Polyhydroxyalkanoates, Hyaluronic acid (HA) Alginate, Cellulose, Chitosan (CS), Collagen Gelatin, Gelatin nanofiber (GNF) Polydimethylsiloxane (PDMS) Fibroin, Hydrogel, Poly-glutamic acid [$(C_5H_7NO_3)_n$]
Biocomposite material	GO-PCL PLLA/PCL/GNF/AuNP PCL/silk fibroin (PCL/SF) PCL/SF/Au(SiO_2) Polysulphone-modified MWCNT Alginate-nano-silica hydrogel Chitosan hydrogel reinforced HAP nanorod (CS-HAPNr) AuNP-collagen Chitosan-collagen-HAP (Cs/Col/HAP) HAP/poly(lactic-co-glycolic acid)

Figure 1 illustrates a multifaceted approach that combines cell therapy, bioactive materials, and biomolecules or external stimulation to treat musculoskeletal diseases affecting bone and muscle through regenerative medicine.

External stimulation, along with signaling molecules, aptamers, peptides, and growth factors can modulate immune response and promote tissue growth essential in regeneration. External stimuli like light, temperature, ultrasound, topography, magnetic fields, or electrical stimulation can activate intracellular signaling cascades, augment intrinsic tissue cell growth, and mitigate regeneration under natural conditions [78]. Therefore, a synergistic strategy that integrates innovative biomaterials, growth factors, and external stimulation offers a comprehensive approach to tissue engineering, effectively addressing patients' needs for treating musculoskeletal conditions. Nevertheless, to treat musculoskeletal disorders, optimizing the stimuli responsiveness and refining currently available therapeutic methods can leverage the effectiveness of therapeutic approaches.

To enhance the effectiveness of stem cell therapy in treating cardiovascular diseases, Chen et al. [35] incorporated a silica–iron oxide (silica-Fe_3O_4) nanoparticle, a superparamagnetic mesocellular foam to increase cell viability and sustained release of cargo in human MSCs. Alternatively, the higher porosity and surface area of mesoporous silica nanoparticles, metallic oxide nanoparticles, HAPs and titanium dioxide (TiO_2) nanotubes have enabled applications in drug delivery, imaging, osteoblast differentiation, and bone regeneration [79]. Surface coating of superparamagnetic iron oxide nanoparticles with silica was shown to reduce the long-term toxicity effects and improve biocompatibility via enhancement of resistance to lysosomal acidity. The Fe_3O_4/SiO_2 nanoparticles were surface-passivated with amino- and sulfonate-silanes, and toxicity studies were performed on A549 and HeLa cell lines [80]. Lactate dehydrogenase (LDH) assay, which measures cell membrane integrity, showed high levels of membrane damage in bare nanoparticles, while the nanoparticles that were surface coated showed a marginal increase in LDH levels, suggesting the role of surface passivation on cell viability and membrane damage. TiO_2 nanoparticles have been shown to cause genotoxic effects by inducing oxidative DNA damage, and excessive production of ROS can result in strand breakage after 24 h of exposure [81,82]. Data indicated that a decrease in mitochondrial dehydrogenase activity in human lymphocytes, caused by TiO_2 nanoparticles, could increase oxidative stress in cells and serve as mediators for TiO_2 genotoxicity [83]. Genotoxicity studies performed using a comet assay agreed with cytotoxicity measurements, with bare TiO_2 nanoparticles exhibiting increased DNA damage [81,83].

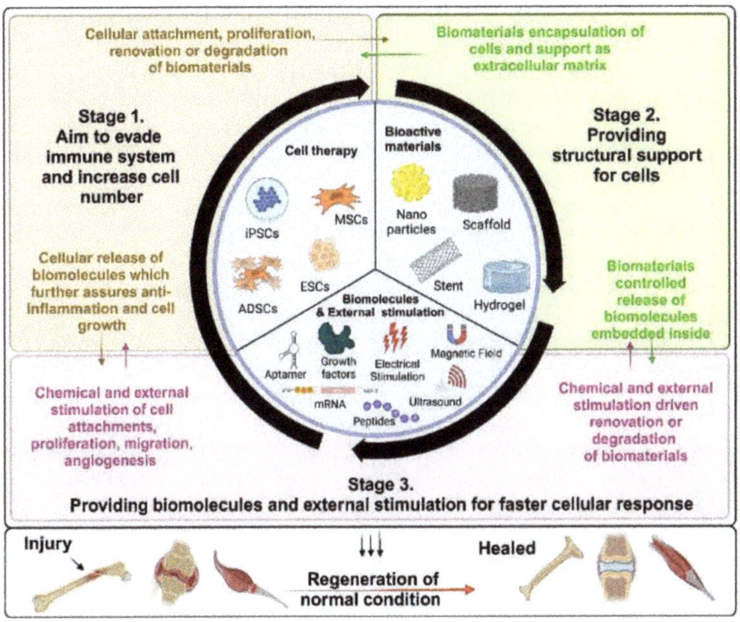

Figure 1. An integrative approach that combines bioactive materials, biomolecules and external stimulation, and cell therapy in musculoskeletal disorders. The figure is reproduced from ref. [78], an open access article under the CC BY-NC-ND license, https://creativecommons.org/licenses/by-nc-nd/4.0/ (accessed on 1 November 2024).

Studies suggested that TiO_2 nanoparticles are prone to accumulation in the alimentary tract, central nervous system, liver, kidney, pulmonary tract, and reproductive system and form agglomerates with other mammalian cells; TiO_2 particles less than 25 nm can accumulate in the spleen, lungs, and kidney while particles of around 80 nm can accumulate in the liver [84]. Since these nanoparticles have the potency to cross the blood–brain barrier, they can induce apoptosis in the hippocampal neurons [84]. TiO_2 nanoparticles of certain sizes can induce oxidative stress, toxic effects, DNA strand breakage, and chromosome damage, leading to elevated expression of inflammatory cytokines and increased occurrence of apoptosis [85]. Studies on LA-9 mouse fibroblast cell lines suggested TiO_2 nanoparticles could induce intracellular ROS production at 150 and 250 μg/mL concentrations after exposure for 24 h [86]. Furthermore, cells exposed to 150 μg/mL concentration exhibited an increased percentage of apoptotic cells when compared to negative control [86]. Cytotoxicity measurements of 3–600 μg/mL of homogeneous and weakly aggregated TiO_2 nanoparticle solution showed variations in cell viability at lower concentrations of TiO_2 (3–6 μg/mL) [87]. Cell viability decreased at concentrations > 6 μg/mL, while cells cultured in 600 μg/mL TiO_2 nanoparticles for 48 h showed a 37% decrease in cell viability [87]. After

48 h of exposure to L929 cell lines, significant cytotoxicity was observed. Higher concentrations of TiO_2 nanoparticles increased oxidative stress, leading to increased ROS and LDH levels, and decreased glutathione (GSH) and superoxide dismutase (SOD) levels [87]. Detailed reviews on the detrimental, hazardous effects of TiO_2 nanoparticles, along with toxicity, can be found elsewhere [83,84,88,89].

Friggeri et al. [90] investigated the potential application of graphene oxide–polycaprolactone (GO-PCL) composite material as a multifunctional 3D-printed scaffold in tissue engineering and medical device design. The composite material exhibited antibacterial and adhesive growth properties, highlighting its potential applications in tissue engineering (Figure 2). PCL and PCL-GO composite of 1 wt.% concentration had low cytotoxicity and did not impair the mechanical properties of PCL. The adsorption of solvent molecules by GO reduced bacterial adhesion during surgical procedures. The cellular adhesion on PCL-GO was significantly lower compared to PCL samples. Biocompatibility analysis was performed with VERO, HEK, C2C12, and RAW cell lines that were plated on PCL and PCL-GO scaffolds, and cell viability after 72 h was measured using the bioluminescence method (Figure 2A). PCL-GO scaffolds had lower cellular adhesion for the cell lines compared to PCL and plastic wells, which were used as controls. The cell adhesion for control was higher than for PCL, as shown in Figure 2A. The reduced cell adhesion of the PCL-GO scaffold was related to PCL's hydrophobicity, which limit intermolecular contacts with the surface. The toxicity profile of VERO cells lines conditioned with PCL-GO scaffolds was observed to be lower than those of cells conditioned with PCL after 7 days of scaffold submersion in DMEM (Dulbecco's Modified Eagle's Medium), as shown in Figure 2B. The decreased cell adhesion in PCL-GO lessened biofouling risks in a surgical environment. Controlling the long-term cell adhesion ensured scaffold biodegradability and population over time, which was vital in promoting successful tissue regeneration. Fluorescence images showed visible dead cells (shown in red) that point to a local cytotoxic effect of PCL-GO along the scaffold borders, as indicated by the arrows (Figure 2C,D). PCL and PCL-GO scaffolds were washed with PBS or PBS + ethanol (Figure 2E), and the number of cells attached to the scaffold increased with the number of washes. PCL-GO scaffold showed a five-fold increase in cell attachment compared to PCL. The bioconductivity of PCL-GO scaffolds reached that of PCL after repeated washing with DMEM medium (Figure 2F). Lastly, fluorescence imaging of VERO cell lines grown on DMEM-washed scaffolds did not show the presence of dead cells along the scaffold borders (Figure 2G,H). The study proposed a low-cost, yet versatile strategy in the application of GO-PCL composite material as a 3D-printed scaffold for developing biomedical implants.

The scaffolding function of poly (L-lactic acid) (PLLA)/PCL composite matrix containing gelatin nanofibers (GNFs) and gold nanoparticles (AuNPs) were studied in bone tissue engineering [91]. The solubility properties, coupled with biocompatibility, biodegradability, and low immunogenicity, make gelatin a suitable biopolymer for tissue engineering applications [92]. Furthermore, the Arg-Gly-Asp (RGD) domains in gelatin promote cell attachment and proliferation. The induced toxicity and cellular proliferation of MG-63 cells on the fabricated scaffolds were measured using MTT and LDH assays [91]. PLLA/PCL/GNF/AuNPs (80 ppm) showed the highest cell growth at 72 h (Figure 3). PLLA/PCL/GNF/AuNPs (160 ppm) at 72 h after cell seeding had lower cell proliferation than control, while test groups showed higher cell proliferation than control.

The rat calvarial defect model was used to evaluate bone healing, and the histomorphometric data is provided in Table 2. The poorest bone regeneration was observed in the negative control group. The defect treated with PLA/PCL/GNF/AuNPs (80 ppm) exhibited the highest fibroblast + fibrocyte, osteoblast + osteocyte, and osteon values, which indicated the highest bone regeneration. The observation of osteoclast cells demonstrated greater bone regeneration than bone resorption; bone regeneration was enhanced in defects treated with PLA/PCL/GNF/AuNPs (80 ppm) [91].

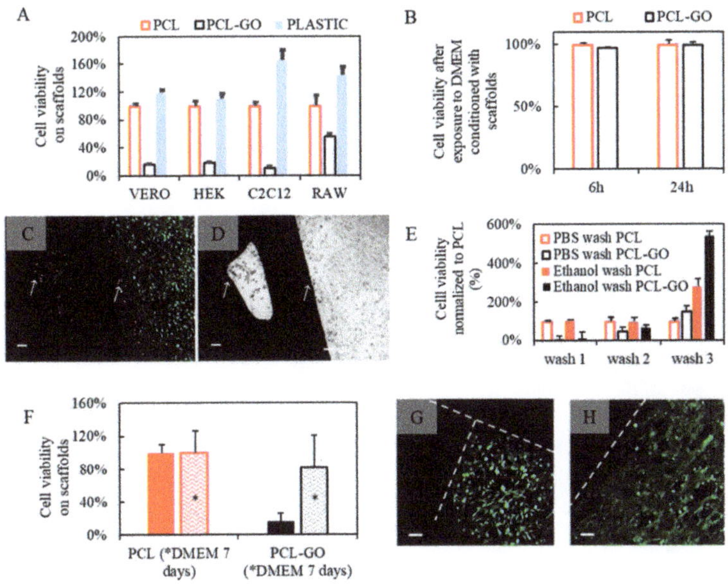

Figure 2. (**A**) Viability of different cell lines grown on scaffolds or plastic wells evaluated using a luminescence assay. (**B**) Toxicity of DMEM exposed to scaffolds for 7 days and used to treat VERO cells seeded on 96-well plates. Toxicity on VERO cells was evaluated after 6 or 24 h of treatment. (**C**) Fluorescence imaging of VERO cells surrounding grids of PCL-GO. Cells have been labeled with calcein (green) to evaluate viable cells and propidium iodide (red) to evaluate dead cells. The brightfield image of the same sample is shown in (**D**). Arrows indicate boundaries between the scaffold and well. (**E**) Evaluation of VERO cell viability on scaffolds washed with PBS or PBS + ethanol. (**F**) Cell viability on the scaffold without DMEM washing and after washing with DMEM, indicated with asterisks (*). (**G**,**H**) Representative fluorescence images of cells surrounding grids (dashed lines) after DMEM washings; no red dead cells were visible. The scale bar is 100 μm. The figure and caption are reproduced from ref. [90]. Copyright © 2024, G. Friggeri, I. Moretti, F. Amato, A. G. Marrani, F. Sciandra, S. G. Colombarolli, A. Vitali, S. Viscuso, A. Augello, L. Cui, G. Perini, M. De Spirito, M. Papi, V. Palmieri. This article is distributed under a Creative Commons Attribution (CC BY) license, http://creativecommons.org/licenses/by/4.0/ (accessed on 1 November 2024).

Table 2. Histomorphometric analysis of bone tissue regeneration in the defects treated with PLA/PCL/GNF, PLA/PCL/GNF/AuNPs (40 ppm), PLA/PCL/GNF/AuNPs (80 ppm), and PLA/PCL/GNF/AuNPs (160 ppm). Reproduced from ref. [91].

	Negative Control	PLA/PCL/GNF	PLA/PCL/GNF/AuNPs		
			40 ppm	80 ppm	160 ppm
Fibroblast + fibrocyte	62.10 ± 3.55	79.42 ± 18.64	93.90 ± 16.03	109.11 ± 22.45	87.09 ± 13.10
Chondroblast + chondrocyte	102.40 ± 5.44	88.50 ± 11.37	53.73 ± 6.74	55.26 ± 4.99	67.32 ± 7.88
Osteoblast + osteocyte	23.25 ± 4.65	51.25 ± 18.09	85.34 ± 16.22	97.60 ± 27.16	76.31 ± 15.05
Osteoclast	3.78 ± 1.66	4.54 ± 1.09	3.31 ± 2.14	2.65 ± 1.28	4.21 ± 0.98
Osteon	3.67 ± 1.27	5.57 ± 1.99	6.93 ± 2.44	7.55 ± 1.69	4.50 ± 1.08

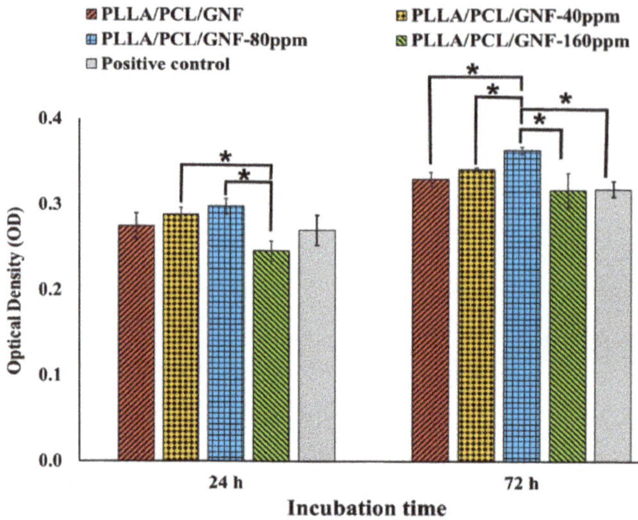

Figure 3. MG-63 cell proliferation on the fabricated scaffolds was measured using an MTT assay at 24 and 72 h after cell seeding. Control: Tissue Culture Plate (TCP). Data represented as mean ± SD, $n = 3$. * $p < 0.05$ (obtained by one-way ANOVA). The Figure is reproduced from ref. [91]. This article is licensed under a Creative Commons Attribution 4.0 International License, http://creativecommons.org/licenses/by/4.0/ (accessed on 1 November 2024).

The in vivo data indicated defects treated with PLA/PCL/GNF/AuNPs (80 ppm) induced higher neo-bone formation. AuNPs (80 ppm) incorporated in the scaffolds did not interrupt the normal functioning of cells. PLA/PCL/GNF/AuNP (80 ppm) scaffold induced the highest bone regeneration, and the incorporation of AuNPs and GNFs mimicked the native structure of bone and stimulated bone healing.

Nanofibrous composite scaffolds of PCL, PCL/silk fibroin (SF), and PCL/SF/Au(SiO_2) were fabricated to study bone tissue regeneration of human MSCs (hMSCs) [93]. The nanofibrous scaffolds had high porosity and exhibited a uniformly connected fibrous structure with polydispersed Au(SiO_2) nanoparticles on the scaffold surface. Table 3 shows the fiber diameter, pore size, % porosity, and tensile strength of PCL, PCL/SF, and PCL/SF/Au(SiO_2) electrospun nanofibrous scaffolds.

Table 3. Characterization of nanofibrous scaffolds of PCL, PCL/SF, and PCL/SF/Au(SiO_2). Reproduced from ref. [93].

Nanofibrous Constructs	Fiber Diameter (nm)	Pore Size (μm)	Porosity (%)	Tensile Strength (MPa)
PCL	215 ± 32.12	1.45 ± 0.26	88 ± 4.3	7.63
PCL/SF	164 ± 18.65	2.12 ± 0.31	92 ± 6.3	11.67
PCL/SF/Au(SiO_2)	172 ± 24.22	2.35 ±0.22	90 ± 7.5	12.11

The scaffolds have a fiber diameter in the range of 164 ± 18.65 to 215 ± 32.12 nm, as shown in Table 3. The pore sizes of PCL, PCL/SF, and PCL/SF/Au(SiO_2) are 1.45, 2.12, and 2.35 μm, respectively. The porosity values of PCL, PCL/SF, and PCL/SF/Au(SiO_2) scaffolds are 88 ± 4.3%, 92 ± 6.3%, and 90 ± 7.5%, respectively, which are within the desirable range for bone tissue regeneration. The porosity of 88% in PCL was below the optimal requirement of 90%. Inclusion of Au(SiO_2) to the scaffold led to increased porosity, as shown in Table 3. Cell proliferation studies of hMSCs on TCP, PCL, PCL/SF, and PCL/SF/Au(SiO_2) nanofibrous scaffolds on days 7, 14, and 21 suggested that hMSCs grown on PCL had a lower proliferation level than PCL/SF and PCL/SF/Au(SiO_2), which was attributed to the absence of active binding sites in PCL. The higher proliferation level ($p < 0.05$) for PCL/SF and PCL/SF/Au(SiO_2) scaffolds contributed to the bioactive SF and Au(SiO_2) that increased the hydrophilicity essential for cell adhesion [93]. The Au(SiO_2) stimulated cell growth and tissue formation, while SF and Au(SiO_2) stimulated the proliferation of hMSCs without inducing any toxicity.

Composite membranes based on polysulphone-modified multi-walled carbon nanotubes and short carbon fibers were synthesized to elucidate the physicochemical, mechanical, and biological properties [94]. The in vitro results indicated that the membranes were biocompatible when in contact with MG-63 cells. The interaction mechanism between the cells and the composites was influenced by the porosity of the membranes, the presence of carbon additives on the membrane surface, and surface chemistry. Sithole et al. [95] developed a novel 3D-printed biomaterial scaffold to guide host cell growth for bone tissue regeneration. To investigate the potential for bone tissue regeneration/formation, 3D-printed scaffolds were implanted in nasal bone defects of New Zealand white rabbits. Osteoblast-like MG63 cells were used to culture the 3D printed scaffold for 1, 3, and 7 days, respectively. The energy-dispersive X-ray (EDX) analysis showed the presence of Cl, Ca, Na, Si, and P in the mineralized 3D-printed biomaterial scaffold immersed in simulated fluid (SBF) at pH 7.4 for 7 days. On the other hand, the EDX analysis of non-mineralized control scaffold showed the presence of Na, Si, and Cl. This indicated that the biomineral scaffolds had essential biomineralization capabilities crucial for biological growth. The scanning electron microscopy (SEM) images of osteoblast-like MG63 cell adhesion onto 3D-printed biomaterial scaffolds and the scaffold without cells (control) is depicted in Figure 4. The MG63 cells were successfully seeded onto the 3D-printed scaffold, and the rough surface was suitable for cell adhesion to the surface of the implanted devices. In vitro release kinetics experiments were performed to elucidate the performance of the biomaterial scaffold and evaluate the release of BMP-7 protein/growth factor from the biomaterial scaffold in a way that mimicked natural bone fracture healing [95]. For this, the scaffold was loaded with 40.45 ng/mL of BMP-7, and a BMP-7 ELISA Kit was used to study the release profile. BMP-7 was found to adhere to the biomaterial scaffolds surface leading to early burst release. Cell viability measurements further supported the biocompatibility of 3D-printed biomaterial scaffolds, as osteoblast-like MG63 cells were observed to be viable for 7 days.

A biodegradable nanochannel (BNC) membrane was developed through hierarchical cell assembly of 3D manganese dioxide (MnO_2) nanostructures for applications in tissue engineering, cell assembly, and disease modeling, featuring rapid and controllable degradation under biological conditions [96]. As discussed in this study, rapid biodegradation rates of one to three orders of magnitude higher than current polymeric materials make MnO_2 a desirable nanomaterial for membrane construction. The BNC membrane enabled the generation of robust scaffold-free cell sheets and provided a versatile cell assembly-based approach. The cell patterning was visualized using two cell lines labeled with green fluorescence protein (GFP, Cell A) and red fluorescence protein (RFP, Cell B). A densely packed "core–shell" co-assembly of Cells A and B was observed with Cell A surrounded by Cell B, confirming high precision in the patterning process. To recapitulate chronic inflammation in the human skin wound healing process, multicellular patterning of human monocyte-derived macrophages and human fibroblasts was carried out using BNC membrane, placed on a bio-inert cellulose membrane to prevent the tissue from folding. The migration rate of fibroblasts was much slower ($\approx$1 mm/day) compared to conventional scratch assays. The BNC membrane was shown to be suitable in murine diabetic wound healing and chronic skin inflammation. Both in vivo models and in vitro experiments demonstrated the therapeutic efficacy of the BNC cell assembly technique for heterogeneous cell patterning across various cell types. This study by Yang et al. [96] demonstrated a proof-of-concept application of BNC membrane-based 3D cell sheet assembly, achieved in a scaffold-free manner, which is suitable for treating wound healing.

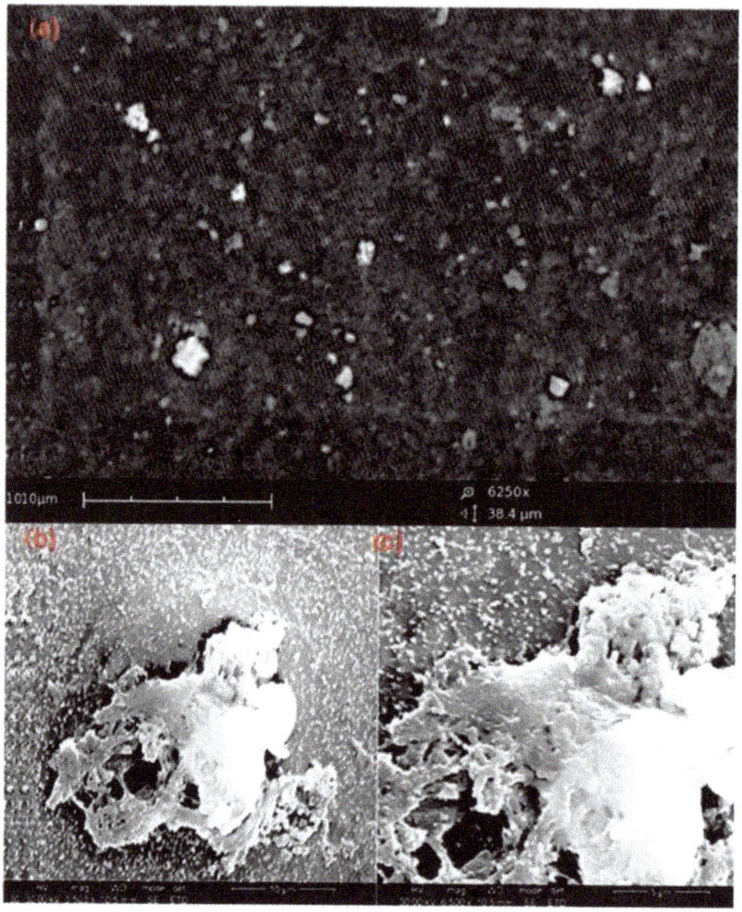

Figure 4. The SEM images on 3D-printed biomaterial scaffolds seeded with osteoblast-like MG-63 cells and the scaffold without cells (control). (**a**) Reference scaffold or scaffold without the cells, (**b**) day 7 cell adhered on the surface of the printed scaffold, (**c**) magnified image of day 7 of cell adhered to the printed scaffold surface. Reprinted with permission from ref. [95]. This article is an open access article distributed under the terms and conditions of the Creative Commons Attribution (CC BY) license, https://creativecommons.org/licenses/by/4.0/ (accessed on 1 November 2024).

To improve the selective differentiation and transplantation of human patient-derived neural stem cells (NSCs), Yang et al. [97] developed a 3D biodegradable hybrid inorganic (BHI) nanoscaffold for stem cell-based tissue engineering and therapy. A biodegradable MnO_2 hybrid scaffold was fabricated, and the interaction of 2D MnO_2 nanosheets with laminin, an ECM protein was investigated. Laminin exhibited a 7.5-fold increase in adsorption on MnO_2 compared to glass (control) and polymer substrates. This was attributed to the polar-π and electrostatic interactions between the biomolecules and MnO_2 nanosheets. The neuronal differentiation of stem cells using MnO_2 was studied using human-induced pluripotent stem cell (hiPSC)-derived NSCs as a model system. A 43% increase in neuronal differentiation and 11-fold increase in neurite outgrowth was observed, due to improved laminin binding and focal adhesion-related pathways. This study validated the potential of hybrid MnO_2 nanoscaffold in improving stem cell survival and inducing neuronal differentiation in vivo.

3. Nanotechnology Strategies in Regenerative Medicine and Wound Healing

The application of nanotechnology in tissue engineering and stem cell biology has fostered new frontiers in research [98]. The novel design principles and characterization techniques, coupled with tunable intrinsic properties and functional specificity, can stimulate the integration of nanomaterials in regenerative medicine therapies [66]. A basic understanding of cell topography is crucial, particularly in terms of how the surrounding environment influences cellular behavior. This includes insights into physical interactions such as cell adhesion, proliferation with scaffold materials, and differentiation. These factors play a significant role in regulating signaling behavior and cellular fate [66,99]. Brovkina and Dashinimaev reviewed regenerative medicine approaches in diabetes mellitus [100]. Li et al. [101] reviewed the application of nanotechnology (nanomedicine, nanomedicine carriers, nanocontrast agents, and nanosensors) in diabetes management and diabetic regenerative medicine. Nanomaterial scaffolds that mimic the natural ECM enhance cell–scaffold interactions, promote the formation of implanted/transplanted tissue-like structures, and facilitate adhesion, oxygen exchange, and nutrient transport [101]. In the treatment of diabetes, nanoparticles were shown to protect insulin from degradation within the gastrointestinal environment and modulate insulin release in response to changes in glucose concentration [102–104]. Polymeric nanoparticles, liposomes, and lipid nanoparticles are the main types of nanoparticles employed in diabetes treatment [70]. These nanoparticles can deliver drugs to the target site, cross the intestinal epithelial barrier while shielding the drugs from environmental conditions, increase drug adsorption, and improve receptor-mediated endocytosis [105,106]. The application of inorganic nanoparticles and polymer nanocomposite materials in early diagnosis and diabetes regenerative medicine has been reviewed elsewhere [102,107–109].

3.1. Engineered Nanomaterials

Engineered nanomaterials or precisely engineered nano–bio interfaces can mimic the natural ECM and facilitate stem cell proliferation, migration, and differentiation into desired cell types [110,111], bone and cartilage repair, and nerve regeneration [111–113]. At the molecular level, engineered nanomaterials can be tailored by modulating the surface properties to provide localized, sustained release of biological molecules and increase therapeutic efficacy [114]. Przekora reviewed the application of engineered biomaterials in bone and cartilage regeneration for long-lasting antimicrobial protection [115]. Wu et al. developed nanotopography patterns in the forms of nano pillars, -holes, and -grills on polycarbonate surfaces to study their effects on human MSC chondrogenesis [116]. Using nanoimprinting technique, nanopillar, nanohole, and nanogrill patterns were constructed to provide topological cues in cartilage collagen fibril orientation. The nanotopographies were patterned on PCL film, and chondroitin sulfate was immobilized on the surface to provide a chondro-inductive biochemical cue for MSC chondrogenesis [117,118]. MSCs on nanogrill and non-patterned surfaces showed an increased cell proliferation rate with a significantly higher DNA amount at day 14 compared to day 1. The initial proliferation on the nanogrill was similar to the that of the non-patterned surface but stopped by day 14. Comparatively, MSCs on nanopillar and nanohole surfaces underwent negligible proliferation; the DNA amount throughout the culture period was not significantly higher than at day 1. The different nanopatterned topography affected the proliferation profile of MSCs. Significantly higher upregulation of hyaline chondrogenic markers and type II collagen was detected on nanopillar and nanohole surfaces, while the expression of this marker was delayed and was at lower levels on the nanogrill surface. This study demonstrated the effect of nano-topography on MSC chondrogenic differentiation; the nanotopographical pattern triggered changes in MSC morphology and cytoskeletal structure, thereby affecting cell aggregation and differentiation.

Atomic layer deposition (ALD) is a cutting-edge nanofabrication technique used in engineering nano-bio interfaces, understanding nano–bio interactions, and developing defect-free and pinhole-free hetero-structured sensing electrodes and layered nanofilms [119,120].

ALD mediates the deposition of thin films on the surface of substrates to form a homogenous coating [121,122]. Bechelany et al. [121] reviewed the synthesis of nanostructured materials using the ALD method for applications in sensing (using nanopore technology) and gas separation membrane technology in water treatment. Pure titanium and Ti-6Al-4V alloy are two biomedical implants widely used in dental and orthopedic applications due to their biocompatibility, superior mechanical properties, and corrosion resistance. However, these implant materials suffer wear and tribocorrosion that releases ions and metallic particles, causing implant loosening and loss in stability [123,124]. TiO_2 coating using ALD improved the mechanical properties, interfacial biocompatibility, water resistance, and cell adherence for high cell–surface interactions [125]. Capek et al. [125] investigated the cellular response of nanotubular oxides of Ti and Ti-6Al-4V alloy to improve cell growth on implants using the ALD technique. Surfaces with ultrathin 1c TiO_2 ALD coatings had a thickness of 0.055 nm and did not result in any morphological changes. Cell proliferation studies of MG-63 cells on ALD-coated Ti-6Al-4V foils showed the growth of elongated MG-63 cells on AM-TiAlV coated with 1c TiO_2 ALD in comparison with uncoated AM-TiAlV by approximately 15%. After 48 h and 72 h, a 2.5-fold and 4-fold increase in the number of cells cultured on AM-TiAlV + 1c TiO_2 was observed. The increased cell proliferation on 1c TiO_2 ALD coating was attributed to the chemistry of the surface and wettability whereby the ultrathin TiO_2 coating lessened the negative cytotoxicity effect of poisonous elements, such as F and V, in biomedical Ti-6Al-4V alloys.

ALD at low temperatures improved the mechanical durability, wetting, and optical properties of TiO_2/SiO_2 nanoparticle films on glass and polycarbonate substrates with precise control of thickness [126]. ALD was utilized for conformal coating of Al_2O_3 oxide thin films, which is a well-defined ALD system on TiO_2/SiO_2 nanoparticle layer-by-layer (LbL) films as a function of ALD-cycle number. The surface morphology of LbL films was not affected by the ALD (thickness of ~1 nm) compared to ~22 nm for silica nanoparticles used as the outermost layer. The contact angle of water on TiO_2/SiO_2 films on glass increased with the number of ALD cycles, and impurity levels were ~0.1–1 atom %. The superhydrophilicity (contact angle $< 5°$) was restored in the films, and the films were superhydrophilic even after 30 days. Priming polycarbonate substrates with Al_2O_3 ALD coatings before LbL assembly significantly improved the uniformity of TiO_2/SiO_2 films, and ALD-deposited Al_2O_3 showed a surface potential of 20.6 ± 4 mV.

ALD was used for the uniform coating of nanostructured TiO_2 on commercial Ti substrates using tetrakis(dimethylamino)titanium in combination with H_2O or ozone as an oxidant [125]. The ALD-coated Ti-TiO_2 sample thin films were studied on three different types of bacteria (gram-positive, gram-negative, and antibiotic-resistant). In vitro bacterial results indicated that Ti-TiO_2 at 160 °C significantly inhibited the adhesion and growth of *S. aureus*, *E. coli*, and MRSA bacteria (exceeding 80%) compared to Ti control. The osteoblast adhesion on Ti-TiO_2 samples was significantly higher than Ti control after 4 h of culture. The enhanced adhesion and proliferation of osteoblast cells were facilitated by the increased surface roughness and surface energy of the TiO_2-coated samples, which promoted cell–surface interactions between the osteoblasts and the surface [125].

Room-temperature ALD technique was used to coat collagen membranes with MgO [127] for improved osteointegration. This method enhanced the biocompatibility of collagen and allowed for the deposition of ultra-thin layers of metal oxides that featured uniform thickness and composition, ensuring a conformal, pinhole-free coating [128]. SEM micrographs of uncoated collagen (Col), collagen after 200 ALD cycles of MgO (Col-MgO200), and collagen after 500 ALD cycles of MgO (Col-MgO500) showed regular patterns on the Col-MgO500 sample. As the number of ALD cycles increased, collagen fibers became thicker, and striations in the individual fibers were no longer visible on the coated samples. Atomic force microscopy demonstrated a wavy topography of the samples due to the fibrous and wavy topography of the collagen membrane. The MgO-coated collagen membrane was non-toxic to human cells, displayed improved antimicrobial activity against bacterial biofilms, and facilitated the controlled release of Mg ions [128]. ALD was used for coating commercially obtained nanoporous alumina membranes with platinum or zinc oxide [129]. An ALD of 8 nm of platinum was coated onto a 20 nm pore size nanoporous alumina membrane, which had high porosity and monodispersed pore size. The pores of PEGylated platinum-coated nanoporous alumina membranes were free of fouling. No protein adsorption, fibrin networks, or platelet aggregation were observed on the membrane surface, suggesting that ALD can serve as a scalable, cost-effective method.

3.2. Carbon Nanomaterials

The design and application of carbon nanotube-based nano–electron mechanical systems (NEMS) in medical nanorobots was reviewed by Popov et al. [130]. Carbon nanotube-based NEMS actuation was achieved using a nanomanipulator, an electric field acting on encapsulated ions, and a magnetic field acting on the magnetic core of the piston. The movable core of the nanotube's wall was polarized by the electric field [130]. The electric field could move metal atoms within the nanotube, transforming them into ions. Wang et al. [131] fabricated 3D carbon nanotube-based composite scaffolds for applications in cartilage repair and tissue engineering. The structure and morphology of carboxylated single-walled carbon nanotube (SWNT-COOH) composite scaffolds were characterized using SEM. The pore size and compression strength of the composite scaffold was modulated by changing the SWNT-COOH concentration. The compression strength increased, while the size of the pore opening decreased with increasing weight percent of SWNT-COOHs. The BMSC proliferation on the composite scaffold suggested that at 7 days, the number of BMSCs in the 2.0 wt. % group was significantly higher than in the 0, 0.5, and 1.0 wt. % groups. However, there was no obvious difference in the number of BMSCs between the three groups. The restorative effects of SWNT-COOH scaffolds were studied using a rabbit model of articular cartilage defects (Figure 5).

In the control group, the cartilage defects were hollow, and reddish granulation tissue at the bottom surface of defects was observed 4 weeks after implantation. The 0 wt. %, 0.5 wt. %, 1.0 wt. %, and 2.0 wt. % groups suggested better filling of defects with no degradation in the composite scaffolds. Repair of cartilage defects was enhanced at 8 weeks after implantation, and the hollows in the control group reduced by half. Except for the control and 2.0 wt. % groups, cartilage defects were fully replaced with repaired tissue at 12 weeks. The cartilage defects in the 0 wt. % group were filled with regenerative cartilage, as shown in Figure 5D. The defects were covered with a smooth, white, semi-transparent cartilage in the 0.5 wt. % group, indicating that this group was superior in repairing cartilage defects.

The chondrogenic potential of multi-walled carbon nanotubes (MWCNT) and hyaluronic acid (HA) was investigated for 3D bioprinting of cartilage constructs [132]. A live/dead assay was performed on the bio-printed constructs after 1 day, 14 days, and 21 days to determine the viability of hMSC-AT cells in the 3D scaffolds. HA-enriched constructs showed a decline in cell viability. The control medium also showed a constant decrease in cell viability, while the differentiation medium exhibited a profound decline in cell viability, despite having the highest transcriptional activity. The MWCNT constructs did not show any decline in cell viability, indicating its protective effect on cells. This was due to the resemblance of CNT to collagen fibrils, which had a stimulating effect on the cells. Scaffolds supplemented with MWCNTs and HA showed the highest cell viability. MWCNTs had a favorable influence on cell viability, while the inclusion of HA had a negative impact on the integrity of constructs. Live/dead analysis indicated that the viability of cells decreased over time for scaffolds without additives (control and differentiation medium), and with HA.

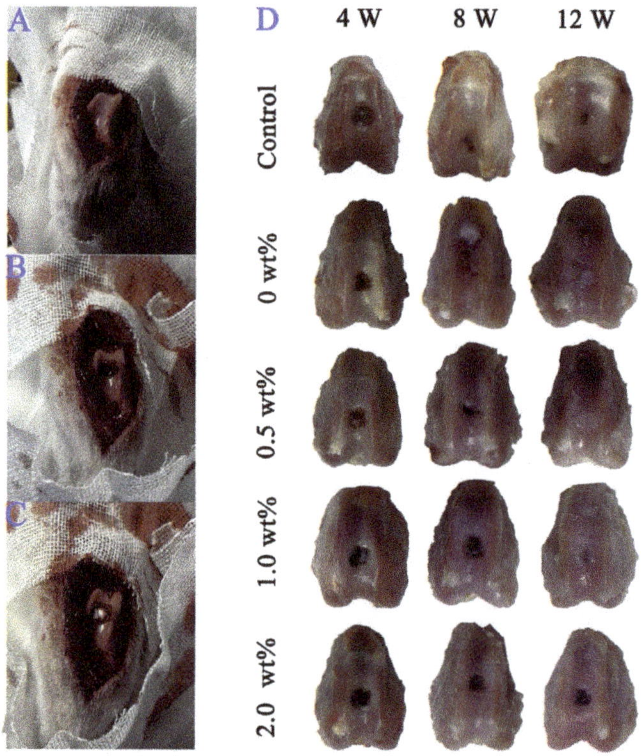

Figure 5. A rabbit model depicting articular cartilage defects and gross observation of defects in the allotted time after operation. (**A**) Smooth and glossy surface of normal articular cartilage. (**B**) Rabbit model of articular cartilage defects. (**C**) Rabbit model of articular cartilage defects filled with composite scaffolds. (**D**) Cartilage defects in control, 0 wt. %, 0.5 wt. %, 1.0 wt. %, and 2.0 wt. % groups at 4 weeks, 8 weeks, and 12 weeks of implantation. Reprinted with permission from ref. [131] from Elsevier.

Polyhydroxylated fullerene (fullerol) was studied for the osteogenic differentiation of human ADSCs and the effect of fullerol on osteogenic differentiation was determined independently and in osteogenic medium [133]. The expression of osteogenic markers was elevated in the presence of a high dose of fullerol. Fullerol inhibited the expression of adipogenic and osteogenic markers in a mouse bone marrow stromal cell line. Fullerol was shown to be nontoxic toward human ADSCs at doses up to 10 µM. Pre-treatment and co-treatment of fullerol enhanced the osteogenic capacity of human ADSCs in culture, and fullerol stimulated gene expression of FoxO1 and its target genes Runx2 and SOD2. As osteogenesis and adipogenesis are regarded to be closely correlated with the capacity of the progenitor cells during bone regeneration, fullerol was shown to inhibit the differentiation of marrow stromal cells and OP9 preadipocytes into adipocytes due to its scavenging activity against intracellular ROS. This study suggested that fullerol could serve as a novel pharmaceutical candidate to produce augmented bone regeneration.

CNT-coated and textured polydimethylsiloxane (PDMS) scaffolds were studied to support the growth of chondrocytes–cells that produce and maintain cartilage [134]. The COOH functionalization led to improved cell adhesion. In this study, four different scaffold architectures, namely, flat (F) and textured (T) PDMS (F-PDMS and T-PDMS), were coated with MWCNT-COOH (F-PDMS-MWCNT and T-PDMS-MWCNT). The addition of MWCNTs atop PDMS produced an isotropic fibrous architecture and enhanced roughness that mimicked the structure in native cartilage. The chondrocytes were observed to form clusters that reduced cellular contact with the substrate. The nano-level roughness increased by ~270% and cell adhesion improved with the addition of MWCNT coating on PDMS scaffolds.

Graphene oxide (GO) and reduced graphene oxide (rGO) mats were examined for their potential to induce neuronic differentiation of ADSCs, which can serve as alternatives to MSCs, demonstrating their ability to differentiate into neuron-like cells [135]. The study demonstrated that graphene nanomaterial-based mats served as suitable stem cell culture platforms for the neurogenic differentiation of ADSCs and GO accelerated the differentiation of ADSCs into neuron-like cells. Neuron-like cells displayed typical neuronal morphology, and the growth, proliferation, and neurogenic differentiation rates of ADSCs seeded on GO mats were superior to those on other substrates.

PCL-based nanocomposite scaffolds with GO nanoshells and two osteogenic drugs, dexamethasone and simvastatin (PCL/GO-Dex and PCL/GO-Sim), were fabricated to examine the osteogenic differentiation, cell viability, and biomineralization on MSCs [136]. The dexamethasone releases from PCL/Dex and PCL/GO-Dex were 69.61 and 49.97%, while the simvastatin releases from PCL/Sim and PCL/GO-Sim were 84.02 and 99.27%, respectively. The presence of GO nanosheets led to less drug being encapsulated in scaffolds, which may be due to the π-π stacking interactions between the drug and GO. PCL/GO-Dex nanocomposite had a higher ability to induce differentiation of MSCs to bone cells. While cell viability increased on PCL/GO and PCL/Dex scaffolds compared to PCL after 7 days, there was no significant change observed after 14 days.

Protein and protease inhibitors were synthesized to improve the therapeutic efficacy of carbon nanofiber-based scaffolds in chronic wound fractures, bone defect repair, and bone marrow-induced inflammation prevention [137]. The nanofibers exhibited superior biocompatibility, cell adhesion, and cell proliferation, and the C-reactive protein was released continuously from the nanofibers for more than four weeks. Doxycycline drug loading of 10% and 15% was considered to investigate the release behavior from the nanoscaffold [137]. For 10% doxycycline drug loading, carbon nanofibers containing only doxycycline released 24% of doxycycline after two weeks of culture. The dual drug nanofibers released 5% of doxycycline when the drug loading of doxycycline was 15%. An increase in doxycycline drug loading from 10 to 15% increased the release rate from the dual drug nanofibers. This was correlated to changes in the hydrophilic properties of the fibers. Shen et al. [138] reported the inhibitory effect of carbon black (Printex 90) on the osteogenesis of bone marrow-derived MSCs and the role of mitochondria in carbon black Printex 90-induced suppression of MSC osteogenesis. Printex 90 dosage of 0.03 μg/mL or below had no detrimental effect on cell viability. For osteogenic differentiation, Printex 90 was added to the osteogenic induction medium, replacing the regular culture medium once MSCs seeded in the culture plates reached 60% confluence [138]. Printex 90 caused a downregulation of osteoblastic markers, reduced the activity of alkaline phosphatase (ALP), and led to poor mineralization of osteogenically induced MSCs.

It is well established that pristine carbon nanomaterials such as carbon nanotubes and graphene-based materials are cytotoxic due to their poor biocompatibility, low solubility in aqueous solvents, and inherent hydrophobicity. Surface functionalization via covalent and noncovalent approaches can improve the biocompatibility, rendering these nanomaterials suitable for biomedical applications including, but not limited to, bone-tissue engineering and regenerative medicine [139]. Ongoing research efforts are focused on reducing cytotoxicity and improving biocompatibility for the successful integration of carbon nanomaterials in future clinical applications [140]. The factors contributing to the toxicity of CNTs include oxidative stress, alterations in protein synthesis, malignant transformation, DNA damage and mutation, interstitial fibrosis, and changes in mitochondrial activity [141–143]. The interaction of carbon nanoparticles with intracellular organelles can lead to an increased production of reactive oxygen species (ROS) and trigger signaling cascades associated with inflammation [140]. Surface modification of GO with biomolecules like proteins, polymers, and/or nucleic acids have been shown to reduce the cytotoxicity of GO [144]. Eivazzadeh-Keihan et al. [145] reviewed approaches to lowering the cytotoxicity of GO, CNT, carbon dots, nanodiamond, and fullerene to enhance the applicability of carbon nanomaterials in artificial bone tissue production and regenerative medicine. The cellular toxicological studies of carbon nanomaterials in different cell types, including macrophages, epithelial cells, and lymphocytes has been reviewed elsewhere [143,146,147].

3.3. Bioactive Glass–Ceramic Nanoparticles and Nano-Silica Hydrogels

Bioactive glass is a group of inorganic bioactive materials composed of Na_2O, CaO, P_2O_5, and SiO_2 that can improve the vascularization, osteoblast adhesion, restoration, and reconstruction of mineralized hard tissues, and differentiation of MSCs, making them suitable candidates for clinical applications [148–151]. Bioactive glass was the first biomaterial, discovered by Larry L. Hench in 1969, that could stimulate osteogenesis and bond with host tissues [152,153]. Bioactive glass–ceramics are classified as Class A bioactive materials due to their unique ability to support osteoconduction, which refers to the growth of bone along the surface of the implant. They also facilitate osteoinduction, meaning that the materials can activate and recruit osteoprogenitor cells, promoting bone growth directly on its surface [154,155]. Over the years, research has increasingly examined the potential of bioactive glass for regenerating cardiac, lung, and peripheral nerve tissues, restoring the musculoskeletal system, and promoting wound healing. Gerhardt and Boccaccini reviewed the recent developments of bioactive glass and glass–ceramic scaffolds in tissue engineering [154].

A sol–gel-derived bioactive glass nanoparticle (BGN) in a 62 SiO_2–34.5 CaO–3.2 P_2O_5 (mol %) composition was fabricated, and the structural, morphological, and elemental properties were characterized using Fourier transform infrared spectroscopy, X-ray diffraction analysis, scanning electron microscopy, elemental analysis, transmission electron microscopy, and solid-state nuclear magnetic resonance [156]. The particle size of the BGN is provided in Table 4.

Table 4. Particle size distribution of BGN synthesized using different fabrication techniques based on TEM image analysis. Reproduced from ref. [156].

	M1-P1	M1-P2	M2-P1	M2-P2 A	M2-P2 B	M2-P2 C
Particle size (nm)	438 ± 17	425 ± 17	86 ± 14	70 ± 13 193 ± 51 495 ± 12	18 ± 2	18 ± 5

TEM images revealed dense, monodispersed, and spherical particles of ~400 nm diameter (Table 4). M2-P2 A protocol yield an average particle size of ~70 nm (52%), ~193 nm (39%), and ~495 nm (9%), respectively, with decreasing H_2O concentration. As depicted in Table 4, stirring time changed the particle size of BGN to ~20 nm (M2-P2 B and M2-P2 C). The osteoconductive behavior of BGN was assessed via in vitro biomineralization studies. For method 1 (M1) BGNs, the vibration peaks that confirmed the presence of a calcium phosphate phase after immersion in SBF were significantly lower than the spectra of method 2 (M2) BGNs. BGNs fabricated using the M2 protocol exhibited half the network connectivity than M1 BGNs protocols due to higher calcium incorporation and underwent a faster bioactive response. The details of the fabrication process are provided in the associated reference [156].

A 3D porous bioactive glass–ceramic scaffold coated with HA-fatty acid was characterized for the first time by De Luca et al. [148] as bone tissue-engineering material. To measure the bioactivity, HA-fatty acid conjugates were deposited on the glass-derived scaffolds. The in vitro bioactivity tests showed a moderate increase in the solution pH from 7.40 to 7.56 after 1 week, which was associated with the ion release from the surface of the scaffold to the simulated body fluid (SBF). The scaffolds soaked for 48 h in SBF indicated the formation of a new phase composed of calcium phosphate globular agglomerates.

The coating of HA-fatty acid conjugate had no detrimental impact on the apatite-forming ability of the scaffolds [148]. These novel bioactive glass–ceramic scaffold biomaterials hold promise for bone repair due to their apatite-forming and tissue regenerative potential.

AuNPs/Au and Ag NPs-containing bioactive glass in alginate–pullulan composite (Alg-Pll-BGAuSP and Alg-Pll-BGAuNCs) were studied in vivo and in vitro on fibroblast and osteoblast cell lines for bone and tissue development, osteoconductivity, and bioactivity [157]. The composites exhibited a viability for the human fibroblast cell line (BJ CRL 2522TM) between 91 and 110% after 24 h. The highest viability of ~110% was obtained in presence of Alg-Pll-βTCP/HA, and Alg-Pll-BGAuSP composite (104%) had the most beneficial effects on bone proliferation [157]. Bioactive Si−Ca−P−Mo glass–ceramic nanoparticle (BBGN) with inlaid molybdate nanocrystals was synthesized for post-surgical melanoma therapy and skin reconstruction [158]. BBGN showed strong antioxidant activity, high photothermal qualities, low toxicity, and efficient infection therapy, and photothermal temperature could be achieved by changing the ratio of Mo^{4+} to Mo^{6+}. The antibacterial efficiency was studied against *Staphylococcus aureus* (*S. aureus*), *Escherichia coli* (*E. coli*), and MRSA in vitro. The three bacterial strains were not inhibited after contact with BBGN-Mo (B-M) for 12 h. The antibacterial effect for the control+ group was <10%. B-0 M with 10 min of 808 nm laser irradiation (B-0M+) demonstrated a 32.7% kill ratio for *S. aureus*, a 47.6% kill ratio for *E. coli*, and a 46.8% kill ratio for MRSA. About 98.9% of *S. aureus*, *E. coli*, and MRSA were killed after being cultured with B-15M+ groups. B-15 M glass–ceramic nanomaterials, due to their free d-electrons, presence of oxygen vacancies, and bioactive composition, demonstrated enhanced photothermal antioxidant, antibacterial, and anti-inflammatory properties, useful for post-surgical bone-infected wounds, inhibition of tumor recurrence, and tissue regeneration [158].

Khatami et al. reported the osteogenic properties of alginate-nano-silica hydrogels enriched with collagen and gelatin on human osteoblast-like MG-63 cells [159]. The osteogenic potential was measured by alkaline phosphatase activity. The addition of collagen and gelatin increased the cell proliferation rate. Biochemical analysis and Alizarin red S staining showed collagen-induced osteogenesis by induction of alkaline phosphatase and matrix mineralization. Silver-doped and gold-doped quaternary 46S6 bioactive glass (BG-Ag and BG-Au) was synthesized, and the cytotoxic effect was studied for potential bone regeneration application [160]. The BG was doped with 0.006 and 0.01 ppm concentrations of AgNPs (AuNPs) to yield BG-Ag1 (BG-Au1) and BG-Ag2 (BG-Au2) composites. Zeta potential and DLS measurements of BG and the nanocomposites showed a reduction in the hydrodynamic particle size of BG-Ag nanocomposites compared to pure BG (Table 5). BG-Ag2 had the lowest polydispersity index (PDI) of 0.262. An increase in zeta potential of the nanocomposites compared to BG was attributed to surface adhesion of Ag^{n+}. Upon incorporation of AuNPs, a decrease in particle size of BG and increase in zeta potential was attributed to the surface adhesion of Au^{n+}, which rendered the surface less negative in charge.

Table 5. Dynamic light scattering (DLS), polydispersity index (PDI), and zeta potential of bioglass and metal–bioglass nanocomposite materials. Reproduced from ref. [160].

Sample	DLS, H_D (nm)	PDI	Zeta Potential (mV)
BG	626 ± 70.51	0.899	−28.9 ± 7.64
BG-Ag1	33.16 ± 3.65	0.853	−21.7 ± 7.55
BG-Ag2	85.81 ± 12.77	0.262	−24.5 ± 4.69
BG-Au1	59.19 ± 10.03	0.789	−12.6 ± 4.85
BG-Au2	92.54 ± 13.80	0.789	−16.5 ± 5.81

The proliferation of osteosarcoma MG-63 cells with different concentrations of BG, BG-Ag1, BG-Ag2, BG-Au1, and BG-Au2 at 0, 2, 4, 6, 8, and 10 µg/mL dosage was considered. Cell proliferation increased even at low concentrations compared to the control group. At a dosage of 6 µg ml^{-1}, the cell proliferation was 141.25%. With increase in dosage to 8 and

10 μg ml^{-1}, the proliferation declined to 128.40% and 128.61%, respectively. This affirmed the biocompatibility of BG and metal-doped BG nanohybrids with MG63 cells [160].

Hydrogels based on mesoporous silica nanoparticles (MSNs), due to their porous homogeneous structure, biocompatibility, high surface area, and tunable core and shell chemistry, serve as ideal candidates in regenerative medicine, as they can mimic the ECM of native tissues [161]. A MSN-PEG nanocomposite hydrogel formed by dynamic covalent crosslinking of thiol surface-functionalized MSNs with thiol-modified polyethylene glycol was developed [161]. The mechanical and self-healing properties of these 3D networks were investigated via macroscopic self-healing and rheological recovery tests. A 7 wt. % of MSN concentration did not influence the self-healing behavior, and recovery of MSN7-PEG hydrogel decreased after applying 500% strain, primarily due to the higher MSN content. The degradation behavior of the hydrogels was measured in PBS buffer or GSH (glutathione). In the presence of GSH, which is a reducing agent, the disulfide bonds in the nanocomposite hydrogels were prone to degradation. MSN2-PEG showed ~80% loss of mass in presence of 300 μM GSH but no significant mass loss in PBS. However, MSN-PEG hydrogels exhibited slow degradation in 300 μM GSH in MSN-PEG nanocomposites, which was due to increased crosslinking density of the MSNs within the polymeric network.

3.4. Calcium Phosphates

Calcium phosphate (CaP)-based biomaterials are widely researched for cell attachment, proliferation, and differentiation in bone defect repair due to their resemblance to the chemical composition of natural minerals [162]. Various CaP-based biomaterials have been utilized in bone regeneration, including tricalcium phosphate, HAP, amorphous calcium phosphate, octacalcium phosphate, dicalcium phosphate anhydrous, dibasic calcium phosphate dihydrate, and tetracalcium phosphate [163]. HAP is the most thermally stable crystalline phase of CaP used in the form of ceramics or composite scaffolds with polymers [162]. Zhi et al. [164] synthesized a biomimetic and injectable hydrogel scaffold based on nano-HAP, collagen (Col), and chitosan (Chi) as workable systemic minimally invasive scaffolds in bone defect repair and bone tissue engineering. The results from SEM, TEM, and XPS suggested that Chi/HA/Col exhibited surface properties akin to physiological bone, and the peaks in the FT-IR spectrum corresponded to those found in natural bone. A mixture of palladium (Pd) and platinum (Pt) nanoparticles, called PAPLAL, was studied for aging-related skin pathologies in mice [165]. PAPLAL did not cause any morphological abnormalities, such as cell infiltration, deposition or cellular damage in mouse skin and exhibited potent antioxidant activity (attributed to the effects of Pt nanoparticles) and attenuated aging-related skin pathologies in vivo.

Non-woven multifunctional nano- and microfiber scaffolds were fabricated using a block copolymer of PCL and poly(lactic acid) (PLA) (PL-b-CL) as a matrix and HAP as a functional agent, dissolved in a binary solvent mixture to produce electrospun scaffolds with porous fibers [166]. The nanofibrous scaffolds were shown to mimic the ECM of bone. As an in vitro model of mature osteoblasts, SaOS-2, an osteosarcoma cell line, displayed osteoblastic capacity to deposit a mineralization-competent ECM. The cells grown on scaffolds with HAP exhibited higher metabolic activity than pristine PL-b-CL. Cell adhesion results suggested that the cells adhered and proliferated, with no substantial cell loss during the experimental period. Cell attachment was influenced by the physical and chemical properties of the scaffolds, namely, functional groups, solubility, crystallinity, electric charge, surface energy, surface roughness, topography, pore size, and stiffness. This was the first study to demonstrate that scaffolds created from a PCL and PLA block copolymer with embedded HAP NPs could enhance the potential of scaffolds for bone tissue engineering applications [166].

Chitosan (CS) hydrogel reinforced with hydroxyapatite nanorod (HAPNr) composite materials was developed for biomedical hydrogels and articular cartilage regeneration [167]. HAPNr incorporated into the CS matrix demonstrated improved mechanical properties, with the best results obtained for CS/1.5HAPNr composites. The cytotoxicity of the composite hydrogel towards L929 cell lines was measured in vitro. Phytohaemagglutinin (PHA)-treated composite had the highest cell viability of 112% and CS/1.5 HAPNr showed better cell proliferation (96 ± 0.41%) than pristine CS hydrogel (90 ± 0.63%) due to the release of Ca^{2+} and PO_4^{3-} ions from HAP that helped anchor CS hydrogel. L929 mouse fibroblast cells exhibited a fusiform morphology, and most cells spread and proliferated on the culture plate. The hydrogels exhibited no apparent cytotoxicity towards L929 cell lines, and the cells proliferated uniformly on the culture plate and exhibited a fusiform morphology. The CS/1.5HAPNr composite hydrogel may serve as a potential biomaterial for tissue engineering cartilage regeneration.

A biphasic cartilage and bone-like scaffold combined with HAP was developed for osteochondral tissue engineering [168]. Three types of self-assembled biphasic artificial cartilage–HAP conjugates from cartilage components, hyaluronic acid (HA), aggrecan (AG), and type II collagen were implanted in vitro into an osteochondral defect in rat knee joints, bone, and cartilage tissues (Figure 6A). The SEM micrographs revealed a fibrous 3D reticulate structure (Figure 6B). Incubation of the chondrocytes with AG/HA/collagen complex for 1 or 2 weeks showed the chondrocytes to be uniformly distributed on the scaffold. On the surface of the self-organized scaffolds, cytoplasmatic extensions denominated the phylopodia, hinting that the chondrocytes were alive after the 2-week incubation (Figure 6C).

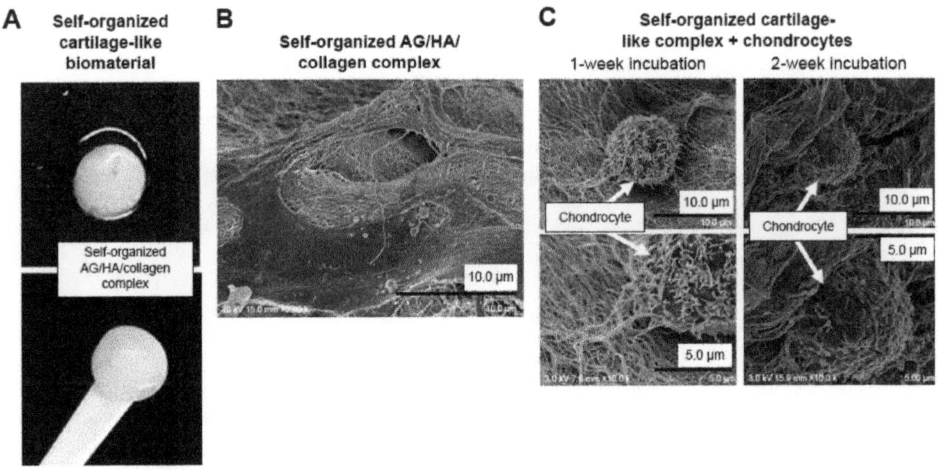

Figure 6. Self-organized cartilage-like scaffold formed from HA, AG, and type II collagen for articular cartilage tissue engineering. (**A**) Macroscopic image of self-organized cartilage-like scaffold. (**B**) Self-organized AG/HA/collagen complex as observed by SEM. (**C**) After 1- and 2-week incubation of chondrocytes with the self-organized AG/HA/collagen complex, chondrocytes were present on the scaffold of fibers forming the complex. Reprinted with permission from ref. [168]. This work is published and licensed by Dove Medical Press Limited. The full terms of this license are available at https://www.dovepress.com/terms.php and incorporate the Creative Commons Attribution–Non-Commercial (unported, v3.0) License (http://creativecommons.org/licenses/by-nc/3.0/, accessed on 1 November 2024).

To investigate the role of scaffold implantation in knee cartilage defects, in the control right knee joints, cartilage defects were filled with scar tissue and clefts were present between the scar and healthy surrounding tissue. In the left knee joints, cartilage defects were filled with repair tissue similar to the surrounding cartilage tissue. In knee joints without implant (sham-operated), fibrous scar tissue with hyperplasia of spindle-shaped fibroblastic cells was observed with no regeneration of cartilage tissue in 4 or 8 weeks. In left knee joints with implanted scaffolds, a stratiform cartilage-like structure with the formation of cartilage cavities and the growth of chondrocytes was observed in the implantation hole, pointing to cartilage regeneration. The osteochondral tissue regeneration rate was 33% in the AG/HA/collagen group, 50% in AG/HA/collagen complex–HAP powder (40 nm size), 75% in AG/HA/collagen complex–HAP powder (5 μm size), and 100% in the AG/HA/collagen complex–HAP block conjugate-implanted group [168]. Cartilage-like scaffold without conjugated HAP was insufficient to regenerate damaged subchondral bone tissue. The self-organized cartilage-like biomaterial conjugated with HAP not only repaired articular cartilage defects but also the degenerated subchondral bone in osteochondral defects.

Nanocrystalline hydroxyapatite (nHAP) was utilized to demonstrate how the differentiation of hMSCs was affected by nHAP loading and induced by bone-like scaffolds [169]. The nHAP dispersed in polymeric composites at >22 wt. % loading induced osteoblast differentiation in vitro via the PiT1 and Erk1/2 signaling pathways. When nHAP-poly(thioketal urethane) nHAP-PTKUR scaffolds were implanted in two anatomic sites in rats, namely, the femoral diaphysis and femoral metaphysis, new cartilage was formed in the absence of nHAP at the diaphyseal but not at the metaphyseal sites. New bone formation increased with nHAP loading at metaphyseal but not diaphyseal sites. Thus, nHAP composite scaffolds induced osteogenic differentiation of adherent endogenous progenitor cells in a dose-dependent manner. nHAP-PTKUR scaffolds promoted intramembranous and endochondral ossification dependence on material formulation and anatomic sites, indicating the importance of these factors as key regulators in assessing overall outcomes and the efficacy of novel biomaterials for bone regeneration.

Huang et al. [170] developed a gelatin/HAP-based hydrogel material using microextrusion 3D bioprinting and enzymatic cross-linking to promote the proliferation of human umbilical cord blood-derived mesenchymal stem cells (hUCB-MSCs) and repair of knee cartilage defects. To produce bioink for 3D-printed gelatin scaffolds, 10% of gelatin with 5% volume weight of HAP provided the best gelation kinetics and rheological properties. Two types of scaffolds, namely, 10% of gelatin and 10% of gelatin with 5% volume weight of HAP, was used in the study. SEM images showed porosity rates of 85.26 ± 2.09% for the gelatin scaffold and 81.29 ± 2.05% for the gelatin/HAP scaffold. The cytotoxicity was measured by culturing hUCB-MSCs in the extracted solutions of the scaffolds (Figure 7A). The extract solutions supported a high cell viability of >75%, as shown in Figure 7B. A continuous increase in cell population confirmed that the extracted solutions of scaffold-facilitated cell proliferation and migration of hUCB-MSCs (Figure 7C). The 3D printing of gelatin scaffolds assisted by enzymatic cross-linking and HAP doping strengthened the ability of the gelatin scaffold to enhance chondrogenic differentiation of stem cells and regeneration of bone and cartilage.

Biocompatible nanocomposite materials based on chitosan, HAP, and riboflavin have shown high antioxidant activity and inhibitory effect on *Staphylococcus aureus* compared to pure chitosan, useful in cell adhesion and the tissue regeneration process [171]. Riboflavin (RF) increased the antioxidant activity of Cs/HAP/RF composite material compared to pure Cs and HAP–doped Cs and reduced the inhibitory effect on *Staphylococcus aureus*. MTT assay provided insights into the effect of different concentrations of RF on cell metabolic activity of mouse embryonic fibroblasts (NIH3T3) and human osteosarcoma (U-2OS) cell lines. Cell migration/mobility response of NIH3T3 cell lines were accessed at different concentrations of riboflavin for 48 h of exposure. The scratch closure of NIH3T3 cells was close to 100%, suggesting riboflavin to have no negative effect on cell migration; rather, it activated cell proliferation process [171].

Chitosan–collagen–HAP (Cs/Col/HAP) composites are bioactive, biocompatible, and multifunctional biomaterials that find application in tissue engineering [172]. SEM micrographs of the composite membrane showed the Cs membrane to align as fibers with pore sizes of 80 nm to 2 μm. Incorporation of collagen changed the morphology of the Cs membrane; at low collagen content, pleats were formed, and areas surrounded by fibers suggested that collagen regions were blended into the chitosan matrix. At high collagen content, a uniform morphology was observed, characteristic of polymer blends. The cell viability was measured by MTS for buccal fat pad (hBFP)-MSCs. The membranes with the highest collagen and/or HAP content showed the highest cell adhesion values for in vitro-cultured hBFP-MSC. The membranes with low collagen and/or low HAP content showed similar cell viability compared to pure chitosan membranes. The largest cell viability was obtained when both compounds were present in the chitosan membrane with the highest collagen and HAP (Cs/0.75 collagen/0.75 HAP) content, showing 75% higher viability than pure chitosan. This study proposed that membranes obtained by combining chitosan with collagen and HAP enhanced cell adhesion ability, compared to the chitosan membranes alone.

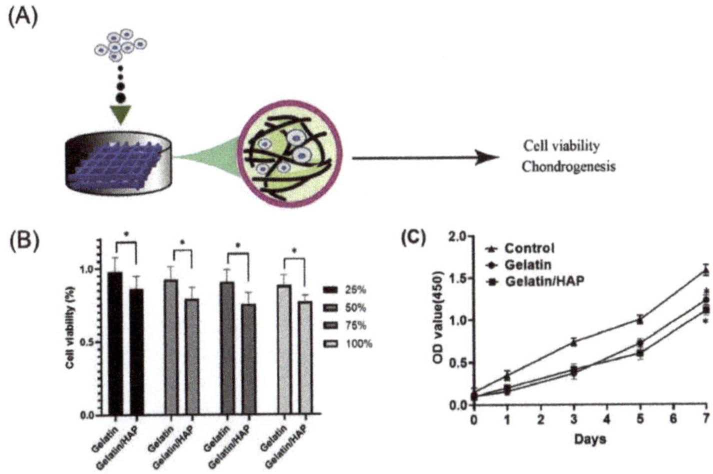

Figure 7. Proliferation and induction of hUCB-MSCs on scaffolds. (**A**) Schematic illustration of the experimental design of differentiation of human umbilical cord blood-derived mesenchymal stem cells (hUCB-MSCs). (**B**) Cytotoxicity measurement of the scaffold extraction solutions. *, $p < 0.05$. (**C**) The proliferation of hUCB-MSCs cells on the scaffold material shows a continuous increase in cell population. Reprinted with permission from ref. [170] from Royal Society of Chemistry.

A partially crystallized CaP and dicalcium phosphate anhydrate particle, integrated with black phosphorous nanosheets, was used to prepare calcium phosphate bone cements (CPCs) for potential application for bone regeneration [173]. The physicochemical, photothermal, biodegradability, cytotoxicity, and osteoinductive properties of CPCs were investigated to develop an innovative bone defect repair material. SEM images were analyzed to observe the adhesion of MC3T3-E1 cells on the surface of bone cement (BP/CPC) after co-culture for 1 and 24 h. At 1 h, cell morphology on the CPC surface was spherical, but on the 100 BP/CPC surface, the morphology was irregular. On the 200 BP/CPC surface, the morphology of cells was similar to that of 100 BP/CPC, and there were more pseudopodia. After 24 h of coculture, cells on the three bone cements were long and spindle-shaped, with the appearance of plate pseudopodia and filamentous pseudopodia. The BP/CPC successfully promoted cell adhesion in the early stages and demonstrated its biocompatibility. The addition of black phosphorus enhanced the strength of the bone cement. BP/CPC induced cell growth and showed very low to no toxicity for MC3T3-E1 cells. Experiments on osteoblast development showed that BP/CPC under near-infrared conditions accelerated the mineralization of bone cement and promoted the upregulation of osteogenic marker proteins in cells.

3.5. Magnesium Oxide Alloys

Magnesium (Mg) alloys can stimulate bone formation, increase new bone quality, promote the growth of bone tissues, and induce bone cell activation [174–176]. Degradable Mg alloys are of particular interest, as Mg can degrade under physiological conditions and eliminate the requirement for secondary surgical removal [127,177]. Mg-based implants are also efficient in stabilizing fractures and aiding healing process [178]. Liu et al. [175] studied the osteogenesis, angiogenesis, and antibacterial properties of Mg-Cu alloys, namely, Mg-0.05Cu, Mg-0.2Cu, and Mg-0.5Cu for orthopedic applications. Cell viability of pre-osteoblasts (MC3T3-E1) and endothelial cells (HUVECs) towards Mg-Cu alloys and pure Mg extraction media was studied for 24 h of incubation. Pre-osteoblast viability of Mg-0.03Cu and Mg-0.19Cu alloys were >100%, whereas Mg-0.57Cu had a slightly smaller value than pure Mg and other alloys. Increasing Cu concentrations in the Mg-Cu alloy decreased HUVECs viability from 125 to 75%. The results suggested that Mg-0.03Cu and Mg-0.19Cu stimulate the growth of MC3T3-E1 and HUVECs cells. Cell proliferation of MC3T3-E1 and HUVECs in the extraction media of Mg-Cu alloys and pure Mg was studied. On day 1, the control groups, pure Mg and its alloys did not show any difference in cell number in MC3T3-E1 cells. Mg-Cu alloys indicated less cell proliferation after 3 and 5 days. Cu depressed MC3T3-E1 cell proliferation. In HUVECs, Mg-0.03Cu showed the most cell proliferation after 1, 3, and 5 days, while Mg-0.57Cu exhibited the opposite response. Cell adhesion studies on cytoskeletons of MC3T3-E1 cells after 12 h of immersion in the extracts revealed spreading and superior filopodia extension for pure Mg and Mg-Cu alloys when compared to control. Compared to pure Mg and Mg-0.57Cu, MC3T3-E1 cells cultured in Mg-0.03Cu and Mg-0.19Cu extracts had more plump focal adhesion via well-organized F-actin stress fibers. The aggregation of HUVECs diminished with increasing Cu concentration, and cell status deteriorated slightly. This indicated that Mg-0.03Cu and Mg-0.19Cu extracts are favorable for initial attachment and spreading of HUVECs and MC3T3-E1 cells. The combined effects of high alkalinity and copper release from biodegradable Mg-Cu alloys, along with their enhanced mechanical and antibacterial properties compared to pure Mg, make these alloys suitable for stimulating osteogenesis and angiogenesis in orthopedic applications.

Saha et al. [127] reported the conformal deposition of MgO films on collagen using ALD technique at room temperature. This was done to improve the osteointegration and reinforce the mechanical strength and biocompatibility of collagen. Biocompatibility analysis of the membranes did not reveal any foreign body reactions in Col (control), Col-MgO200, and Col-MgO500 experimental samples. The Col-MgO200 and Col-MgO500 groups showed higher lymphocyte count compared to Col ($p < 0.05$), and coated groups showed similar results (Col-MgO200 = Col-MgO500; $p > 0.05$). Data from inflammatory profiles and blood cell counts suggested the following order: Col-MgO500 > Col-MgO200 > Blood clot—BC (negative control) > Col groups. BC had the lowest number of lymphocytes, while Col-MgO200 had the highest number (Col-MgO200 > Col-MgO500 = Col > BC) at 14 days, and higher inflammatory cells were found for Col-MgO200 group at 28 days as well. Histology characteristics of bone healing at 7-, 14-, and 28-day periods indicated that BC forms few newly formed bones until 28 days. Col-MgO200 and Col-MgO500 groups facilitated osteopromotive properties with increased osteoblastic activity, and Col-Mg200 and 500 showed good compatibility in the subcutaneous tissues. This study hinted at the need for pre-clinical assessments to confirm membrane compatibility and bone reconstruction.

Li et al. [176] reported the post-degradation effect of pure Mg, Mg-1Y, Mg-5Al, and Mg-2Ca alloys for the differentiation, proliferation, and gene expression of human mesenchymal stem cells (hMSCs). Mg^{2+} ions increased cell proliferation, and a maximum concentration of approximately 8.0×10^{-4} M was favorable for ATP production, above which ATP production decreased. Mg-2Ca alloy had minimal effect on osteogenic differentiation, with Mg-1Y and pure Mg having a superior effect on the proliferation and differentiation of hMSCs.

An immunomodulatory multicellular scaffold based on manganese silicate (MnS) nanoparticles with tendon/bone-related cells was fabricated for integrated regenerative recovery of tendon-to-bone injury [179]. The MnS nanoparticles were shown to enhance the specific differentiation of multicellular scaffolds via regulating macrophages. This was attributed to the secretion of PGE2 in macrophages induced by Mn ions. The tendon stem/progenitor cells (TSPCs) and BMSCs were distributed in a layered manner and the multicellular MnS-based nanoparticle scaffolds were implanted into macrophage-depleted rats to reveal the role of immunomodulatory processes in the differentiation of scaffolds. The Mn ions stimulated macrophages to secrete more PGE2, and a notable enhancement in the secretion of PGE2 in macrophages suggested that Mn ions could induce the secretion of PGE2 in macrophages to enhance the tenogenic differentiation of TSPCs and osteogenic differentiation of BMSCs.

3.6. Gold Nanoparticles

Gold nanoparticles (AuNPs) proffer tunable size, surface functionalization, unique physiochemical properties, and surface plasmon resonance with light at a wavelength of 520 nm [180], making it a suitable entrant in biomedicine, drug delivery, and imaging [181, 182]. Using coarse-grained molecular dynamics, Gupta and Rai [183] provided molecular level insights on the permeation free energy of dodecanethiol-coated neutral hydrophobic AuNPs of 2–5 nm dimension via a model skin lipid membrane (Figure 8). Charged AuNPs were adsorbed on the bilayer headgroup, while neutral hydrophobic AuNPs disrupted the bilayer and penetrated inside. AuNPs of 2 nm did not cause any structural changes in the bilayer. However, larger AuNPs, ranging from 3 to 5 nm, significantly altered the structure and packing of the bilayer. Some lipids from the bottom leaflet formed a hump-like structure to accommodate these bigger nanoparticles. The permeability was maximum for neutral 2 nm AuNP and minimum for 3 nm cationic AuNP. AuNPs created vacancies in the bilayer interior and induced undulation that increased with nanoparticle size. The permeation-free energy calculation suggested a very small barrier for neutral hydrophobic AuNP at the head group of the bilayer, while a free energy minimum was observed for charged AuNPs. Neutral hydrophobic AuNPs created more disruptions in the bilayer, which reduced the local and overall order parameter of the bilayers. The smaller neutral hydrophobic AuNP was more permeable, and charged AuNPs experienced resistance to permeation in the bilayer interior.

Lin et al. [181] investigated the biocompatibility and cell viability of AuNPs towards MSCs by analyzing the optimal AuNP concentration for tissue engineering efficacy. The % cell viability of MSCs at different AuNP concentrations of 1.25, 2.5, 5, and 10 ppm were investigated at 24, 48, and 72 h, respectively (Figure 9A). AuNPs at 1.25 and 2.5 ppm stimulated the lowest intracellular ROS generation, while AuNP at 10 ppm induced the highest production of ROS at time points of 24 and 48 h (Figure 9B,C). Fluorescence images of MSC cell adhesion stimulated by the AuNP solutions showed MSCs' morphology to be filopodia and lamellipodia in both the 1.25 and 2.5 ppm AuNPs, demonstrating improved cell adhesion and migration capacity. AuNPs at concentrations of 1.25 and 2.5 ppm demonstrated better biocompatibility. The AuNPs fabricated with polymeric biomaterials exhibited superior biosafety and reduced toxicity beneficial in clinical applications.

AuNPs enhanced the differentiation of osteoblastic MC3T3-E1 cells and antimycin A-induced mitochondrial dysfunction, as demonstrated by an MTT assay [184]. The MC3T3-E1 pre-osteoblastic cell line is widely used as a cell culture model for osteoblast differentiation. A significant increase in cell growth was revealed by the treatment of 5, 10, and 20 nm AuNPs at the concentrations of 1~10, 2~10, and 1~10 μg/mL, respectively, which suggested that AuNPs did not induce any cytotoxicity and stimulated osteoblast growth. AuNPs with a diameter of 10 nm considerably promoted osteogenesis through activating earlier osteogenic differentiation. Thus, AuNP increased the bone matrix by stimulating ALP activity, collagen synthesis, and osteocalcin synthesis in osteoblastic MC3T3-E1 cells. AuNPs (a) had a protective effect against oxidative stress-induced dysfunction in osteoblasts, (b) had a stimulating effect on the early differentiation stages of osteoblasts, and (c) protected osteoblasts from antimycin A-induced cell damage.

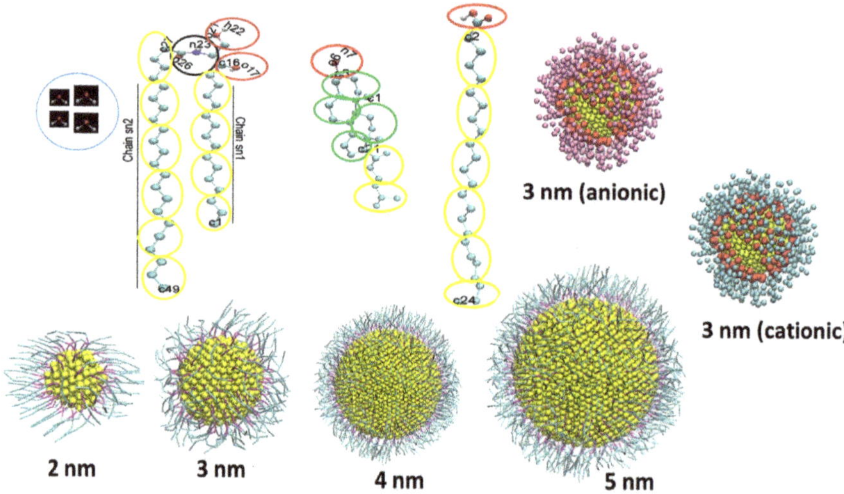

Figure 8. Coarse-grained molecular structures of long-chain ceramides, cholesterol, and fatty acids of skin lipid matrix, water, and thiol-coated AuNPs used in the simulations. Reproduced from ref. [183]. This work is licensed under a Creative Commons Attribution 4.0 International License, http://creativecommons.org/licenses/by/4.0/ (accessed on 1 November 2024).

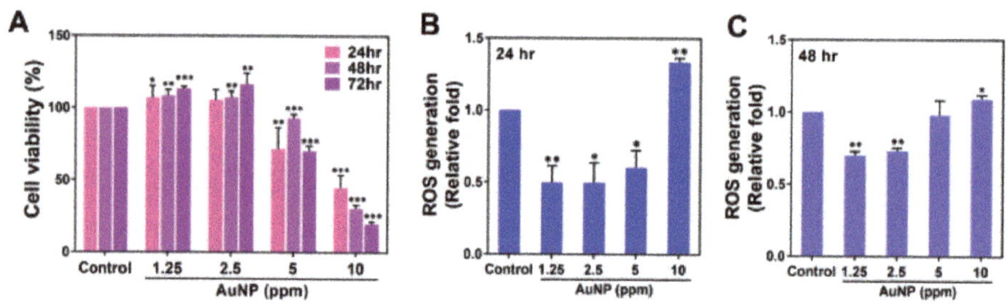

Figure 9. The cell viability and ROS generation of MSCs with different concentrations of AuNPs. (**A**) The cell viability of MSCs treated with AuNPs indicated that AuNP 1.25 and 2.5 ppm treatments enhanced MSC proliferation at 48 and 72 h, while the AuNP 5 and 10 ppm treatments had significantly lower cell viability. (**B**,**C**) Intracellular ROS generation in MSCs was detected using the FACS method after 24 and 48 h of treatment. AuNP 1.25 and 2.5 ppm induced lower ROS production at both time points. The results are displayed as mean ± SD (n = 3). * $p < 0.05$, ** $p < 0.01$, *** $p < 0.001$: compared to the control group (TCPS). Reproduced from ref. [181]. Licensee MDPI, Basel, Switzerland. This article is an open access article distributed under the terms and conditions of the Creative Commons Attribution (CC BY) license, https://creativecommons.org/licenses/by/4.0/ (accessed on 1 November 2024).

In soft tissue augmentation, AuNP-collagen constructs (AuNP-CC) were applied as soft tissue fillers to investigate the biodegradability and local tissue reaction of AuNP-collagen fillers compared to commercially available HA and cross-linked collagen fillers [185]. Two conjugates were constructed: AuNP-CC1, fabricated with the addition of 10% of pre-polymerized porcine Type I collagen; and AuNP-CC2, fabricated with the addition of 3% glycerin. The tissue response (neovascularization and granulation tissue) scores for the cohorts were investigated at 1, 3, and 6 months. Cross-linked collagen, AuNP-CC1, and AuNP-CC2 demonstrated slight to mild neovascularization and scored in the mild to moderate range for granulation tissue. The AuNP-CC formulations scored in the slight to mild range for granulation tissue. By 6 months, the cross-linked collagen demonstrated mild to moderate neovascularization and was in the mild range for granulation tissue. The AuNP-CC formulations scored slight to mild in each category. Cross-linked collagen showed increased tissue response scores over time. AuNP-CC2 had a similar response with increasing tissue response scores; however, the irritancy scores were more stable over time. These responses indicated active resorption of the constructs along with active remodeling [185].

The biomedical applications of AuNPs have been extended to wound-healing process [186,187]. Poomrattanangoon and Pissuwan [188] synthesized AuNPs coated with Collagen-I (Collagen-I@AuNPs) and studied the efficacy towards human skin fibroblast (HSF) cells for wound closure, cell proliferation, pro-inflammatory responses, growth factor generation, and in vitro wound healing. AuNPs and Collagen-I@AuNPs at 3, 5, 10, 15, and 25 $\mu g\ mL^{-1}$ concentrations were applied to HSF cells, and cell viability was measured for 24 h. Attenuated total reflectance (ATR)-FTIR measurements confirmed a change in the structure of Collagen-I and AuNPs resulting from the interaction between Collagen-I and AuNPs. For HSF cells treated with Collagen-I@AuNPs, cell viability was slightly higher than cells treated with AuNPs (Figure 10). The % cell viabilities of HSF cells after treating with 3, 5, 10, 15, and 25 $\mu g\ mL^{-1}$ of Collagen-I@AuNPs were 100.38 ± 3.51, 101.87 ± 2.40, 101.43 ± 2.32, 97.18 ± 2.25, and $97.40 \pm 1.18\%$, respectively (Figure 10b). HSF cells treated with AuNPs at concentrations of 15 and 25 $\mu g\ mL^{-1}$ showed a significant reduction in cell viability (95.08 ± 0.68 and $96.18 \pm 0.56\%$) compared to control that suggested that AuNPs and Collagen-I@AuNPs are non-cytotoxic.

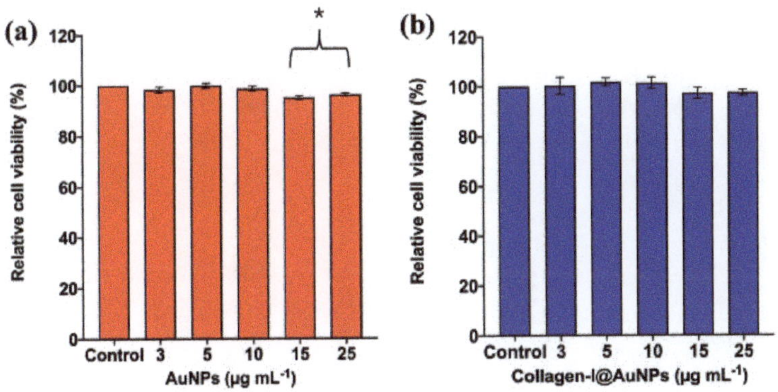

Figure 10. Relative cell viability of HSF cells treated with (**a**) AuNPs and (**b**) Collagen-I@AuNPs. The asterisk indicates a significant difference in cell viability compared to untreated HSF cells (control; $p < 0.05$). Statistical analysis was performed using Tukey_Kramer ($n \geq 7$). Reprinted with permission from ref. [188] from Elsevier. This article is under a Creative Commons license, http://creativecommons.org/licenses/by-nc-nd/4.0/ (accessed on 1 November 2024).

To study the effects of AuNPs and Collagen-I@AuNPs on wound closure, in vitro measurements using a wound scratch assay were performed [188]. The % wound closures of scratched HSF cells treated with 3 $\mu g\ mL^{-1}$ AuNPs and Collagen-I@AuNPs for 24, 48, and 72 h are shown in Figure 11a, and images of cell migration are shown in Figure 11b. The scratched HSF cells treated with 3 $\mu g\ mL^{-1}$ AuNPs and Collagen-I@AuNPs had a higher wound closure percentage than untreated control cells. After 24 h treatment of scratched HSF cells treated with AuNPs and Collagen-I@AuNPs, the percentages of wound closure were $\sim 57.16 \pm 1.31$ and $\sim 67.67 \pm 1.67\%$, respectively (Figure 11a). Collagen-I@AuNPs accelerated the healing process of scratched HSF cells compared to AuNPs. After 48 h, the percentage of wound closure in scratched HSF cells treated with AuNPs ($91.01 \pm 2.71\%$)

and Collagen-I@AuNPs (95.19 ± 1.67%) increased compared to 24 h (Figure 11a). The percentage of wound closure in untreated cells increased to 59.96 ± 2.66%. Complete closure in scratched HSF cells treated with AuNPs or Collagen-I@AuNPs was observed for 72 h of treatment. This implied that Collagen-I protein was vital in enhancing wound closure, and enhanced wound closure of scratched HSF cells treated with AuNPs may be attributed to the antioxidative properties of AuNPs vital in cell migration and healing [188].

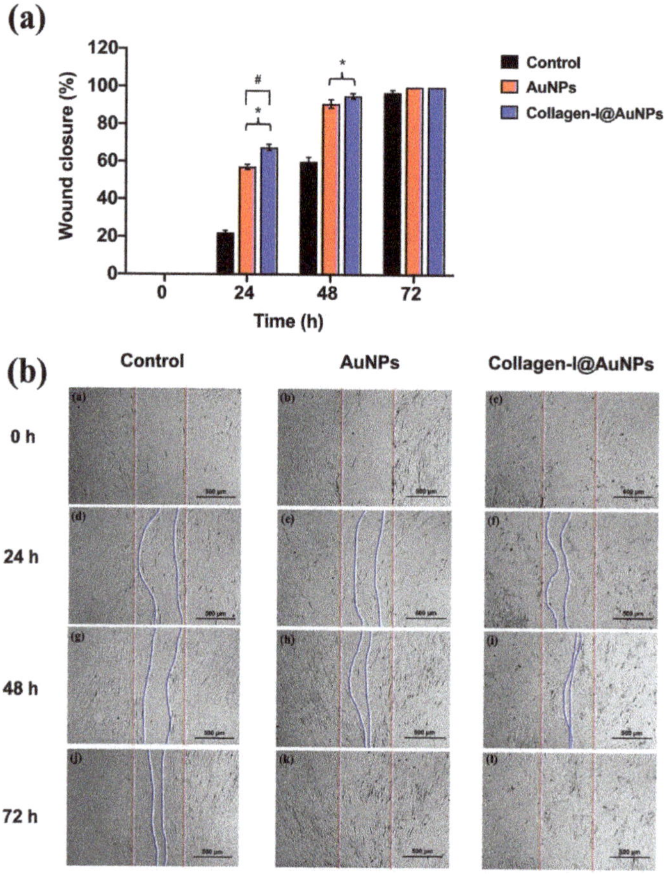

Figure 11. (**a**) Percentage of wound closure of scratched HSF cells treated with 3 μg mL^{-1} AuNPs and Collagen-I@AuNPs for 24, 48, and 72 h. * Significant difference in wound closure percentage at $p < 0.05$ compared with control cells (untreated HSF cells) at 24 and 48 h, respectively. # Significant difference in wound closure percentage of HSF cells treated with AuNPs and Collagen-I@AuNPs for 24 h, at $p < 0.05$. Statistical analysis was performed using the Tukey–Kramer test ($n \geq 11$). (**b**) Cell migration images of scratched HSF cells treated with 3 μg mL^{-1}AuNPs and Collagen-I@AuNPs for 24 h, 48 h, and 72 h. The panels (**a**–**l**) denote the cell migration images of control, AuNPs, and Collagen-I@AuNPs at 0, 24, 48, and 72 h, respectively. The figure and caption are reprinted with permission from ref. [188] from Elsevier. This article is under a Creative Commons license, http://creativecommons.org/licenses/by-nc-nd/4.0/ (accessed on 1 November 2024).

While the therapeutic and biomedical applications of AuNPs are well established, research has shown that the cytotoxicity of AuNPs is influenced by composition, size, shape, surface charge, and cell type [189]. The size of AuNPs was shown to impact ubiquitin proteasome system (UPS) such as the ubiquitin-specific protease (USP) and ubiquitin carboxyl-terminal hydrolase L1 (UCHL-1) and impact the proliferation of A549 cells [189]. A study on the size-dependent cytotoxicity of polyethyleneglycol (PEG)-coated gold nanoparticles (AuNPs) in mice found that particles sized between 10 and 60 nm had harmful effects, such as changes in cell shape, inhibition of cell proliferation, and DNA mutations [190]. In contrast, particles sized between 5 and 30 nm showed no signs of toxicity. Negatively charged (citrate-capped) and positively charged (cysteamine-capped) gold nanoparticles (AuNPs) induced cell death in a dose-dependent manner [191]. The negatively charged AuNPs caused a gradual process of cell death, while the positively charged AuNPs led to abrupt destruction of MDA-MB-231 cells. This rapid cell death was associated with increased phosphorylation of histone H3 at serine 10, indicating mitotic catastrophe [191]. Anti-proliferative activity and cytotoxicity of AuNPs on MCF7 human breast cancer cells was reported by Paramasivam et al. [192].

3.7. Biomimetic Natural Biomaterials

Biomimetic biomaterials are classified as scaffolding materials that provide the structural basis for cells to adhere, migrate, and proliferate [193]. They mimic the characteristics of the natural ECM and are currently used in various applications, including implantable materials, drug delivery systems, orthopedics, and musculoskeletal tissue engineering [193–195]. Liu et al. reviewed recent advances in the fabrication, functionality, and potential applications of biomimetic natural biomaterials [196]. Ma reviewed the importance and broad implications of current biomimetic materials and novel processing technologies in tissue engineering and regeneration [197]. Biomimetic natural biomaterials are combined with tissue engineering applications due to their non-toxicity, non-genotoxicity, and non-teratogenicity effects to healthy native tissue. Polylactic acid, polyhydroxyalkanoates and their derivatives, HA, alginate, cellulose, and chitosan, as well as collagen, gelatin, fibroin, poly-glutamic acids, and antimicrobial peptides, are the commonly used biomimetic natural biomaterials [194]. The development of biomimetic natural biomaterials that replicate the biological and physico-chemical characteristics of the natural ECM integrates additional functionality and therapeutic effects, as illustrated in Figure 12.

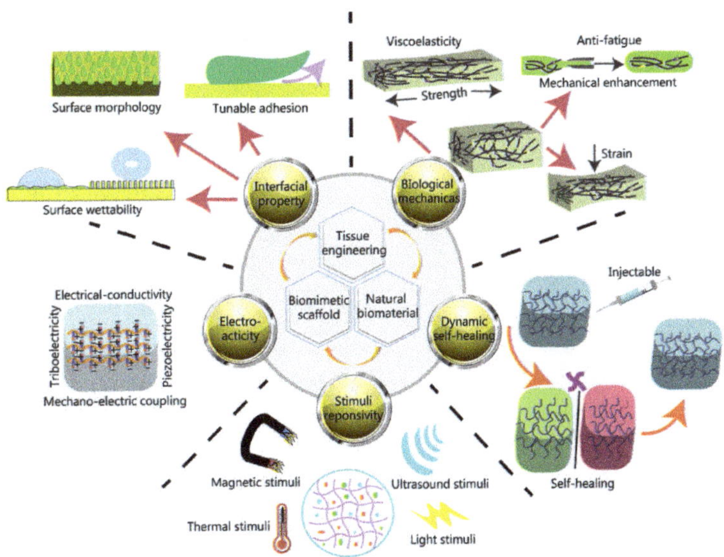

Figure 12. The innovative strategies involved in the rational design of biomimetic natural biomaterials that can mimic the ECM of tissues. Reproduced from ref. [196]. This article is licensed under a Creative Commons Attribution 4.0 International License, http://creativecommons.org/licenses/by/4.0/ (accessed on 1 November 2024).

4. Challenges in the Use of Inorganic Biomaterials in Regenerative Medicine

Clinical application of inorganic biomaterials in regenerative medicine confronts challenges, including, but not limited to, long-term fate in biocompatibility and bioaccumulation, nanoparticle/nanomaterial toxicity, better assessment tools, optimization of the mechanical properties, stem cell fate, and so on. The toxicity, carcinogenicity, bioaccumulation, and teratogenicity of nanoparticles are dose- and exposure-dependent [75]. Advances in materials science and engineering technology for the design and application of biomimetic 3D-printed degradable scaffolds have paved the way for pioneering research strategies in regenerative therapy [198]. For example, a biocompatible, high-performance semicrystalline polymer, poly (ether-ether-ketone) (PEEK), named bone cement, was one of the most prominent candidates in bone implants approved by the FDA [199,200]. Since PEEK is chemically inert and has poor integration, incorporating bioactive metals such as strontium or hydroxyapatite could stimulate cell differentiation [200,201]. Thus, the design and development of 'immune interactive' biomaterials can minimize immune response [202], such as mediating macrophage polarization, decreasing body rejection, and governing the outcome of tissue engineering and regenerative therapy [200].

Lele et al. [203] performed a detailed study of engineered biomaterials for different clinical trials and specific diseases. Ophthalmic disease had the highest number of clinical trials, followed by diseases of blood vessels and the oral cavity (Figure 13). Clinical trials suggested that synthetic polymers can be used in the fabrication of contact lenses in the treatment of ophthalmic diseases, while natural polymers can be used in the treatment of skin diseases. Dental applications utilized autologous biomaterials, ceramics, and composites. Metals and combination materials, such as polymers coated on metals, are ubiquitous for treating coronary artery disease.

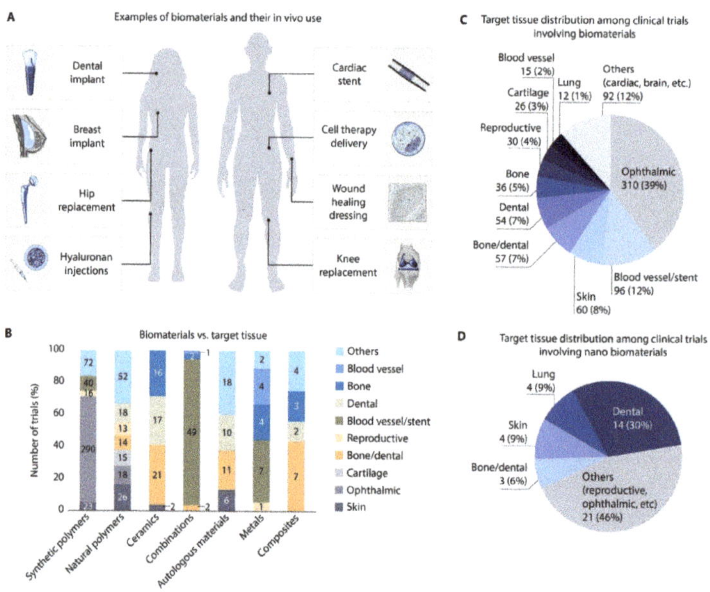

Figure 13. Clinical trials involving biomaterials in diseases and various target tissues. (**A**) The in vivo application of biomaterial in diseases and target tissues. The distribution of clinical trials involving biomaterials in (**B**) disease categories and (**C**) various target tissues. (**D**) The distribution of clinical trials involving nanobiomaterials in different target tissues. Reproduced from ref. [203]. This is an open-access article distributed under the terms of the Creative Commons Attribution-NonCommercial license, https://creativecommons.org/licenses/by-nc/4.0/ (accessed on 1 November 2024).

Tissue-engineered constructs have been successfully utilized in the repair of skin, bladder, bones, and cartilage [7]. Engineered tissue constructs can be assembled from primary cells or specific tissue culture lines: fibroblasts, endothelial cells, or cardiac muscle [204]. However, tissue constructs may not be able to seamlessly mimic the biological tissue due to differences in structure, organization, and lower cell density [204]. This gives rise to functional differences. Tissue engineering constructs also require a continued blood supply for long-term stability following implantation [205]. Thus, a continuing goal is to develop engineered tissue constructs with mechanical, structural, and functional properties that mimic biological tissue function and provide cues to diverse intercellular interactions.

Chronic wound healing is another area of regenerative medicine, currently undergoing clinical trials. Advanced wound care approaches, such as stem cells, biomaterials, and innovative treatment techniques are integrated to promote tissue regeneration and healing. However, there have been limited clinical case studies for bone tissue engineering using scaffolds and osteogenic cells, pointing to the slow advances in clinical tissue engineering [9,206]. Tissue rejection, challenges in surgical reconstruction, and lack of donor tissues are some of the major drawbacks that may impair the integration of inorganic biomaterials with natural tissues or organs to drive cell differentiation. Further, scalability, cost, regulatory issues, and uptake are other factors that largely limit the clinical translation of complex structures [207,208]. Also, ethical concerns concerning the utilization of tissue cells and human embryonic stem cells cannot be sidelined. Thus, clinical translation of regenerative medicine requires the development of appropriate guidelines for the safe and effective delivery of regenerative medicine strategies. The involvement of start-up corporations and enterprises is essential for streamlining the development of innovative therapies, and the speedy and effective delivery of regenerative medicine to physicians and patients [209]. These advances are pivotal in disseminating cutting-edge research and facilitating clinical applications. By bridging the gap between laboratory research and practical implementation, we can ensure that advancements in regenerative medicine reach those in need more swiftly and effectively.

5. Conclusions

Regenerative medicine and tissue engineering integrate engineering, materials science, clinical translation, stem cell science, and biology to promote the regeneration of diseased tissues and the formation of new tissues and organ systems. Advances in regenerative engineering have significantly improved the management of traumatic patients requiring organ replacement and/or repair, while also promoting the development of biological substitutes for diseased or damaged tissues. These innovations offer renewed hope for augmenting patient recovery and long-term outcomes in severe injury, trauma, or chronic situations. The review offers a detailed overview of the recent advancements, developments, and applications of inorganic nanoparticles and biomaterials in tissue engineering and regenerative therapy. At the intersection of biomaterial science and nanotechnology, inorganic biomaterials are extensively being studied for diagnostic, imaging, therapeutic delivery applications, and regenerative medicine. Development of inorganic biomaterials to stimulate new cells and regenerative tissue involves a detailed understanding of (1) fabrication techniques, (2) development of composition, morphology, and functionality, and (3) optimizing the mechanical and bioactive properties to obtain the desired function. Technical advances have facilitated the fabrication of engineered tissue constructs and tailor-made grafts that mimic natural tissue. Given the complexity of tissue engineering procedures, interdisciplinary collaboration between engineers, scientists, biologists, and clinical specialists is much sought to guide novel technological advancements in regenerative medicine. This review also discusses the challenges and opportunities of combining regenerative medicine strategies with inorganic-based nanoparticles and biomaterials. The growing application of synergistic strategies in regenerative medicine offers a promising opportunity to integrate advanced techniques, applications, and prospects for use in clinical trials and modern healthcare.

Funding: This research received no external funding.

Conflicts of Interest: The author has no conflicts of interest.

References

1. Mao, A.S.; Mooney, D.J. Regenerative Medicine: Current Therapies and Future Directions. *Proc. Natl. Acad. Sci. USA* **2015**, *112*, 14452. [CrossRef] [PubMed]
2. Mason, C.; Dunnill, P. A Brief Definition of Regenerative Medicine. *Regen. Med.* **2008**, *3*, 1. [CrossRef] [PubMed]
3. Jin, Y.; Li, S.; Yu, Q.; Chen, T.; Liu, D. Application of Stem Cells in Regeneration Medicine. *MedComm* **2023**, *4*, e291. [CrossRef] [PubMed]
4. McKinley, K.L.; Longaker, M.T.; Naik, S. Emerging Frontiers in Regenerative Medicine. *Science* **2023**, *380*, 796. [CrossRef]
5. Bianchi, E.; Vigani, B.; Viseras, C.; Ferrari, F.; Rossi, S.; Sandri, G. Inorganic Nanomaterials in Tissue Engineering. *Pharmaceutics* **2022**, *14*, 1127. [CrossRef]
6. Salgado, A.J.; Oliveira, J.M.; Martins, A.; Teixeira, F.G.; Silva, N.A.; Neves, N.M.; Sousa, N.; Reis, R.L. Chapter One—Tissue Engineering and Regenerative Medicine: Past, Present, and Future. *Int. Rev. Neurobiol. Acad. Press* **2013**, *108*, 1. [CrossRef]
7. O'Brien, F.J. Biomaterials & Scaffolds For Tissue Engineering. *Mater. Today* **2011**, *14*, 88. [CrossRef]
8. Vishwakarma, A.; Sharpe, P.; Shi, S.; Ramalingam, M. Chapter 1—An Introduction to Stem Cell Biology and Tissue Engineering. In *Stem Cell Biology and Tissue Engineering in Dental Sciences*; Academic Press: Cambridge, MA, USA, 2015; Volume 1. [CrossRef]
9. Ikada, Y. Challenges in Tissue Engineering. *J. Royal Soc. Interface* **2006**, *3*, 589. [CrossRef]
10. Wang, Y.; Zhang, W.; Yao, Q. Copper-based Biomaterials for Bone and Cartilage Tissue Engineering. *J. Orthop. Translat.* **2021**, *29*, 60. [CrossRef]
11. Yun, Y.R.; Won, J.E.; Jeon, E.; Lee, S.; Kang, W.; Jo, H.; Jang, J.H.; Shin, U.S.; Kim, H.W. Fibroblast Growth Factors: Biology, Function, and Application For Tissue Regeneration. *J. Tissue Eng.* **2010**, 218142. [CrossRef]
12. Sheikh, Z.; Javaid, M.A.; Hamdan, N.; Hashmi, R. Bone Regeneration Using Bone Morphogenetic Proteins and Various Biomaterial Carriers. *Materials* **2015**, *8*, 1778. [CrossRef] [PubMed]
13. Katagiri, T.; Watabe, T. Bone Morphogenetic Proteins. *Cold Spring Harb. Perspect. Biol.* **2016**, *8*, a021899. [CrossRef] [PubMed]
14. Lademann, F.; Hofbauer, L.C.; Rauner, M. The Bone Morphogenetic Protein Pathway: The Osteoclastic Perspective. *Front. Cell Dev. Biol.* **2020**, *8*, 586031. [CrossRef] [PubMed]
15. Kulikauskas, M.R.; Shaka, X.; Bautch, V.L. The Versatility and Paradox of BMP Signaling in Endothelial Cell Behaviors and Blood Vessel Function. *Cell Mol. Life Sci.* **2022**, *79*, 77. [CrossRef]
16. Zakrzewski, W.; Dobrzyński, M.; Szymonowicz, M.; Rybak, Z. Stem Cells: Past, Present, and Future. *Stem Cell Res. Ther.* **2019**, *10*, 68. [CrossRef]
17. Kolios, G.; Moodley, Y. Introduction to Stem Cells and Regenerative Medicine. *Respiration* **2013**, *85*, 3. [CrossRef]
18. Abdal Dayem, A.M.; Saadeldin, I.M.; Zhang, S. Editorial: Crosslinking ROS signaling and stem cells. *Front. Cell Dev. Biol.* **2022**, *10*, 1101802. [CrossRef]
19. Amini, A.R.; Laurencin, C.T.; Nukavarapu, S.P. Bone Tissue Engineering: Recent Advances and Challenges. *Crit. Rev. Biomed. Eng.* **2012**, *40*, 363. [CrossRef]
20. Zhang, C.P.; Fu, X.B. Therapeutic Potential of Stem Cells in Skin Repair and Regeneration. *Chin. J. Traumatol.* **2008**, *11*, 209. [CrossRef]
21. Zhang, W.; Huang, X. Stem Cell-Based Drug Delivery Strategy For Skin Regeneration and Wound Healing: Potential Clinical Applications. *Inflamm. Regen.* **2023**, *43*, 33. [CrossRef]
22. Duscher, D.; Barrera, J.; Wong, V.W.; Maan, Z.N.; Whittam, A.J.; Januszyk, M.; Gurtner, G.C. Stem Cells in Wound Healing: The Future of Regenerative Medicine? A Mini-Review. *Gerontology* **2016**, *62*, 216. [CrossRef] [PubMed]
23. Hoang, D.M.; Pham, P.T.; Bach, T.Q.; Ngo, A.T.; Nguyen, Q.T.; Phan, T.T.; Nguyen, G.H.; Le, P.T.; Hoang, V.T.; Forsyth, N.R.; et al. Stem Cell-Based Therapy For Human Diseases. *Signal Transduct. Target. Ther.* **2022**, *7*, 272. [CrossRef] [PubMed]
24. Rajabzadeh, N.; Fathi, E.; Farahzadi, R. Stem Cell-Based Regenerative Medicine. *Stem Cell Investig.* **2019**, *6*, 19. [CrossRef] [PubMed]
25. Tabansky, I.; Stern, J.N.H. Basics of Stem Cell Biology as Applied to the Brain. In *Stem Cells in Neuroendocrinology*; Pfaff, D., Christen, Y., Eds.; Springer: Cham, Switzerland, 2016; pp. 11–24.
26. Harris, D.T.; Badowski, M.; Ahmad, N.; Gaballa, M.A. The potential of Cord Blood Stem Cells for use in Regenerative Medicine. *Expert Opin. Biol. Ther.* **2007**, *7*, 1311. [CrossRef] [PubMed]
27. Devi, S.; Bongale, A.M.; Tefera, M.A.; Dixit, P.; Bhanap, P. Fresh Umbilical Cord Blood-A Source of Multipotent Stem Cells, Collection, Banking, Cryopreservation, and Ethical Concerns. *Life* **2023**, *13*, 1794. [CrossRef]
28. Sanchez-Petitto, G.; Rezvani, K.; Daher, M.; Rafei, H.; Kebriaei, P.; Shpall, E.J.; Olson, A. Umbilical Cord Blood Transplantation: Connecting Its Origin to Its Future. *Stem Cells Transl. Med.* **2023**, *12*, 55. [CrossRef]
29. Barker, J.N.; Byam, C.E.; Kernan, N.A.; Lee, S.S.; Hawke, R.M.; Doshi, K.A.; Wells, D.S.; Heller, G.; Papadopoulos, E.B.; Scaradavou, A.; et al. Availability of Cord Blood Extends Allogeneic Hematopoietic Stem Cell Transplant Access to Racial and Ethnic Minorities. *Biol. Blood Marrow Transpl.* **2010**, *16*, 1541. [CrossRef]

30. Gragert, L.; Eapen, M.; Williams, E.; Freeman, J.; Spellman, S.; Baitty, R.; Hartzman, R.; Rizzo, J.D.; Horowitz, M.; Confer, D.; et al. HLA Match Likelihoods For Hematopoietic Stem-Cell Grafts in the U.S. Registry. *N. Engl. J. Med.* **2014**, *371*, 339. [CrossRef]
31. Weiss, M.L.; Troyer, D.L. Stem Cells in the Umbilical Cord. *Stem Cell Rev.* **2006**, *2*, 155. [CrossRef]
32. Kurtzberg, J.; Lyerly, A.D.; Sugarman, J. Untying the Gordian Knot: Policies, Practices, and Ethical Issues Related to Banking of Umbilical Cord Blood. *J. Clin. Investig.* **2005**, *115*, 2592. [CrossRef]
33. Hoover-Plow, J.; Gong, Y. Challenges For Heart Disease Stem Cell Therapy. *Vasc. Health Risk Manag.* **2012**, *8*, 99. [CrossRef] [PubMed]
34. Mohr, A.; Zwacka, R. The Future of Mesenchymal Stem Cell-Based Therapeutic Approaches For Cancer—From Cells to Ghosts. *Cancer Lett.* **2018**, *414*, 239. [CrossRef] [PubMed]
35. Chen, F.; Zhao, E.R.; Hableel, G.; Hu, T.; Kim, T.; Li, J.; Gonzalez-Pech, N.I.; Cheng, D.J.; Lemaster, J.E.; Xie, Y.; et al. Increasing the Efficacy of Stem Cell Therapy via Triple-Function Inorganic Nanoparticles. *ACS Nano* **2019**, *13*, 6605. [CrossRef]
36. Nourian Dehkordi, A.; Mirahmadi Babaheydari, F.; Chehelgerdi, M.; Raeisi Dehkordi, S. Skin Tissue Engineering: Wound Healing Based on Stem-Cell-Based Therapeutic Strategies. *Stem Cell Res. Ther.* **2019**, *10*, 111. [CrossRef] [PubMed]
37. Butler, K.L.; Goverman, J.; Ma, H.; Fischman, A.; Yu, Y.M.; Bilodeau, M.; Rad, A.M.; Bonab, A.A.; Tompkins, R.G.; Fagan, S.P. Stem Cells and Burns: Review and Therapeutic Implications. *J. Burn Care Res.* **2010**, *31*, 874. [CrossRef] [PubMed]
38. Chu, G.Y.; Chen, Y.F.; Chen, H.Y.; Chan, M.H.; Gau, C.S.; Weng, S.M. Stem Cell Therapy on Skin: Mechanisms, Recent Advances and Drug Reviewing Issues. *J. Food Drug Anal.* **2018**, *26*, 14. [CrossRef]
39. Ojeh, N.; Pastar, I.; Tomic-Canic, M.; Stojadinovic, O. Cells in Skin Regeneration, Wound Healing, and Their Clinical Applications. *Int. J. Mol. Sci.* **2015**, *16*, 25476. [CrossRef]
40. Díaz-García, D.; Filipová, A.; Garza-Veloz, I.; Martinez-Fierro, M.L. A Beginner's Introduction to Skin Stem Cells and Wound Healing. *Int. J. Mol. Sci.* **2021**, *22*, 11030. [CrossRef]
41. Gurtner, G.C.; Werner, S.; Barrandon, Y.; Longaker, M.T. Wound Repair and Regeneration. *Nature* **2008**, *453*, 314. [CrossRef]
42. Clevers, H. Wnt/Beta-Catenin Signaling in Development and Disease. *Cell* **2006**, *127*, 469. [CrossRef]
43. Logan, C.Y.; Nusse, R. The Wnt Signaling Pathway in Development and Disease. *Annu. Rev. Cell Dev. Biol.* **2004**, *20*, 781. [CrossRef] [PubMed]
44. Holland, J.D.; Klaus, A.; Garratt, A.N.; Birchmeier, W. Wnt Signaling in Stem and Cancer Stem Cells. *Curr. Opin. Cell Biol.* **2013**, *25*, 254. [CrossRef] [PubMed]
45. Tottoli, E.M.; Dorati, R.; Genta, I.; Chiesa, E.; Pisani, S.; Conti, B. Skin Wound Healing Process and New Emerging Technologies for Skin Wound Care and Regeneration. *Pharmaceutics* **2020**, *12*, 735. [CrossRef] [PubMed]
46. Khaled, M.M.; Ibrahium, A.M.; Abdelgalil, A.I.; El-Saied, M.A.; El-Bably, S.H. Regenerative Strategies in Treatment of Peripheral Nerve Injuries in Different Animal Models. *Tissue Eng. Regen. Med.* **2023**, *20*, 839. [CrossRef]
47. Frausin, S.; Viventi, S.; Verga Falzacappa, L.; Quattromani, M.J.; Leanza, G.; Tommasini, A.; Valencic, E. Wharton's Jelly Derived Mesenchymal Stromal Cells: Biological Properties, Induction of Neuronal Phenotype and Current Applications in Neurodegeneration Research. *Acta Histochem.* **2015**, *117*, 329. [CrossRef]
48. Seyed-Forootan, K.; Karimi, H.; Jafarian, A.-A.; Seyed-Forootan, N.-S.; Ravari, F.K.; Karimi, A.-M. Nerve Regeneration and Stem Cells. *Biomed. J. Sci.Tech. Res.* **2019**, *18*, 13540. [CrossRef]
49. Pan, M.; Wang, X.; Chen, Y.; Cao, S.; Wen, J.; Wu, G.; Li, Y.; Li, L.; Qian, C.; Qin, Z.; et al. Tissue Engineering With Peripheral Blood-Derived Mesenchymal Stem Cells Promotes the Regeneration of Injured Peripheral Nerves. *Exp. Neurol.* **2017**, *292*, 92. [CrossRef]
50. Zhang, R.C.; Du, W.Q.; Zhang, J.Y.; Yu, S.X.; Lu, F.Z.; Ding, H.M.; Cheng, Y.B.; Ren, C.; Geng, D.Q. Mesenchymal Stem Cell Treatment For Peripheral Nerve Injury: A Narrative Review. *Neural Regen. Res.* **2021**, *16*, 2170. [CrossRef]
51. Yi, S.; Zhang, Y.; Gu, X.; Huang, L.; Zhang, K.; Qian, T.; Gu, X. Application of Stem Cells in Peripheral Nerve Regeneration. *Burns. Trauma* **2020**, *8*, tkaa002. [CrossRef]
52. Zeng, C.W.; Zhang, C.L. Neuronal Regeneration After Injury: A New Perspective on Gene Therapy. *Front. Neurosci.* **2023**, *17*, 1181816. [CrossRef]
53. Zang, X.; Song, J.; Li, Y.; Han, Y. Targeting Necroptosis as an Alternative Strategy in Tumor Treatment: From Drugs to Nanoparticles. *J. Control Release* **2022**, *349*, 213. [CrossRef] [PubMed]
54. Spirescu, V.A.; Chircov, C.; Grumezescu, A.M.; Vasile, B.Ș.; Andronescu, E. Inorganic Nanoparticles and Composite Films for Antimicrobial Therapies. *Int. J. Mol. Sci.* **2021**, *22*, 4595. [CrossRef] [PubMed]
55. Kim, D.; Kim, J.; Park, Y.I.; Lee, N.; Hyeon, T. Recent Development of Inorganic Nanoparticles for Biomedical Imaging. *ACS Cent. Sci.* **2018**, *4*, 324. [CrossRef] [PubMed]
56. Alshammari, B.H.; Lashin, M.M.; Mahmood, M.A.; Al-Mubaddel, F.S.; Ilyas, N.; Rahman, N.; Sohail, M.; Khan, A.; Abdullaev, S.S.; Khan, R. Organic and Inorganic Nanomaterials: Fabrication, Properties and Applications. *RSC Adv.* **2023**, *13*, 13735. [CrossRef]
57. Giljohann, D.A.; Seferos, D.S.; Daniel, W.L.; Massich, M.D.; Patel, P.C.; Mirkin, C.A. Gold Nanoparticles for Biology and Medicine. *Angew. Chem. Int. Ed. Engl.* **2010**, *49*, 3280. [CrossRef]
58. Ma, X.; Luan, Z.; Li, J. Inorganic Nanoparticles-Based Systems in Biomedical Applications of Stem Cells: Opportunities and Challenges. *Int. J. Nanomed.* **2023**, *18*, 143. [CrossRef]
59. Zheng, Y.; Wu, J.; Zhu, Y.; Wu, C. Inorganic-Based Biomaterials for Rapid Hemostasis and Wound Healing. *Chem. Sci.* **2022**, *14*, 29. [CrossRef]

60. Ma, J.; Wu, C. Bioactive Inorganic Particles-Based Biomaterials for Skin Tissue Engineering. *Exploration* **2022**, *2*, 20210083. [CrossRef]
61. Dalisson, B.; Barralet, J. Bioinorganics and Wound Healing. *Adv. Healthc. Mater.* **2019**, *8*, e1900764. [CrossRef]
62. Anselmo, A.C.; Mitragotri, S. A Review of Clinical Translation of Inorganic Nanoparticles. *AAPS J.* **2015**, *17*, 1041. [CrossRef]
63. Zheng, X.; Zhang, P.; Fu, Z.; Meng, S.; Dai, L.; Yang, H. Applications of Nanomaterials in Tissue Engineering. *RSC Adv.* **2021**, *11*, 19041. [CrossRef] [PubMed]
64. Brokesh, A.M.; Gaharwar, A.K. Inorganic Biomaterials for Regenerative Medicine. *ACS Appl. Mater. Interfaces* **2020**, *12*, 5319. [CrossRef] [PubMed]
65. Gao, Y.; Wang, X.; Fan, C. Advances in Graphene-Based 2D Materials for Tendon, Nerve, Bone/Cartilage Regeneration and Biomedicine. *iScience* **2024**, *27*, 110214. [CrossRef] [PubMed]
66. Arora, P.; Sindhu, A.; Dilbaghi, N.; Chaudhury, A.; Rajakumar, G.; Rahuman, A.A. Nano-Regenerative Medicine Towards Clinical Outcome of Stem Cell and Tissue Engineering in Humans. *J. Cell Mol. Med.* **2012**, *16*, 1991. [CrossRef]
67. Kim, Y.-H.; Vijayavenkataraman, S.; Cidonio, G. Biomaterials and Scaffolds for Tissue Engineering and Regenerative Medicine. *BMC Methods* **2024**, *1*, 2. [CrossRef]
68. Chan, B.P.; Leong, K.W. Scaffolding in Tissue Engineering: General Approaches and Tissue-Specific Considerations. *Eur. Spine J.* **2008**, *17* (Suppl. S4), 467. [CrossRef]
69. Krishani, M.; Shin, W.Y.; Suhaimi, H.; Sambudi, N.S. Development of Scaffolds from Bio-Based Natural Materials for Tissue Regeneration Applications: A Review. *Gels* **2023**, *9*, 100. [CrossRef]
70. Mitchell, M.J.; Billingsley, M.M.; Haley, R.M.; Wechsler, M.E.; Peppas, N.A.; Langer, R. Engineering Precision Nanoparticles for Drug Delivery. *Nat. Rev. Drug Discov.* **2021**, *20*, 101. [CrossRef]
71. Raju, G.S.R.; Dariya, B.; Mungamuri, S.K.; Chalikonda, G.; Kang, S.M.; Khan, I.N.; Sushma, P.S.; Nagaraju, G.P.; Pavitra, E.; Han, Y.K. Nanomaterials Multifunctional Behavior For Enlightened Cancer Therapeutics. *Semin. Cancer Biol.* **2021**, *69*, 178. [CrossRef]
72. Zheng, J.J.; Jiang, X.C.; Li, Y.S.; Gao, J.Q. Inorganic Nanoparticle-Integrated Mesenchymal Stem Cells: A Potential Biological Agent For Multifaceted Applications. *MedComm* **2023**, *4*, e313. [CrossRef]
73. Devin, J.E.; Attawia, M.A.; Laurencin, C.T. Three-Dimensional Degradable Porous Polymer-Ceramic Matrices For Use in Bone Repair. *J. Biomater. Sci. Polym. Ed.* **1996**, *7*, 661. [CrossRef] [PubMed]
74. Blokhuis, T.J.; Arts, J.J. Bioactive and Osteoinductive Bone Graft Substitutes: Definitions, Facts and Myths. *Injury* **2011**, *42*, S26. [CrossRef] [PubMed]
75. Hasan, A.; Morshed, M.; Memic, A.; Hassan, S.; Webster, T.J.; Marei, H.E. Nanoparticles in Tissue Engineering: Applications, Challenges and Prospects. *Int. J. Nanomed.* **2018**, *13*, 5637. [CrossRef] [PubMed]
76. Antony, S.A.; Anju, T.R.; Thomas, B. Nature-Inspired Biomimetic Polymeric Materials and Their Applications. In *Handbook of Biopolymers*; Thomas, S., AR, A., Jose Chirayil, C., Thomas, B., Eds.; Springer: Singapore, 2022. [CrossRef]
77. Kučuk, N.; Primožič, M.; Knez, Ž.; Leitgeb, M. Sustainable Biodegradable Biopolymer-Based Nanoparticles for Healthcare Applications. *Int. J. Mol. Sci.* **2023**, *24*, 3188. [CrossRef]
78. Park, S.; Rahaman, K.A.; Kim, Y.C.; Jeon, H.; Han, H.S. Fostering Tissue Engineering and Regenerative Medicine to Treat Musculoskeletal Disorders in Bone and Muscle. *Bioact. Mater.* **2024**, *40*, 345. [CrossRef]
79. Zhang, Q.; Xiao, L.; Xiao, Y. Porous Nanomaterials Targeting Autophagy in Bone Regeneration. *Pharmaceutics* **2021**, *13*, 1572. [CrossRef]
80. Malvindi, M.A.; De Matteis, V.; Galeone, A.; Brunetti, V.; Anyfantis, G.C.; Athanassiou, A.; Cingolani, R.; Pompa, P.P. Toxicity Assessment of Silica Coated Iron Oxide Nanoparticles and Biocompatibility Improvement by Surface Engineering. *PLoS ONE* **2014**, *9*, e85835. [CrossRef]
81. Di Bucchianico, S.; Cappellini, F.; Le Bihanic, F.; Zhang, Y.; Dreij, K.; Karlsson, H.L. Genotoxicity of TiO_2 Nanoparticles Assessed by Mini-Gel Comet Assay and Micronucleus Scoring with Flow Cytometry. *Mutagenesis* **2017**, *32*, 127. [CrossRef]
82. Santonastaso, M.; Mottola, F.; Colacurci, N.; Iovine, C.; Pacifico, S.; Cammarota, M.; Cesaroni, F.; Rocco, L. In vitro Genotoxic Effects of Titanium Dioxide Nanoparticles (n-TiO_2) in Human Sperm Cells. *Mol. Reprod. Dev.* **2019**, *86*, 1369. [CrossRef]
83. Chen, T.; Yan, J.; Li, Y. Genotoxicity of Titanium Dioxide Nanoparticles. *J. Food Drug Anal.* **2014**, *22*, 95. [CrossRef]
84. Manzoor, Q.; Sajid, A.; Ali, Z.; Nazir, A.; Sajid, A.; Imtiaz, F.; Iqbal, S.; Younas, U.; Arif, H.; Iqbal, M. Toxicity Spectrum and Detrimental Effects of Titanium Dioxide Nanoparticles as an Emerging Pollutant: A Review. *Desalination Water Treat.* **2024**, *317*, 100025. [CrossRef]
85. Saquib, Q.; Al-Khedhairy, A.A.; Siddiqui, M.A.; Abou-Tarboush, F.M.; Azam, A.; Musarrat, J. Titanium Dioxide Nanoparticles Induced Cytotoxicity, Oxidative Stress and DNA Damage in Human Amnion Epithelial (WISH) Cells. *Toxicol. Vitro* **2012**, *26*, 351. [CrossRef] [PubMed]
86. Fattori, A.C.; Brassolatti, P.; Feitosa, K.A.; Gonçalves, M.P.; Correia, R.D.; Albuquerque, Y.R.; Rodolpho, J.M.; Luna, G.L.; Bernardi, J.C.; Zucolotto, V.; et al. Titanium Dioxide Nanoparticle (TiO_2 NP) Induces Toxic Effects on LA-9 Mouse Fibroblast Cell Line. *Cell Physiol. Biochem.* **2023**, *57*, 63. [CrossRef] [PubMed]
87. Jin, C.Y.; Zhu, B.S.; Wang, X.F.; Lu, Q.H. Cytotoxicity of Titanium Dioxide Nanoparticles in Mouse Fibroblast Cells. *Chem. Res. Toxicol.* **2008**, *21*, 1871. [CrossRef]

88. Afşar, O.; Oltulu, Ç. Evaluation of the Cytotoxic Effect of Titanium Dioxide Nanoparticles in Human Embryonic Lung Cells. *Turk. J. Med. Sci.* **2023**, *53*, 1648. [CrossRef]
89. Gojznikar, J.; Zdravković, B.; Vidak, M.; Leskošek, B.; Ferk, P. TiO_2 Nanoparticles and Their Effects on Eukaryotic Cells: A Double-Edged Sword. *Int. J. Mol. Sci.* **2022**, *23*, 12353. [CrossRef]
90. Friggeri, G.; Moretti, I.; Amato, F.; Marrani, A.G.; Sciandra, F.; Colombarolli, S.G.; Vitali, A.; Viscuso, S.; Augello, A.; Cui, L.; et al. Multifunctional Scaffolds For Biomedical Applications: Crafting Versatile Solutions with Polycaprolactone Enriched by Graphene Oxide. *APL Bioengr.* **2024**, *8*, 016115. [CrossRef]
91. Samadian, H.; Khastar, H.; Ehterami, A.; Salehi, M. Bioengineered 3D Nanocomposite Based on Gold Nanoparticles and Gelatin Nanofibers for Bone Regeneration: In Vitro and In Vivo Study. *Sci. Rep.* **2021**, *11*, 13877. [CrossRef]
92. Samadian, H.; Zamiri, S.; Ehterami, A.; Farzamfar, S.; Vaez, A.; Khastar, H.; Alam, M.; Ai, A.; Derakhshankhah, H.; Allahyari, Z.; et al. Electrospun Cellulose Acetate/Gelatin Nanofibrous Wound Dressing Containing Berberine for Diabetic Foot Ulcer Healing: In Vitro and In Vivo Studies. *Sci. Rep.* **2020**, *10*, 8312. [CrossRef]
93. Gandhimathi, C.; Quek, Y.J.; Ezhilarasu, H.; Ramakrishna, S.; Bay, B.H.; Srinivasan, D.K. Osteogenic Differentiation of Mesenchymal Stem Cells with Silica-Coated Gold Nanoparticles for Bone Tissue Engineering. *Int. J. Mol. Sci.* **2019**, *20*, 5135. [CrossRef]
94. Wedel-Grzenda, A.; Fraczek-Szczypta, A.; Terrones, M.; Elias, A.L.; Lekka, M.; Menaszek, E.; Blazewicz, S. Polysulphone composite membranes modified with two types of carbon additives as a potential material for bone tissue regeneration. *Bull. Mater. Sci.* **2017**, *40*, 201. [CrossRef]
95. Sithole, M.N.; Kumar, P.; Du Toit, L.C.; Erlwanger, K.H.; Ubanako, P.N.; Choonara, Y.E. A 3D-Printed Biomaterial Scaffold Reinforced with Inorganic Fillers for Bone Tissue Engineering: In Vitro Assessment and In Vivo Animal Studies. *Int. J. Mol. Sci.* **2023**, *24*, 7611. [CrossRef] [PubMed]
96. Yang, L.; Rathnam, C.; Hou, Y.; Patel, M.; Cai, L.; Lee, K.-B. Rapid and Controllable Multilayer Cell Sheet Assembly via Biodegradable Nanochannel Membranes. *Adv. Funct. Mater.* **2024**, 2403367. [CrossRef]
97. Yang, L.; Chueng, S.D.; Li, Y.; Patel, M.; Rathnam, C.; Dey, G.; Wang, L.; Cai, L.; Lee, K.B. A Biodegradable Hybrid Inorganic Nanoscaffold for Advanced Stem Cell Therapy. *Nat. Commun.* **2018**, *9*, 3147. [CrossRef]
98. Baydar, S.Y.; Ay, H.F.; Cakir, R. Frontiers of Stem Cell Engineering For Nanotechnology-Mediated Drug Delivery Systems. *ADMET DMPK* **2024**, *12*, 225. [CrossRef]
99. Yang, Y.; Leong, K.W. Nanoscale Surfacing For Regenerative Medicine. *WIREs Nanomed. Nanobiotechnol.* **2010**, *2*, 478. [CrossRef]
100. Brovkina, O.; Dashinimaev, E. Advances and Complications of Regenerative Medicine in Diabetes Therapy. *PeerJ* **2020**, *8*, e9746. [CrossRef]
101. Li, D.; Liu, Y.; Wu, N. Application Progress of Nanotechnology in Regenerative Medicine of Diabetes Mellitus. *Diabetes Res. Clin. Pract.* **2022**, *190*, 109966. [CrossRef]
102. Andreadi, A.; Lodeserto, P.; Todaro, F.; Meloni, M.; Romano, M.; Minasi, A.; Bellia, A.; Lauro, D. Nanomedicine in the Treatment of Diabetes. *Int. J. Mol. Sci.* **2024**, *25*, 7028. [CrossRef]
103. Siwakoti, P.; Rennie, C.; Huang, Y.; Li, J.J.; Tuch, B.E.; McClements, L.; Xu, X. Challenges with Cell-based Therapies for Type 1 Diabetes Mellitus. *Stem Cell. Rev. Rep.* **2023**, *19*, 601. [CrossRef]
104. Yao, Y.; Ji, K.; Wang, Y.; Gu, Z.; Wang, J. Materials and Carriers Development for Glucose-Responsive Insulin. *Acc. Mater. Res.* **2022**, *3*, 960. [CrossRef]
105. Bakhru, S.H.; Furtado, S.; Morello, A.P.; Mathiowitz, E. Oral Delivery of Proteins by Biodegradable Nanoparticles. *Adv. Drug Deliv. Rev.* **2013**, *65*, 811. [CrossRef] [PubMed]
106. Rabanel, J.M.; Aoun, V.; Elkin, I.; Mokhtar, M.; Hildgen, P. Drug-Loaded Nanocarriers- Passive Targeting and Crossing of Biological Barriers. *Curr. Med. Chem.* **2012**, *19*, 3070. [CrossRef] [PubMed]
107. Veiseh, O.; Tang, B.C.; Whitehead, K.A.; Anderson, D.G.; Langer, R. Managing Diabetes with Nanomedicine: Challenges and Opportunities. *Nat. Rev. Drug Discov.* **2015**, *14*, 45. [CrossRef] [PubMed]
108. Sharma, R.; Borah, S.J.; Bhawna; Kumar, S.; Gupta, A.; Kumari, V.; Kumar, R.; Dubeyg, K.K.; Kumar, V. Emerging Trends in Nano-Based Antidiabetic Therapeutics: A Path to Effective Diabetes Management. *Mater. Adv.* **2023**, *4*, 3091. [CrossRef]
109. Liu, Y.; Zeng, S.; Ji, W.; Yao, H.; Lin, L.; Cui, H.; Santos, H.A.; Pan, G. Emerging Theranostic Nanomaterials in Diabetes and Its Complications. *Adv. Sci.* **2022**, *9*, e2102466. [CrossRef]
110. Sonomoto, K.; Yamaoka, K.; Kaneko, H.; Yamagata, K.; Sakata, K.; Zhang, X.; Kondo, M.; Zenke, Y.; Sabanai, K.; Nakayamada, S.; et al. Spontaneous Differentiation of Human Mesenchymal Stem Cells on Poly-Lactic-Co-Glycolic Acid Nano-Fiber Scaffold. *PLoS ONE* **2016**, *11*, e0153231. [CrossRef]
111. Umer, A.; Ghouri, M.D.; Muyizere, T.; Aqib, R.M.; Muhaymin, A.; Cai, R.; Chen, C. Engineered Nano-Bio Interfaces for Stem Cell Therapy. *Precis. Chem.* **2023**, *1*, 341. [CrossRef]
112. Teo, B.K.; Wong, S.T.; Lim, C.K.; Kung, T.Y.; Yap, C.H.; Ramagopal, Y.; Romer, L.H.; Yim, E.K. Nanotopography Modulates Mechanotransduction of Stem Cells and Induces Differentiation Through Focal Adhesion Kinase. *ACS Nano* **2013**, *7*, 4785. [CrossRef]
113. Ludin, A.; Gur-Cohen, S.; Golan, K.; Kaufmann, K.B.; Itkin, T.; Medaglia, C.; Lu, X.J.; Ledergor, G.; Kollet, O.; Lapidot, T. Reactive Oxygen Species Regulate Hematopoietic Stem Cell Self-Renewal, Migration and Development, as well as their Bone Marrow Microenvironment. *Antioxid. Redox Signal.* **2014**, *21*, 1605. [CrossRef]

114. Wang, C.; Liu, D.; Zhang, C.; Sun, J.; Feng, W.; Liang, X.J.; Wang, S.; Zhang, J. Defect-Related Luminescent Hydroxyapatite-Enhanced Osteogenic Differentiation of Bone Mesenchymal Stem Cells Via an ATP-Induced cAMP/PKA Pathway. *ACS Appl. Mater. Interfaces* **2016**, *8*, 11262. [CrossRef] [PubMed]
115. Przekora, A. Current Trends in Fabrication of Biomaterials for Bone and Cartilage Regeneration: Materials Modifications and Biophysical Stimulations. *Int. J. Mol. Sci.* **2019**, *20*, 435. [CrossRef]
116. Wu, Y.N.; Law, J.B.; He, A.Y.; Low, H.Y.; Hui, J.H.; Lim, C.T.; Yang, Z.; Lee, E.H. Substrate Topography Determines the Fate of Chondrogenesis from Human Mesenchymal Stem Cells Resulting in Specific Cartilage Phenotype Formation. *Nanomedicine.* **2014**, *10*, 1507. [CrossRef] [PubMed]
117. Wu, Y.N.; Yang, Z.; Hui, J.H.; Ouyang, H.W.; Lee, E.H. Cartilaginous ECM Component-Modification of the Micro-Bead Culture System For Chondrogenic Differentiation of Mesenchymal Stem Cells. *Biomaterials* **2007**, *28*, 4056. [CrossRef] [PubMed]
118. Varghese, S.; Hwang, N.S.; Canver, A.C.; Theprungsirikul, P.; Lin, D.W.; Elisseeff, J. Chondroitin Sulfate Based Niches For Chondrogenic Differentiation of Mesenchymal Stem Cells. *Matrix Biol.* **2008**, *27*, 12. [CrossRef] [PubMed]
119. Xu, H.; Akbari, M.K.; Kumar, S.; Verpoort, F.; Zhuiykov, S. Atomic layer deposition—State-of-the-art approach to nanoscale hetero-interfacial engineering of chemical sensors electrodes: A review. *Sens. Actuators B Chem.* **2021**, *331*, 129403. [CrossRef]
120. Liu, M.; Li, X.; Karuturi, S.K.; Tok, A.I.; Fan, H.J. Atomic Layer Deposition for Nanofabrication and Interface Engineering. *Nanoscale* **2012**, *4*, 1522–1528. [CrossRef]
121. Bechelany, M.; Balme, S.; Miele, P. Atomic Layer Deposition of Biobased Nanostructured Interfaces for Energy, Environmental and Health Applications. *Pure Appl. Chem.* **2015**, *87*, 751. [CrossRef]
122. Bishal, A.K.; Sukotjo, C.; Takoudis, C.G. Room Temperature TiO_2 Atomic Layer Deposition on Collagen Membrane from a Titanium Alkylamide Precursor. *J. Vacuum Sci. Technol. A.* **2017**, *35*, 01B134. [CrossRef]
123. Campoccia, D.; Montanaro, L.; Arciola, C.R. The Significance of Infection Related to Orthopedic Devices and Issues of Antibiotic Resistance. *Biomaterials* **2006**, *27*, 2331. [CrossRef]
124. Chung, C.J.; Long, H.Y. Systematic Strontium Substitution in Hydroxyapatite Coatings on Titanium via Micro-Arc Treatment and their Osteoblast/Osteoclast Responses. *Acta Biomater.* **2011**, *7*, 4081. [CrossRef] [PubMed]
125. Capek, J.; Sepúlveda, M.; Bacova, J.; Rodriguez-Pereira, J.; Zazpe, R.; Cicmancova, V.; Nyvltova, P.; Handl, J.; Knotek, P.; Baishya, K.; et al. Ultrathin TiO_2 Coatings via Atomic Layer Deposition Strongly Improve Cellular Interactions on Planar and Nanotubular Biomedical Ti Substrates. *ACS Appl. Mater. Interfaces* **2024**, *16*, 5627. [CrossRef] [PubMed]
126. Dafinone, M.I.; Feng, G.; Brugarolas, T.; Tettey, K.E.; Lee, D. Mechanical Reinforcement of Nanoparticle Thin Films Using Atomic Layer Deposition. *ACS Nano* **2011**, *5*, 5078. [CrossRef] [PubMed]
127. Saha, S.; Costa, R.C.; Silva, M.C.; Fonseca-Santos, J.M.; Chen, L.; Phakatkar, A.H.; Bhatia, H.; Faverani, L.P.; Barão, V.A.; Shokuhfar, T.; et al. Collagen Membrane Functionalized with Magnesium Oxide via Room-Temperature Atomic Layer Deposition Promotes Osteopromotive and Antimicrobial Properties. *Bioact. Mater.* **2023**, *30*, 46. [CrossRef]
128. Johnson, R.W.; Hultqvist, A.; Bent, S.F. A Brief Review of Atomic Layer Deposition: From Fundamentals to Applications. *Mater. Today* **2014**, *07*, 236–246. [CrossRef]
129. Narayan, R.J.; Adiga, S.P.; Pellin, M.J.; Curtiss, L.A.; Hryn, A.J.; Stafslien, S.; Chisholm, B.; Shih, C.C.; Shih, C.M.; Lin, S.J.; et al. Atomic Layer Deposition-Based Functionalization of Materials for Medical and Environmental Health Applications. *Philos. Trans. A Math Phys. Eng. Sci.* **2010**, *368*, 2033. [CrossRef]
130. Popov, A.M.; Lozovik, Y.E.; Fiorito, S.; Yahia, L. Biocompatibility and Applications of Carbon Nanotubes in Medical Nanorobots. *Int. J. Nanomed.* **2007**, *2*, 361.
131. Wang, J.; Huang, C.; Wang, Y.; Chen, Y.; Ding, Z.; Yang, C.; Chen, L. Exploration of the Single-Walled Carbon Nanotubes' Influence for Cartilage Repair. *Colloids Surf. A Physicochem. Eng. Asp.* **2020**, *606*, 125520. [CrossRef]
132. Szymański, T.; Semba, J.A.; Mieloch, A.A.; Cywoniuk, P.; Kempa, M.; Rybka, J.D. Hyaluronic Acid and Multiwalled Carbon Nanotubes as Bioink Additives for Cartilage Tissue Engineering. *Sci. Rep.* **2023**, *13*, 646. [CrossRef]
133. Yang, X.; Li, C.J.; Wan, Y.; Smith, P.; Shang, G.; Cui, Q. Antioxidative Fullerol Promotes Osteogenesis of Human Adipose-Derived Stem Cells. *Int. J. Nanomed.* **2014**, *9*, 4023. [CrossRef]
134. Elídóttir, K.L.; Scott, L.; Lewis, R.; Jurewicz, I. Biomimetic Approach to Articular Cartilage Tissue Engineering Using Carbon Nanotube-Coated and Textured Polydimethylsiloxane Scaffolds. *Ann. N. Y. Acad. Sci.* **2022**, *1513*, 48. [CrossRef] [PubMed]
135. Feng, Z.Q.; Yan, K.; Shi, C.; Xu, X.; Wang, T.; Li, R.; Dong, W.; Zheng, J. Neurogenic Differentiation of Adipose Derived Stem Cells on Graphene-Based Mat. *Mater. Sci. Eng. C Mater. Biol. Appl.* **2018**, *90*, 685. [CrossRef] [PubMed]
136. Rostami, F.; Tamjid, E.; Behmanesh, M. Drug-Eluting PCL/Graphene Oxide Nanocomposite Scaffolds For Enhanced Osteogenic Differentiation of Mesenchymal Stem Cells. *Mater. Sci. Eng. C Mater. Biol. Appl.* **2020**, *115*, 111102. [CrossRef]
137. Wei, S.; Zou, L.; Hu, X.; Li, Q. Carbon Nanomaterials in the Treatment of Infectious Bone Defects and Wound Scars after Wushu Fractures. *J. Chem.* **2020**, 2094273. [CrossRef]
138. Shen, Y.; Wu, L.; Qin, D.; Xia, Y.; Zhou, Z.; Zhang, X.; Wu, X. Carbon Black Suppresses the Osteogenesis of Mesenchymal Stem Cells: The Role of Mitochondria. *Part Fibre Toxicol.* **2018**, *15*, 16. [CrossRef]
139. Zheng, S.; Tian, Y.; Ouyang, J.; Shen, Y.; Wang, X.; Luan, J. Carbon Nanomaterials for Drug Delivery and Tissue Engineering. *Front. Chem.* **2022**, *10*, 990362. [CrossRef]
140. Holmannova, D.; Borsky, P.; Svadlakova, T.; Borska, L.; Fiala, Z. Carbon Nanoparticles and Their Biomedical Applications. *Appl. Sci.* **2022**, *12*, 7865. [CrossRef]

141. Li, L.; Zhu, Z.; Xiao, W.; Li, L. Multi-Walled Carbon Nanotubes Promote Cementoblast Differentiation and Mineralization through the TGF-β/Smad Signaling Pathway. *Int. J. Mol. Sci.* **2015**, *16*, 3188. [CrossRef]
142. Saleemi, M.A.; Hosseini Fouladi, M.; Yong, P.V.C.; Chinna, K.; Palanisamy, N.K.; Wong, E.H. Toxicity of Carbon Nanotubes: Molecular Mechanisms, Signaling Cascades, and Remedies in Biomedical Applications. *Chem. Res. Toxicol.* **2021**, *34*, 24. [CrossRef]
143. Liu, Y.; Zhao, Y.; Sun, B.; Chen, C. Understanding the Toxicity of Carbon Nanotubes. *Acc. Chem. Res.* **2013**, *46*, 702. [CrossRef]
144. Hu, W.; Peng, C.; Lv, M.; Li, X.; Zhang, Y.; Chen, N.; Fan, C.; Huang, Q. Protein Corona-Mediated Mitigation of Cytotoxicity of Graphene Oxide. *ACS Nano* **2011**, *5*, 3693. [CrossRef] [PubMed]
145. Eivazzadeh-Keihan, R.; Maleki, A.; de la Guardia, M.; Bani, M.S.; Chenab, K.K.; Pashazadeh-Panahi, P.; Baradaran, B.; Mokhtarzadeh, A.; Hamblin, M.R. Carbon Based Nanomaterials for Tissue Engineering of Bone: Building New Bone on Small Black Scaffolds: A Review. *J. Adv. Res.* **2019**, *18*, 185. [CrossRef] [PubMed]
146. Yuan, X.; Zhang, X.; Sun, L.; Wei, Y.; Wei, X. Cellular Toxicity and Immunological Effects of Carbon-based Nanomaterials. *Part Fibre Toxicol.* **2019**, *16*, 18. [CrossRef]
147. Peng, Z.; Liu, X.; Zhang, W.; Zeng, Z.; Liu, Z.; Zhang, C.; Liu, Y.; Shao, B.; Liang, Q.; Tang, W.; et al. Advances in the Application, Toxicity and Degradation of Carbon Nanomaterials in Environment: A Review. *Environ. Int.* **2020**, *134*, 105298. [CrossRef] [PubMed]
148. De Luca, S.; Verdoliva, V.; Kargozar, S.; Baino, F. Bioactive Glass-Ceramic Scaffolds Coated with Hyaluronic Acid-Fatty Acid Conjugates: A Feasibility Study. *J. Funct. Biomater.* **2023**, *14*, 26. [CrossRef]
149. Cannio, M.; Bellucci, D.; Roether, J.A.; Boccaccini, D.N.; Cannillo, V. Bioactive Glass Applications: A Literature Review of Human Clinical Trials. *Materials* **2021**, *14*, 5440. [CrossRef]
150. Sergi, R.; Bellucci, D.; Cannillo, V. A Review of Bioactive Glass/Natural Polymer Composites: State of the Art. *Materials* **2020**, *13*, 5560. [CrossRef]
151. El-Rashidy, A.A.; Roether, J.A.; Harhaus, L.; Kneser, U.; Boccaccini, A.R. Regenerating Bone with Bioactive Glass Scaffolds: A Review of in vivo Studies in Bone Defect Models. *Acta Biomater.* **2017**, *62*, 1–28. [CrossRef]
152. Fiume, E.; Barberi, J.; Verné, E.; Baino, F. Bioactive Glasses: From Parent 45S5 Composition to Scaffold-Assisted Tissue-Healing Therapies. *J. Funct. Biomater.* **2018**, *9*, 24. [CrossRef]
153. Hench, L.L. The Story of Bioglass. *J. Mater. Sci. Mater. Med.* **2006**, *17*, 967. [CrossRef]
154. Gerhardt, L.C.; Boccaccini, A.R. Bioactive Glass and Glass-Ceramic Scaffolds for Bone Tissue Engineering. *Materials* **2010**, *3*, 3867. [CrossRef] [PubMed]
155. Jones, J.R.J. Bioactive Ceramics and Glasses. In *Tissue Engineering Using Ceramics and Polymers*, 1st ed.; Boccaccini, A.R., Gough, J., Eds.; Woodhead Publishing Limited CRC Press: Cambridge, UK, 2007; Volume 1, pp. 52–71.
156. Pajares-Chamorro, N.; Chatzistavrou, X. Bioactive Glass Nanoparticles for Tissue Regeneration. *ACS Omega* **2020**, *5*, 12716. [CrossRef] [PubMed]
157. Dreanca, A.; Bogdan, S.; Popescu, A.; Sand, D.; Pall, E.; Astilean, A.N.; Pestean, C.; Toma, C.; Marza, S.; Taulescu, M.; et al. The Evaluation of the Osteopromoting Capabilities of Composites Based on Biopolymers and Gold/Silver Nanoparticles Doped Bioactive Glasses on an Experimental Rat Bone Defect. *Biomed. Mater.* **2023**, *18*. [CrossRef] [PubMed]
158. Niu, W.; Chen, M.; Guo, Y.; Wang, M.; Luo, M.; Cheng, W.; Wang, Y.; Lei, B. A Multifunctional Bioactive Glass-Ceramic Nanodrug for Post-Surgical Infection/Cancer Therapy-Tissue Regeneration. *ACS Nano* **2021**, *15*, 14323. [CrossRef] [PubMed]
159. Khatami, N.; Khoshfetrat, A.B.; Khaksar, M.; Zamani, A.R.N.; Rahbarghazi, R. Collagen-Alginate-Nano-Silica Microspheres Improved the Osteogenic Potential of Human Osteoblast-Like MG-63 Cells. *J. Cell Biochem.* **2019**, *120*, 15069. [CrossRef]
160. Mostafa, A.A.; El-Sayed, M.M.H.; Emam, A.N.; Abd-Rabou, A.A.; Dawood, R.M.; Oudadesse, H. Bioactive Glass Doped with Noble Metal Nanoparticles for Bone Regeneration: In vitro Kinetics and Proliferative Impact on Human Bone Cell Line. *RSC Adv.* **2021**, *11*, 25628. [CrossRef]
161. Zengin, A.; Castro, J.P.O.; Habibovic, P.; van Rijt, S.H. Injectable, Self-Healing Mesoporous Silica Nanocomposite Hydrogels with Improved Mechanical Properties. *Nanoscale* **2021**, *13*, 1144. [CrossRef]
162. Hou, X.; Zhang, L.; Zhou, Z.; Luo, X.; Wang, T.; Zhao, X.; Lu, B.; Chen, F.; Zheng, L. Calcium Phosphate-Based Biomaterials for Bone Repair. *J. Funct. Biomater.* **2022**, *13*, 187. [CrossRef]
163. Xiao, D.; Zhang, J.; Zhang, C.; Barbieri, D.; Yuan, H.; Moroni, L.; Feng, G. The Role of Calcium Phosphate Surface Structure in Osteogenesis and the Mechanisms Involved. *Acta Biomater.* **2020**, *106*, 22. [CrossRef]
164. Zhi, H.; Qingling, F.; Bo, Y.; Songjian, L. Biomimetic Properties of an Injectable Chitosan/Nano-Hydroxyapatite/Collagen Composite. *Mater. Sci. Eng. C* **2011**, *31*, 683. [CrossRef]
165. Shibuya, S.; Ozawa, Y.; Watanabe, K.; Izuo, N.; Toda, T.; Yokote, K.; Shimizu, T. Palladium and Platinum Nanoparticles Attenuate Aging-Like Skin Atrophy via Antioxidant Activity in Mice. *PLoS ONE* **2014**, *9*, e109288. [CrossRef] [PubMed]
166. Leal, F.; Nirwan, V.; Gonçalves, A.M.; Panitschewski, N.; Filová, E.; Fahmi, A.; Costa, P.F. Bio-Inspired Nanoporous Scaffold: Electrospun Hybrid Fibers Based on Self-Assembled Block Copolymer Mineralized with Inorganic Nanoparticles for Bone Tissue Engineering. *Int. J. Polym. Mater. Polym. Biomater.* **2023**, *73*, 1054. [CrossRef]
167. Kumar, B.Y.S.; Isloor, A.M.; Kumar, G.C.M.; Inamuddin Asiri, A.M. Nanohydroxyapatite Reinforced Chitosan Composite Hydrogel with Tunable Mechanical and Biological Properties for Cartilage Regeneration. *Sci. Rep.* **2019**, *9*, 15957. [CrossRef] [PubMed]

168. Kumai, T.; Yui, N.; Yatabe, K.; Sasaki, C.; Fujii, R.; Takenaga, M.; Fujiya, H.; Niki, H.; Yudoh, K. A Novel, Self-Assembled Artificial Cartilage-Hydroxyapatite Conjugate for Combined Articular Cartilage and Subchondral Bone Repair: Histopathological Analysis of Cartilage Tissue Engineering in Rat Knee Joints. *Int. J. Nanomed.* **2019**, *14*, 1283. [CrossRef]
169. Boller, L.A.; Shiels, S.M.; Florian, D.C.; Peck, S.H.; Schoenecker, J.G.; Duvall, C.; Wenke, J.C.; Guelcher, S.A. Effects of Nanocrystalline Hydroxyapatite Concentration and Skeletal Site on Bone and Cartilage Formation in Rats. *Acta Biomater.* **2021**, *130*, 485. [CrossRef]
170. Huang, J.; Huang, Z.; Liang, Y.; Yuan, W.; Bian, L.; Duan, L.; Rong, Z.; Xiong, J.; Wang, D.; Xia, J. 3D Printed Gelatin/Hydroxyapatite Scaffolds for Stem Cell Chondrogenic Differentiation and Articular Cartilage Repair. *Biomater. Sci.* **2021**, *9*, 2620. [CrossRef] [PubMed]
171. Gaweł, J.; Milan, J.; Żebrowski, J.; Płoch, D.; Stefaniuk, I.; Kus-Liśkiewicz, M. Biomaterial Composed of Chitosan, Riboflavin, and Hydroxyapatite for Bone Tissue Regeneration. *Sci. Rep.* **2023**, *13*, 17004. [CrossRef]
172. Becerra, J.; Rodriguez, M.; Leal, D.; Noris-Suarez, K.; Gonzalez, G. Chitosan-Collagen-Hydroxyapatite Membranes for Tissue Engineering. *J. Mater. Sci. Mater. Med.* **2022**, *33*, 18. [CrossRef]
173. Ma, S.; Wei, Y.; Sun, R.; Xu, H.; Liu, X.; Wang, Y.; Liang, Z.; Hu, Y.; Lian, X.; Ma, X.; et al. Calcium Phosphate Bone Cements Incorporated with Black Phosphorus Nanosheets Enhanced Osteogenesis. *ACS Biomater. Sci. Eng.* **2023**, *9*, 292. [CrossRef]
174. Witte, F.; Kaese, V.; Haferkamp, H.; Switzer, E.; Meyer-Lindenberg, A.; Wirth, C.J.; Windhagen, H. In vivo Corrosion of Four Magnesium Alloys and the Associated Bone Response. *Biomaterials* **2005**, *26*, 3557. [CrossRef]
175. Liu, C.; Fu, X.; Pan, H.; Wan, P.; Wang, L.; Tan, L.; Wang, K.; Zhao, Y.; Yang, K.; Chu, P.K. Biodegradable Mg-Cu Alloys with Enhanced Osteogenesis, Angiogenesis, and Long-Lasting Antibacterial Effects. *Sci. Rep.* **2016**, *6*, 27374. [CrossRef] [PubMed]
176. Li, R.W.; Kirkland, N.T.; Truong, J.; Wang, J.; Smith, P.N.; Birbilis, N.; Nisbet, D.R. The Influence of Biodegradable Magnesium Alloys on the Osteogenic Differentiation of Human Mesenchymal Stem Cells. *J. Biomed. Mater. Res. A* **2014**, *102*, 4346. [CrossRef] [PubMed]
177. Gonzalez, J.; Hou, R.Q.; Nidadavolu, E.P.S.; Willumeit-Römer, R.; Feyerabend, F. Magnesium Degradation under Physiological Conditions—Best Practice. *Bioact. Mater.* **2018**, *3*, 174. [CrossRef]
178. Fischerauer, S.F.; Kraus, T.; Wu, X.; Tangl, S.; Sorantin, E.; Hänzi, A.C.; Löffler, J.F.; Uggowitzer, P.J.; Weinberg, A.M. In vivo Degradation Performance of Micro-Arc-Oxidized Magnesium Implants: A Micro-CT Study in Rats. *Acta Biomater.* **2013**, *9*, 5411. [CrossRef]
179. Du, L.; Wu, J.; Han, Y.; Wu, C. Immunomodulatory Multicellular Scaffolds For Tendon-to-Bone Regeneration. *Sci. Adv.* **2024**, *10*, eadk6610. [CrossRef]
180. Oliveira-Filho, G.B.; Atoche-Medrano, J.J.; Aragón, F.F.H.; Mantilla Ochoa, J.C.; Pacheco-Salazar, D.G.; da Silva, S.W.; Coaquira, J.A.H. Core-Shell Au/Fe_3O_4 Nanocomposite Synthesized by Thermal Decomposition Method: Structural, Optical, and Magnetic Properties. *Appl. Surf. Sci.* **2021**, *563*, 150290. [CrossRef]
181. Lin, R.H.; Lee, H.T.; Yeh, C.A.; Yang, Y.C.; Shen, C.C.; Chang, K.B.; Liu, B.S.; Hsieh, H.H.; Wang, H.M.; Hung, H.S. Favorable Biological Performance Regarding the Interaction Between Gold Nanoparticles and Mesenchymal Stem Cells. *Int. J. Mol. Sci.* **2022**, *24*, 5. [CrossRef]
182. Hu, X.; Zhang, Y.; Ding, T.; Liu, J.; Zhao, H. Multifunctional Gold Nanoparticles: A Novel Nanomaterial for Various Medical Applications and Biological Activities. *Front. Bioeng. Biotechnol.* **2020**, *8*, 990. [CrossRef]
183. Gupta, R.; Rai, B. Effect of Size and Surface Charge of Gold Nanoparticles on their Skin Permeability: A Molecular Dynamics Study. *Sci. Rep.* **2017**, *7*, 45292. [CrossRef]
184. Suh, K.S.; Lee, Y.S.; Seo, S.H.; Kim, Y.S.; Choi, E.M. Gold Nanoparticles Attenuates Antimycin A-Induced Mitochondrial Dysfunction in MC3T3-E1 Osteoblastic Cells. *Biol. Trace Elem. Res.* **2013**, *153*, 428. [CrossRef]
185. Grant, S.A.; Zhu, J.; Gootee, J.; Snider, C.L.; Bellrichard, M.; Grant, D.A. Gold Nanoparticle-Collagen Gels for Soft Tissue Augmentation. *Tissue Eng. Part A* **2018**, *24*, 1091. [CrossRef] [PubMed]
186. Nunan, R.; Harding, K.G.; Martin, P. Clinical Challenges of Chronic Wounds: Searching for an Optimal Animal Model to Recapitulate their Complexity. *Dis. Model Mech.* **2014**, *7*, 1205. [CrossRef] [PubMed]
187. Mendes, C.; Thirupathi, A.; Zaccaron, R.P.; Corrêa, M.E.; Bittencourt, J.V.; Casagrande, L.D.; de Lima, A.C.; de Oliveira, L.L.; de Andrade, T.A.; Gu, Y.; et al. Microcurrent and Gold Nanoparticles Combined with Hyaluronic Acid Accelerates Wound Healing. *Antioxidants* **2022**, *11*, 2257. [CrossRef]
188. Poomrattanangoon, S.; Pissuwan, D. Gold Nanoparticles Coated with Collagen-I and Their Wound Healing Activity in Human Skin Fibroblast Cells. *Heliyon* **2024**, *10*, e33302. [CrossRef] [PubMed]
189. Ibrahim, B.; Akere, T.H.; Chakraborty, S.; Valsami-Jones, E.; Ali-Boucetta, H. Gold Nanoparticles Induced Size Dependent Cytotoxicity on Human Alveolar Adenocarcinoma Cells by Inhibiting the Ubiquitin Proteasome System. *Pharmaceutics* **2023**, *15*, 432. [CrossRef]
190. Zhang, X.D.; Wu, D.; Shen, X.; Liu, P.X.; Yang, N.; Zhao, B.; Zhang, H.; Sun, Y.M.; Zhang, L.A.; Fan, F.Y. Size-Dependent in vivo Toxicity of PEG-Coated Gold Nanoparticles. *Int. J. Nanomed.* **2011**, *6*, 2071. [CrossRef]
191. Surapaneni, S.K.; Bashir, S.; Tikoo, K. Gold Nanoparticles-Induced Cytotoxicity in Triple Negative Breast Cancer Involves Different Epigenetic Alterations Depending upon the Surface Charge. *Sci. Rep.* **2018**, *8*, 12295. [CrossRef]
192. Paramasivam, V.; Paulpandian, P.; Venkatachalam, K.; Hussain, S.; Kangal, A.; Al Farraj, D.A.; Elshikh, M.S.; Balaji, P. Cytotoxicity and Antimicrobial Efficiency of Gold (Au) Nanoparticles Formulated by Green Approach using Andrographis Paniculata Leaf Extract. *J. King Saud. Univ. Sci.* **2023**, *35*, 102687. [CrossRef]

193. Bertsch, C.; Maréchal, H.; Gribova, V.; Lévy, B.; Debry, C.; Lavalle, P.; Fath, L. Biomimetic Bilayered Scaffolds for Tissue Engineering: From Current Design Strategies to Medical Applications. *Adv. Healthc. Mater.* **2023**, *12*, e2203115. [CrossRef]
194. Del Bakhshayesh, A.R.; Asadi, N.; Alihemmati, A.; Tayefi Nasrabadi, H.; Montaseri, A.; Davaran, S.; Saghati, S.; Akbarzadeh, A.; Abedelahi, A. An Overview of Advanced Biocompatible and Biomimetic Materials for Creation of Replacement Structures in the Musculoskeletal Systems: Focusing on Cartilage Tissue Engineering. *J. Biol. Eng.* **2019**, *13*, 85. [CrossRef]
195. Patterson, J.; Martino, M.M.; Hubbell, J.A. Biomimetic Materials in Tissue Engineering. *Mater. Today* **2010**, *13*, 14. [CrossRef]
196. Liu, S.; Yu, J.M.; Gan, Y.C.; Qiu, X.Z.; Gao, Z.C.; Wang, H.; Chen, S.X.; Xiong, Y.; Liu, G.H.; Lin, S.E.; et al. Bioomimetic Natural Biomaterials For Tissue Engineering and Regenerative Medicine: New Biosynthesis Methods, Recent Advances, and Emerging Applications. *Mil. Med. Res.* **2023**, *10*, 16. [CrossRef]
197. Ma, P.X. Biomimetic Materials for Tissue Engineering. *Adv. Drug Deliv. Rev.* **2008**, *60*, 184. [CrossRef] [PubMed]
198. Qin, L.; Yang, S.; Zhao, C.; Yang, J.; Li, F.; Xu, Z.; Yang, Y.; Zhou, H.; Li, K.; Xiong, C.; et al. Prospects and Challenges For the Application of Tissue Engineering Technologies in the Treatment of Bone Infections. *Bone Res.* **2024**, *12*, 28. [CrossRef]
199. Senra, M.R.; Marques, M.F.V.; Monteiro, S.N. Poly (Ether-Ether-Ketone) For Biomedical Applications: From Enhancing Bioactivity to Reinforced-Bioactive Composites-An Overview. *Polymers* **2023**, *15*, 373. [CrossRef] [PubMed]
200. Cao, D.; Ding, J. Recent Advances in Regenerative Biomaterials. *Regen. Biomater.* **2022**, *9*, rbac098. [CrossRef]
201. Sun, Y.; Liu, X.; Tan, J.; Lv, D.; Song, W.; Su, R.; Li, L.; Liu, X.; Ouyang, L.; Liao, Y. Strontium Ranelate Incorporated 3D Porous Sulfonated PEEK Simulating MC3T3-E1 Cell Differentiation. *Regen. Biomater.* **2020**, *8*, rbaa043. [CrossRef]
202. Vasconcelos, D.P.; Águas, A.P.; Barbosa, M.A.; Pelegrín, P.; Barbosa, J.N. The Inflammasome in Host Response to Biomaterials: Bridging Inflammation and Tissue Regeneration. *Acta Biomater.* **2019**, *83*, 1. [CrossRef]
203. Lele, M.; Kapur, S.; Hargett, S.; Sureshbabu, N.M.; Gaharwar, A.K. Global Trends in Clinical Trials Involving Engineered Biomaterials. *Sci. Adv.* **2024**, *10*, eabq0997. [CrossRef]
204. Elson, E.L.; Genin, G.M. Tissue Constructs: Platforms for Basic Research and Drug Discovery. *Interface Focus* **2016**, *6*, 20150095. [CrossRef]
205. Hench, L.L.; Thompson, I. Twenty-First Century Challenges for Biomaterials. *J. R. Soc. Interface* **2010**, *7*, S379. [CrossRef] [PubMed]
206. Sharma, P.; Kumar, P.; Sharma, R.; Bhatt, V.D.; Dhot, P.S. Tissue Engineering; Current Status & Futuristic Scope. *J. Med. Life* **2019**, *12*, 225. [CrossRef] [PubMed]
207. Chandra, P.K.; Soker, S.; Atala, A. Chapter 1—Tissue Engineering: Current Status and Future Perspectives. In *Principles of Tissue Engineering*, 5th ed.; Lanza, R., Langer, R., Vacanti, J.P., Atala, A., Eds.; Academic Press: London, UK, 2020; pp. 1–35.
208. Ashammakhi, N.; GhavamiNejad, A.; Tutar, R.; Fricker, A.; Roy, I.; Chatzistavrou, X.; Hoque Apu, E.; Nguyen, K.L.; Ahsan, T.; Pountos, I.; et al. Highlights on Advancing Frontiers in Tissue Engineering. *Tissue Eng. Part B Rev.* **2022**, *28*, 633. [CrossRef] [PubMed]
209. Badylak, S.F.; Nerem, R.M. Progress in Tissue Engineering and Regenerative Medicine. *Proc. Natl. Acad. Sci. USA* **2010**, *107*, 3285. [CrossRef] [PubMed]

inorganics

MDPI

Article

Biodegradable Ca^{2+} Doped Mesoporous Silica Nanoparticles Promote Chemotherapy Synergism with Calcicoptosis and Activate Anti-Tumor Immunity

Chao Liu [1,*], Xiaohui Tang [2] and Gaofei Huang [1]

[1] School of Medical Engineering and Technology, Xinjiang Medical University, Urumqi 830017, China; huanggaofei@xjmu.edu.cn
[2] Core Facility, Xinjiang Medical University, Urumqi 830017, China; tangxh0429@163.com
* Correspondence: cliu20@sohu.com

Abstract: Mesoporous silica nanoparticles (MSNs), an excellent carrier material, have been widely used in tumor therapy as a vector for numerous therapeutic substances to boost therapeutical efficiency and specificity, such as loading them with chemotherapy drugs to improve the efficacy of chemotherapy. Nevertheless, they still face hurdles, such as lack of specificity and poor efficacy of monotherapy. The construction of multifunctional MSNs with excellent therapeutic effects by introducing metal ions has attracted the attention of many researchers. Herein, we demonstrated a calcium doped, chemotherapy drug doxorubicin (Dox) loaded, specific degradation nanoplatform, prepared using the sol–gel method by introducing calcium ions into an MSN framework, which enabled the doped nanoplatform to enhance chemotherapy and activate anti-tumor immune response. As a proof of concept, the doping of Ca^{2+} endowed MSNs with excellent specific degradation and pH responsive drug release, and enabled the synergy of chemotherapy and calcicoptosis. Furthermore, this nanoplatform also effectively elicited immunogenic cell death (ICD) and promoted the maturation of dendritic cells (DCs), realizing the activation of the anti-tumor immune system. The Ca^{2+} doped MSNs (CMSNs), that can activate immune response with specific degradation capability, demonstrate a practical strategy for the effective synergy between chemotherapy and calcicoptosis, providing a new paradigm for promoting chemotherapy-related treatment.

Keywords: mesoporous silica nanoparticles; calcium; calcicoptosis; synergistic treatment; immune activation

1. Introduction

In the past few decades, chemotherapy has played a huge role in tumor treatment, and to this day, chemotherapy remains a first-line choice for the treatment of many advanced malignant tumors [1,2]. However, due to the low specificity, weak cell uptake efficiency and other problems, the clinical efficacy of chemotherapy is not satisfactory [3,4]. With the rapid development of nanotechnology, nanoparticle-based drug delivery systems (DDSs) have shown excellent properties in tumor therapy [5,6]. Loading chemotherapy drugs into nanoparticles can not only improve drug pharmacokinetic characteristics, but also increase drug accumulation in tumor cells and reduce toxic side effects [7]. Among numerous types of nanoparticles, due to their excellent properties such as a high specific surface area and easy functionalization, MSNs are widely utilized as carriers for various chemotherapy agents in tumor treatment research [8,9]. However, MSNs' DDSs encounter some issues, such as unsatisfactory therapeutic effects due to the weak biodegradability of MSNs, as well as poor efficacy of monotherapy. Therefore, improving the specific degradation and granting MSNs a multimodal synergistic therapeutic function are highly desirable goals for enhancing the tumor treatment efficacy of MSNs' DDSs.

To improve the specific degradation ability of MSNs' DDSs, many components with tumor microenvironment (TME) responsiveness have been introduced into the MSNs' framework, which not only could break the stable Si–O bonds in MSNs but also endows MSNs with TME responsive degradability [10–13]. Among various approaches to improve the specific degradation of MSNs, metal ion doping has attracted extensive attention. Utilizing metal ions, acid responsive metal–oxygen bonds can form in MSNs to replace the stable Si–O bonds, and the metal–O bonds in the MSNs' framework break responsively under the stimulation of acidic TME, causing metal ions to escape from MSNs, and leading to the collapse of the MSNs' framework [14,15]. Meanwhile, the degradation also accelerates the release of loaded therapeutic agents [16]. More importantly, while improving the specific degradation of MSNs, the metal ions also have therapeutic and diagnostic functions. For example, the reactive oxygen species (ROS), produced by doped metal ions such as Fe^{3+}, Cu^{2+} and Mn^{4+} through Fenton or Fenton-like reaction, can be utilized for tumor treatment [17–19]. In addition to inducing tumor cell apoptosis through chemodynamic therapy (CDT), metal ions can induce tumor cell death through other means, for example, Fe^{2+} causes lipid peroxidation in tumor cells, leading to the occurrence of ferroptosis [20], and Cu^{2+} can disrupt the tricarboxylic acid cycle, resulting in the aggregation of lipoylated proteins and the loss of iron–sulfur clusters, which eventually leads to cuproptosis [21,22]. Furthermore, metal ions also exhibit an excellent immune activation effect, which could further augment the therapeutic effect [23–25]. Although metal ions such as copper and iron ions enable MSNs to acquire an excellent therapeutic effect, they are also more likely to cause cytotoxicity to normal cells. However, compared to other metal ions, Ca^{2+} can cause tumor cell death specifically, and their cytotoxicity to normal cells can be ignored, because tumor cells are more sensitive to changes in Ca^{2+} concentration than normal cells [26,27]. In this context, Ca^{2+} is a suitable and ideal candidate for improving the tumor treatment performance of MSNs' nanoplatform.

Till now, there has been relatively little research utilizing CMSNs for tumor treatment, and the main focus of these investigations was to improve the degradability of the carrier [28–32]. Although these CMSNs, prepared by different approaches, have outstanding biodegradability and have been successfully used in tumor treatment research, they do not devote more attention to the therapeutic function of Ca^{2+} in tumor treatment, and can only achieve synergistic treatment by loading multiple therapeutic agents. Calcicoptosis, as a newly defined programmed cell death, is triggered by Ca^{2+} overload which leads to mitochondrial dysfunction [33,34]. Therefore, utilizing the Ca^{2+} can realize the specific biodegradation of MSNs and also endow the carrier with therapeutic function. Herein, we developed a multifunctional CMSNs nanoplatform for boosting the therapeutic effect of MSNs' DDSs. Among various methods for MSNs' preparation, the sol–gel method, as a classical and reliable preparation method, has attracted much attention [35,36]. This is because, compared with other methods such as the hydrothermal method, microwave method and self-assembly, the sol–gel method has been widely used for its outstanding stability, low cost and easy operation [37]. Moreover, despite the long-term anticancer immunity present in immunotherapy, the poor immunogenicity of MSNs' DDSs hinders the immune response of the DDSs, and few studies have investigated the performance of Ca^{2+} doped MSNs' DDSs in activating anti-tumor immunity. Thus, it is intriguing to explore whether the Ca^{2+} doped nanoplatform could be used to augment immune response.

In this investigation, a multifunctional CMSNs nanoplatform was constructed using the sol–gel method, and then the chemotherapeutic agent Dox was loaded into the as-prepared CMSNs (CMSNs@Dox). Under the stimulation of the acidic environment within tumor cells, Ca–O bonds in the framework of CMSNs@Dox broke responsively, which facilitated the biodegradation of nanoparticles, and also rendered the responsive release of Dox and Ca^{2+}. The excessive Ca^{2+} within the tumor cells caused mitochondrial damage, leading to calcicoptosis, which then synergistically inhibited tumor cells with the chemotherapy drug Dox. In addition, the competence of CMSNs@Dox to trigger ICD and the maturation of DCs was also investigated, and it was found that CMSNs@Dox effectively elicited ICD and DCs' maturation, demonstrating outstanding immune-activating abilities.

Therefore, our Ca^{2+} doped nanoplatform could strengthen chemotherapy through synergy with calcicoptosis, and it also demonstrated a significant anti-tumor immunity activation.

2. Results and Discussion

2.1. Preparation and Characterization of CMSNs

In this work, CMSNs were prepared using the sol–gel method. The microstructure and particle size of the CMSNs were characterized by transmission electron microscopy (TEM) and, as illustrated in Figure 1a,b, CMSNs possessed a uniform spherical structure with a particle size of approximately 360 nm. To verify the successful doping of Ca^{2+} in CMSNs, TEM mapping was used to characterize the main elements in the CMSNs' framework. As can be seen from Figure 1c, silicon, oxygen and calcium were uniformly distributed throughout the entire nanoparticles, which demonstrated the successful doping of Ca^{2+}. To further certify the elemental composition of CMSNs, an X-ray photoelectron spectroscopy (XPS) analysis was performed. The results of XPS (Figure 1d) confirmed the successful doping of Ca^{2+} in CMSNs, and also indicated the formation of Ca–O bonds in CMSNs, in which a Ca 2p peak originated from Ca–O bonds (Figure 1e). However, there were not any Ca signals in the XPS results of MSNs (Figure S1). The XPS analysis of CMSNs also illustrated that the Ca content in CMSNs was about 3.22%, and the Ca content was also verified by Inductively Coupled Plasma Optical Emission Spectrometer (ICP-OES), which demonstrated that the Ca content per mg of CMSNs was 0.024 mg, similar to the results of XPS. Wide-angle X-ray diffraction (WAXRD) was used to analyze the structure of CMSNs, and as shown in the result (Figure 2c), only an amorphous peak was present near 20°, which indicated that CMSNs had an amorphous structure; this was consistent with other literature [38]. To have a comprehensive knowledge of the prepared CMSNs, nitrogen adsorption–desorption was also conducted. As shown in Figure 2a, CMSNs exhibited a classical type IV curve, which indicated that CMSNs had a typical mesoporous structure. The calculated BET (Brunauer–Emmett–Teller, BET) surface and the BJH (Barrett–Joiner–Halenda, BJH) pore volume were 805.95 m^2/g and 0.77 cm^3/g, respectively. Additionally, the pore distribution concentrated in 2.09 nm (Figure 2b). These characterizations demonstrated that doping with Ca^{2+} did not affect the loading features of CMSNs.

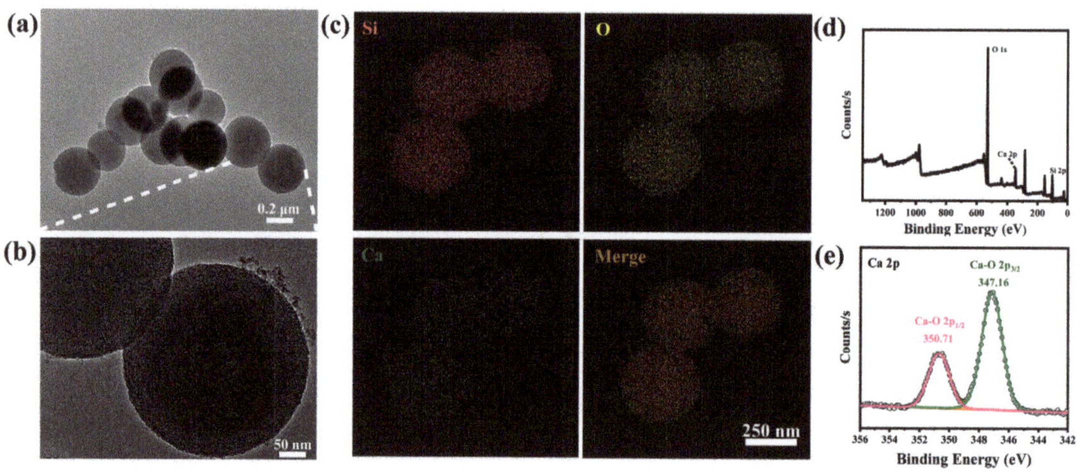

Figure 1. Size and composition characterization. (**a**,**b**) TEM images of CMSNs. (**c**) TEM element mapping of CMSNs. (**d**) Wide scan XPS spectra of CMSNs, (**e**) Ca 2p high solution XPS spectra.

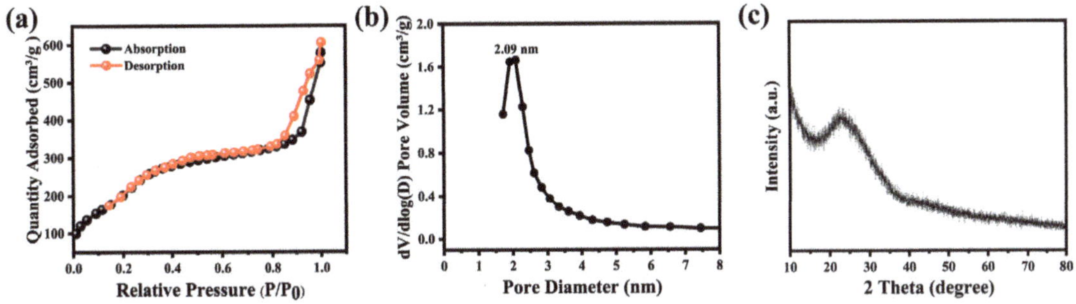

Figure 2. Structural characterization of CMSNs. (**a**) Nitrogen absorption desorption isotherms. (**b**) Pore size distribution of CMSNs. (**c**) WAXRD pattern of CMSNs.

2.2. Biodegradability

Although the above characterizations demonstrated that CMSNs possessed excellent drug loading capability, it is still necessary to circumvent the problem of degradation, which affected the biomedical application of MSNs. Compared with the stable Si–O bonds in MSNs, Ca–O bonds in CMSNs were unstable under acidic conditions, which was due to the cation exchange with H^+ and the cleavage of the Ca–O bond; what is more, the release of Ca^{2+} from Si–O–Ca could produce lots of non-bridging oxygens, which contained a large number of local sites for the nucleophilic attract primarily by OH-, leading to the breakdown of the framework [15,28,39]. The breaking of Ca–O bonds could destroy the structure of CMSNs and accelerate the degradation of CMSNs, characterized by the loss of regular spherical structures. To verify the pH responsive biodegradation behavior of CMSNs, an equal amount of CMSNs were immersed in PBS with different pH values (pH 7.4 and pH 5.6), respectively, and then the morphology of nanoparticles was tracked and observed. As illustrated in Figure 3a, the morphology of the CMSNs showed a changing trend over time when the PBS had a pH of 5.6. At 48 h, the morphology of the nanoparticles has changed from regular spherical to randomly blurred, indicating that the CMSNs possessed outstanding degradability under acidic conditions, which could alleviate the side effects caused by the accumulation of nanoparticles in the body. Considering that the CMSNs did not show significant degradation and structure collapse until 48 h, all cellular anti-tumor evaluations were conducted by co-culturing nanoparticles with cells for 48 h. However, when the CMSNs sample was placed in neutral buffer, it maintained a stable structure all the time (Figure 3b). For MSNs without Ca^{2+} doping, the sample maintained a stable spherical structure under both acidic and neutral conditions (Figure S2), indicating that Ca^{2+} doping could confer a brilliant pH responsive biodegradation ability to traditional MSNs. Of course, it also facilitated the precise release of loaded cargo.

2.3. Dox Loading and pH Responsive Drug Release

In view of the excellent structural characteristics of CMSNs, the chemotherapeutic drug Dox was loaded into the CMSNs, and the pH responsive drug release behavior was investigated. The drug loading efficiency and encapsulation efficiency of the CMSNs were 17.8% and 85%, respectively, which were significantly higher than the MSNs samples without doping, because the metal ions in the doped MSNs skeleton could form coordination complexes with drug molecules [16]. The incorporation of metal ions not only elevated the drug loading ability of CMSNs, but also granted the nanoplatform a pH-triggered drug release due to the excellent degradation ability of CMSNs under an acidic environment. To verify the pH responsive drug release behavior of the nanoplatform, CMSNs@Dox were placed in acidic (pH 5.6) and neutral (pH 7.4) PBS buffer to simulate an intracellular environment and a normal physiological environment. As depicted in Figure 3c, the CMSNs@Dox sample in the neutral buffer showed a slow and insignificant release, while under acidic conditions, the sample showed a more significant release behavior until 48 h, after which the amount of the drug released remained almost unchanged. This was due to the outstanding biodegradability of CMSNs under acidic conditions, and the collapse of the CMSNs' structure facilitated the release of loaded cargo, which was consistent with the results of degradation experiment. Under neutral conditions, only about 15% of the loaded drug was released from CMSNs@Dox after 96 h, while about 37% of the drug was released under acidic conditions, which was twice as much as under neutral conditions. Therefore, CMSNs@Dox possessed a good specific drug release ability, which could effectively improve the precision of treatment and reduce toxic side effects on normal tissues.

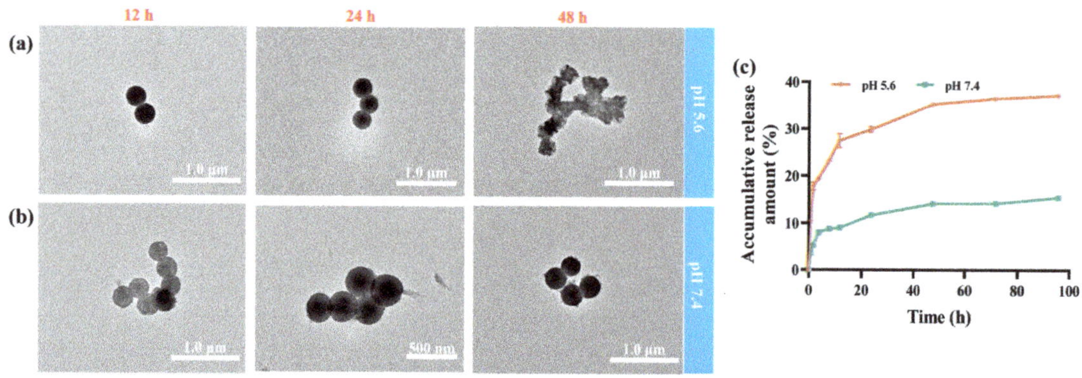

Figure 3. In vitro degradation and drug release of CMSNs. (**a**) TEM images of CMSNs incubated in PBS at (**a**) pH 5.6 and (**b**) pH 7.4 for varied times. (**c**) Drug release profile of CMSNs@Dox at different times in PBS at pH 5.6 and 7.4.

2.4. Calcium Ions Dependent Mitochondrial Dysfunction

The incorporation of Ca^{2+} not only improved the biodegradability of CMSNs, but also endowed CMSNs@Dox with a responsive drug release ability. In addition, the degradation of CMSNs in tumor cells boosted the intracellular Ca^{2+} content, resulting in Ca^{2+} overload intracellularly. Ca^{2+} overload subsequently induced the calcicoptosis characterized by mitochondrial dysfunction (such as decreased mitochondrial membrane potential), thereby making the doped Ca^{2+} possess multiple functions. As shown in Figure 4a, compared with the control and MSNs treated groups, the intracellular Fluo-4 fluorescence intensity was significantly enhanced after CMSNs treatment. Furthermore, the Fluo-4 fluorescence intensity in 4T1 cells treated with different samples was semi-quantitatively by flow cytometry (FCM), and the mean fluorescence intensity (MFI) of Fluo-4 was about three times than the MSNs treatment group (Figure 4b), indicating that the intracellular Ca^{2+} content of 4T1 cells treated with CMSNs was significantly fortified, in other words, CMSNs as a Ca^{2+} reservoir significantly increased the intracellular Ca^{2+} content, facilitating the occurrence of calcicoptosis. In addition, there was no difference in Ca^{2+} signal between the MSNs treatment group and the control group, indicating that CMSNs could certainly enhance the intracellular Ca^{2+} content and cause a Ca^{2+} overload of tumor cells, which provided the prerequisite for calcicoptosis. The mitochondrial membrane potential (MMP) of 4T1 cells treated with various samples was evaluated by JC-1 probe. JC-1 can form aggregates which emit red fluorescence when the MMP is high, while when MMP is low, JC-1 can form monomers emitting a green fluorescence. In other words, when the MMP changed, the ratio of fluorescence intensity between the green fluorescence of monomers and the red fluorescence of aggregates would change [34,40]. As illustrated in Figure 4c, the MMP of cells treated with MSNs for 48 h was not different from that of the control group, indicating that MSNs did not have any impact on mitochondrial function; however, the MMP changed after CMSNs treatment. Moreover, the fluorescence ratio of the JC-1 monomer and polymer was also used to characterize the changes in MMP. As depicted in Figure 4d, different from the control and MSNs treatment group, after treatment with CMSNs, the fluorescence ratio of JC-1 polymer to monomer decreased significantly and was approximately one third of the control and MSNs treatment group (Figure 4d), indicating that CMSNs caused mitochondrial dysfunction, leading to a decrease in MMP. Moreover, 4T1 cells treated with CMSNs also demonstrated apparent early apoptosis (Figure 6), and mitochondrial dysfunction would exhibit significant early apoptosis according to literature [41], which further verified that CMSNs could cause mitochondrial dysfunction and lead to calcicoptosis.

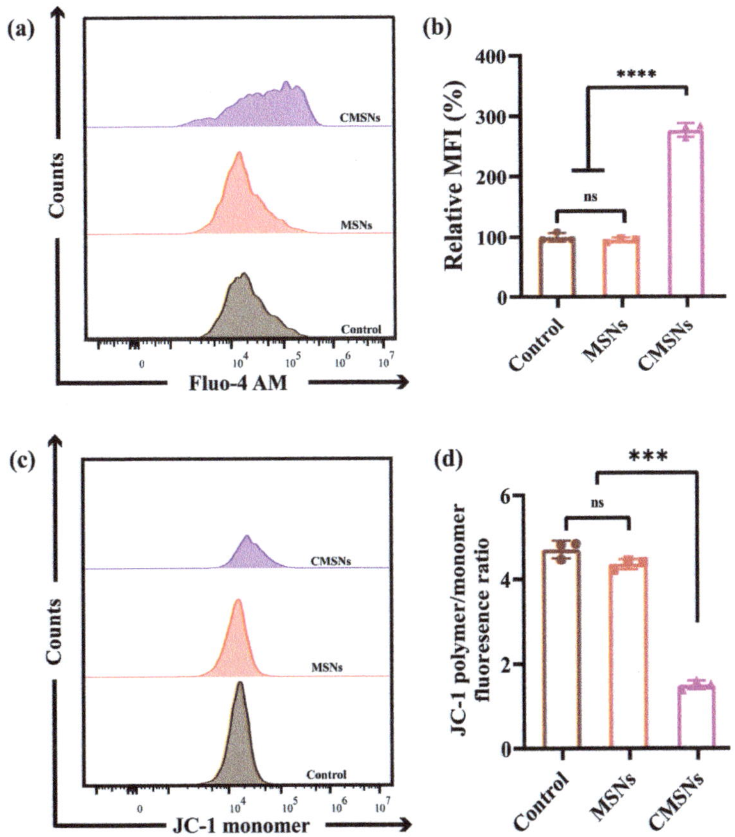

Figure 4. Intracellular Ca^{2+} content and mitochondrial membrane potential analysis. (**a**,**b**) FCM analysis of intracellular Ca^{2+} content including flow cytometric histograms and mean fluorescence intensities (MFI) of Fluo-4 AM. (**c**,**d**) Flow cytometric histogram of JC-1 monomer and the corresponding fluorescence ratio of JC-1 polymer/monomer. Data are shown as mean ± SD, (n = 3, *** $p < 0.001$, **** $p < 0.0001$, ns, not significant).

2.5. Cell Internalization

The uptake of therapeutic agents by tumor cells was a prerequisite for drugs to exert therapeutic effects. Unlike free drugs, nanoparticles that enter cells through endocytosis could improve the uptake efficiency of loaded drugs in tumor cells, thereby enhancing the therapeutic effect. Herein, the uptake behavior of CMSNs was investigated by FCM. As shown in Figure 5a,b, the uptake efficiency of nanoparticles and free Dox in tumor cells increased with time. More importantly, compared with free Dox, tumor cells exhibited higher uptake efficiency towards CMSNs@Dox at any time point, and the MFI of CMSNs@Dox was about five times that of free Dox at 6 h, indicating more chemotherapy drug molecules were internalized by cells with the assistance of CMSNs. Furthermore, CMSNs@Dox could release drugs promptly under the intracellular acidic environment, which would facilitate an enlargement in intracellular drug content and promote the efficacy of chemotherapy.

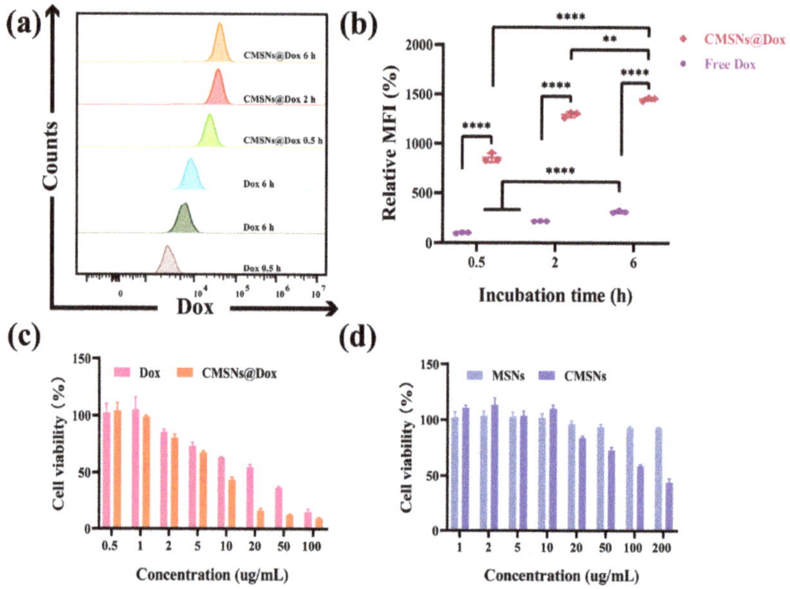

Figure 5. Internalization and in vitro cytotoxicity. (**a**,**b**) FCM profiles and MFI of 4T1 cells co-incubated with CMSNs@Dox and free Dox for 0.5, 2 and 6 h. (**c**,**d**) Cell viability of 4T1 cells after treatment with MSNs, CMSNs, free Dox and CMSNs@Dox for 48 h. Data are shown as mean ± SD, (n = 3, ** $p < 0.01$, **** $p < 0.0001$).

2.6. In Vitro Cytotoxicity and Anti-Tumor Efficiency

To evaluate the anti-tumor effect of CMSNs@Dox, we first investigated the cytotoxicity of MSNs, CMSNs, Dox and CMSNs@Dox against 4T1 cells. After co-incubating 4T1 cells with different samples at 37 °C for 48 h, the cell viability of 4T1 cells was measured by MTT assay. As predicted, CMSNs@Dox showed a superior cell inhibition effect at different concentrations than other treatment groups, and its IC50 value was significantly lower than free Dox (Figure 5c). It is worth noting that, unlike MSNs, which demonstrated excellent biocompatibility, CMSNs exhibited a certain degree of cytotoxicity when the concentration of CMSNs was increased (Figure 5d). This was because a large number of Ca^{2+} were released from CMSNs under the intracellular acidic environment, causing a Ca overload within tumor cells, which further elicited calcicoptosis. Hence, CMSNs@Dox could effectively realize the synergy of chemotherapy and calcicoptosis, which explained the more remarkable anti-tumor effect of CMSNs@Dox compared with free Dox from another perspective. Furthermore, the cytotoxicity of CMSNs and CMSNs@Dox to mouse embryo fibroblasts (3T3 cells) was also conducted, as shown in Figure S7, CMSNs have no cytotoxicity to 3T3 cells when the concentration reaches 200 μg/mL, and the cytotoxicity of CMSNs@Dox to 3T3 cells was significantly higher than that of tumor cells, demonstrating that CMSNs and CMSNs@Dox possess specificity to tumor cells. In a word, the reasons why CMSNs@Dox exhibited a significant anti-tumor effect mainly originated from two aspects: (1) CMSNs@Dox possessed excellent cellular internalization efficiency and outstanding pH responsive drug release ability, leading to an enlargement of intracellular drug accumulation, and (2) CMSNs@Dox enabled chemotherapy in tandem with calcicoptosis.

After the cytotoxicity assay, an Annexin V-FITC/7-AAD apoptosis assay was further utilized to assess the anti-tumor effect of CMSNs@Dox. As can be seen from Figure 6, the apoptosis percentage of 4T1 cells treated by CMSNs@Dox was more significant than the others, which further indicated that CMSNs@Dox posed a superior anti-tumor effect. Consistent with the MTT results, the CMSNs treatment group also showed a significant proportion of cell apoptosis compared with MSNs treatment group. This was because the Ca^{2+} in CMSNs elicited calcicoptosis, which was characterized by mitochondrial dysfunction, and the apparent early apoptosis further indicated the occurrence of calcicoptosis. The above experiments fully demonstrated that CMSNs@Dox could achieve the synergy between calcicoptosis and chemotherapy, and possessed outstanding anti-tumor effects. To demonstrate the synergistic therapeutic effect between chemotherapy and calcicoptosis, the Combination Index (CI) was calculated according to the IC50 value of each treatment, and the obtained CI was 0.116 and less than 1, which indicated that CMSNs@Dox possessed a brilliant synergy ability.

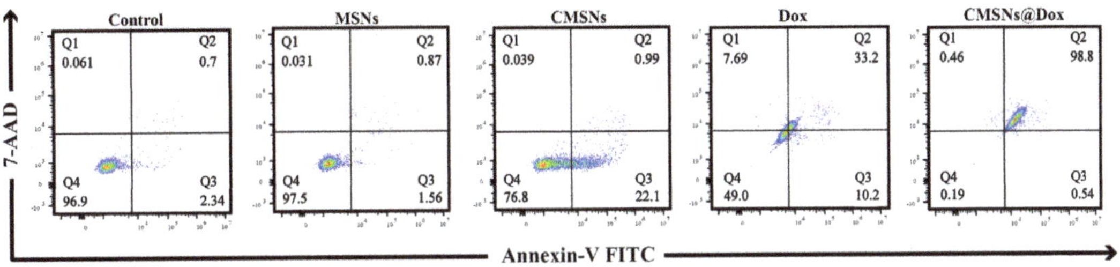

Figure 6. FCM apoptosis analysis of 4T1 cells after treatment with MSNs, CMSNs, Dox and CMSNs@Dox.

2.7. Immunogenic Cell Death and Immune Activation In Vitro

At present, much effort has been paid to the development of ICD-inducing agents due to adaptive immune responses can be driven by ICD [42], and recent investigations have shown that chemotherapy drugs could trigger adaptive immune response by inducing ICD [43]; therefore, it is meaningful to study whether the incorporation of Ca^{2+} could endow the drug delivery vector with the ability to induce ICD, so as to facilitate the CMSNs@Dox nanoplatform to elicit a more significant ICD effect and immune response. As an important indicator of ICD, damage associated molecule patterns (DAMPs) could activate the adaptive immune system by sending a "eat me" signal. Three main DAMPs (ATP, HMGB1, CRT) were evaluated after 4T1 cells were treated with different samples. As shown in Figure 7, compared with the control, MSNs and CMSNs treatment groups, extracellular ATP, HMGB1 and CRT were significantly elevated after free Dox treatment, which verified that chemotherapy could certainly induce ICD. On the other hand, the ATP, HMGB1 and CRT level of 4T1 cells treated with CMSNs was also raised compared with the control and MSNs treatment group, indicating that the Ca^{2+} doped carrier could also elicit ICD, although not as significant as chemotherapy. Among all treatment groups, cells treated with CMSNs@Dox exhibited the highest level of ATP secretion, HMGB1 release and CRT expression, which verified that the synergy of chemotherapy and calcicoptosis elicited ICD more effectively.

ICD is known to enhance anti-tumor immune responses by promoting dendritic cells' maturation through antigen presentation [44]. Therefore, DC2.4 was co-incubated with the 4T1 cells, which were treated with different substances (MSNs, CMSNs, free Dox and CMSNs@Dox) to detect the population of $CD80^+CD86^+$ by FCM analysis. As shown in Figure 8a,b, after MSNs, CMSNs, free Dox and CMSNs@Dox treatment, the percentage of mature DCs was 0.93 ± 0.3%, 31.01 ± 2.52%, 41.1 ± 4.91%, 59.9 ± 4.23%, respectively. Except for the MSNs treatment group, the other groups were significantly higher than the control group. In addition, similar to the free Dox group, the CMSNs treatment group also induced a certain proportion of DCs' maturation, demonstrating that the introduction of Ca^{2+} was essential for enhancing the immunogenicity of the carrier. Certainly, CMSNs@Dox remarkably amplified the percentage of DCs' maturation than the others. The above results demonstrated that the synergy of chemotherapy and calcicoptosis can augment immunogenicity and DCs' maturation, which could activate anti-tumor immunity and alleviate the tumor immunosuppressive microenvironment.

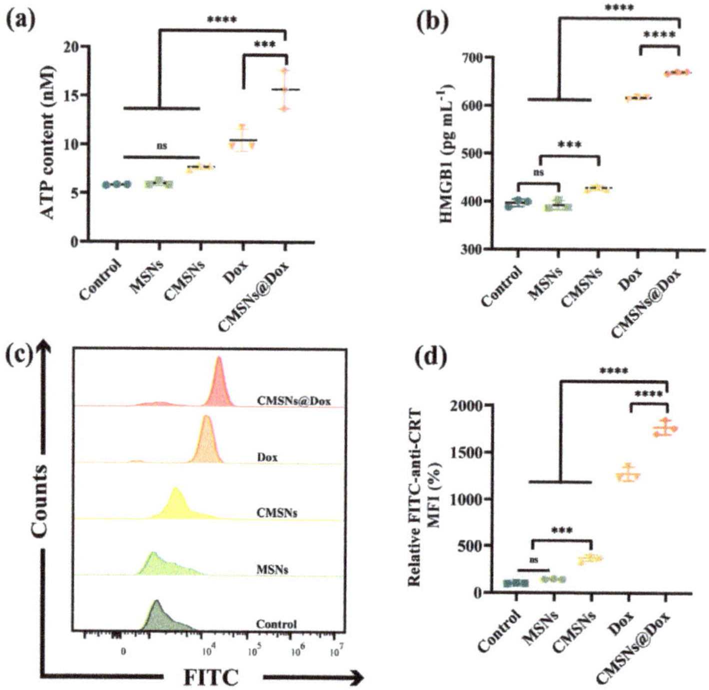

Figure 7. In vitro ICD effect evaluation. (**a**) Extracellular APT and (**b**) HMGB1 level. (**c**,**d**) Expression of CRT on cell surface was analyzed by FCM, the histogram and corresponding MFI. Data are shown as mean ± SD, (n = 3, *** $p < 0.001$, **** $p < 0.0001$, ns, not significant).

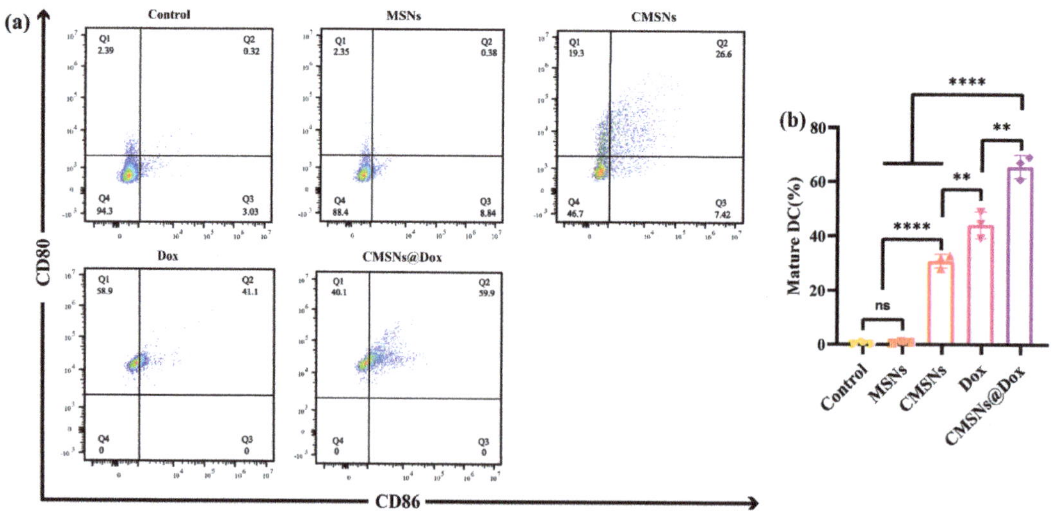

Figure 8. FCM analysis of DCs maturation ratio after treatment with different substances. (**a**) Representative FCM patterns and (**b**) semi-quantification of maturation ratio. Data are shown as mean ± SD, (n = 3, ** $p < 0.01$, **** $p < 0.0001$, ns, not significant).

3. Materials and Methods

3.1. Materials

Tetraethylorthosilicate (TEOS) and Doxorubicin hydrochloride (Dox) were purchased from Aladdin (Shanghai, China). Cetyltrimethylammonium bromide (CTAB) was purchased from J&K Scientific (Beijing, China). Calcium nitrate tetrahydrate (Ca $(NO_3)_2{\cdot}4H_2O$), ammonia ($NH_3{\cdot}H_2O$) and anhydrous ethanol were obtained from Sinopharm (Beijing, China). Roswell Park Memorial Institute (RPMI) 1640, penicillin-streptomycin and Fetal bovine serum (FBS) were purchased from Gibco Life Technologies (Grand Island, NE, USA). The 3-(4,5-dimethylthiazol-2-yl)-2,5-diphenyltetrazolium bromide (MTT) was purchased from Macklin (Shanghai, China). The enhanced mitochondrial JC-1 assay kit, Fluo-4 Calcium Assay kit, and enhanced ATP Assay kit were purchased from Beyotime Biotechnology (Shanghai, China). Annexin V-FITC/7AAD apoptosis detection kit was purchased from Yeasen Biotechnology (Shanghai, China). The FITC anti-Calreticulin antibody was purchased from Stressmarq Bioscience (Victoria, BC, Canada). The high mobility group box 1 (HMGB1) enzyme-linked immunosorbent assay (ELISA) kit was purchased from Wuhan Jiyinmei Biotech (Wuhan, China). All chemicals and reagents are used directly without any purification.

3.2. Synthesis of Virous MSNs

CMSNs were prepared using the classical sol–gel method with a minor modification [38]. A total of 0.08 g CTAB was added to 40 mL mixture of water and anhydrous ethanol (27 mL/13 mL), then 0.5 mL ammonia was added and stirred at 30 °C. After stirring for 30 min, 0.5 mL TEOS was added dropwise to the above mixture. A certain amount of aqueous solution containing 0.22 g Ca $(NO_3)_2{\cdot}4H_2O$ was added to the mixture 30 min after adding TEOS. The reaction was terminated after continuous stirring for 5 h. Nanoparticles were obtained by centrifugation (8000 rpm, 10 min), and washed three times with deionized water and ethanol. The nanoparticles were dried at 60 °C for 12 h and then calcined at 550 °C for 6 h to remove the surfactant CTAB. The preparation procedure of MSNs was almost identical to CMSN, with the difference being that no Ca $(NO_3)_2{\cdot}4H_2O$ added.

3.3. Structural Characterizations

The morphology and particle size of CMSNs were characterized by TEM (HT7700, HITACHI, Chiyoda, Japan). Element characterization was performed using field emission transmission electron microscopy (FETEM) (Tecnai F20, FEI, San Jose, CA, USA) and XPS (Thermofisher escalab 250xi, Thermofisher, Waltham, MA USA). The specific surface area, pore size and pore volume of the nanoparticles were measured through a nitrogen adsorption–desorption isotherm (Micromeritics ASAP2420, Micromeritics, Norcross, GA, USA). The Powder X-ray diffraction (XRD) mode was recorded on an X-ray diffractometer with Cu-Kα radiation (X′perT3 Powder, Almelo, The Netherlands) with a scanning step of 0.02°. Ultraviolet–visible (UV–vis) spectra were recorded using a spectrometer (TU-1901, PUXI, Shanghai, China).

3.4. In Vitro Biodegradation and Stability

An in vitro biodegradation evaluation of CMSNs was conducted by immersing the materials into PBS buffer that simulated the tumor microenvironment and normal physiological environment. Concretely speaking, 10 mg CMSNs and MSNs were dissolved in 30 mL phosphate-buffered saline (PBS) with a pH value of 5.6 (intracellular environment) and 7.4 (normal physiological environment), respectively. The mixture was placed in a shaker (37 °C, 100 rpm), and 1 mL mixture was collected at given time (12 h, 24 h, 48 h), the morphology and structure of the collected samples were observed subsequently by TEM.

3.5. Drug Loading and Release

Drug loading and pH triggered release of samples were carried out according to the published literature [45]. In brief, 15 mg Dox was added to PBS buffer containing 60 mg

CMSNs and stirred at 25 °C for 24 h. The final mixtures were centrifuged (8000 rpm, 10 min) and washed with PBS several times to remove the unloaded drug molecule, and the final product CMSNs@Dox was obtained after lyophilization. At the same time, the absorbance of collected supernatant was measured by a UV–vis spectrophotometer to calculate drug loading efficiency (DL) and encapsulation efficiency (EE).

Drug release was performed by immersing CMSNs@Dox into a PBS buffer that mimics the pH value of tumor cells and the normal physiological environment, respectively. A total of 5 mg CMSNs@Dox was put into 2 mL PBS with pH 5.6 and 7.4, then transferred to a dialysis bag and immersed into 30 mL PBS buffer with corresponding pH value. Finally, the release system was placed in a shaker (37 °C, 100 rpm). At predetermined time points, the same volume of dialysate was taken and promptly supplemented with fresh PBS, an absorption value of dialysate at 480 nm was measured by UV–vis spectrophotometer, and the cumulative drug release was calculated.

3.6. Cell Culture

4T1 mouse breast cells and 3T3 mouse embryo fibroblasts were purchased from the Type Culture Collection of the Chinese Academy of Science (Shanghai, China), 4T1 cells were cultured in RPMI 1640 containing 10% FBS and 1% penicillin-streptomycin in a humidified incubator containing 5% CO_2 at 37 °C, 3T3 cells were cultured in Dulbecco's Modified Eagle Medium containing 10% FBS and 1% penicillin-streptomycin in a humidified incubator containing 5% CO_2 at 37 °C.

3.7. Intracellular Ca^{2+} Detection

Intracellular Ca^{2+} levels were determined by a Fluo-4 Calcium Assay kit. 4T1 cells were inoculated (3×10^5 cells per well) in a 6-well plate and incubated overnight at 37 °C. A medium containing 100 µg/mL CMSNs or MSNs was added and cultured at 37 °C for 48 h. Then, washing three times with PBS, cells were digested by trypsin and washed again. The collected cells were stained for 30 min by Fluo-4 according to the instruction of the Fluo-4 Calcium Assay kit. After washing the stained cells with PBS, cells were resuspended and analyzed by FCM (Cytoflex S, Beckman Coulter, Brea, CA, USA).

3.8. Cell Internalization

4T1 cells were inoculated with 3×10^5 cells per well in a 6-well plate. After incubation for 24 h, the culture media was replaced with the same Dox concentration (5 µg/mL) of free Dox solution and CMSNs@Dox solution. At the predetermined time points (0.5 h, 2 h, 6 h), the culture medium was pipetted and washed with PBS three times, then digested and washed with PBS. FCM analysis was conducted after resuspending the collected cells in 500 µL PBS.

3.9. Mitochondrial Membrane Potential Analysis

4T1 cells (2×10^5 cells per well) were added to a 6-well plate and cultures for 24 h. Then, they were subjected to various treatments including MSNs (100 µg/mL) and CMSNs (100 µg/mL) for 48 h. After washing with PBS three times, cells were digested by trypsin and washed with PBS. The collected cells were stained using a JC-1 assay kit according to the manufacturer's instructions and subsequently analyzed by FCM.

3.10. In Vitro Cytotoxicity

The in vitro cell viability of the 4T1 and 3T3 cells was measured by MTT assay after incubating them with different samples. Cells were seeded in a 96-well plate at a density of 8×10^4 per cell and incubated overnight at 37 °C. The culture medium was replaced with a fresh medium-containing sample (MSNs, CMSNs, Dox, CMSNS@Dox) with different concentrations. After co-incubation for 48 h, MTT working solution (10 µL, 5 mg/mL) was added to each well, and the incubation was continued for 4 h. The media was pipetted from the wells, followed by adding 150 μL DMSO to dissolve formazan. Finally, the cell viability was calculated according to the formazan absorbance at 570 nm.

3.11. Apoptosis

Annexin V-FITC/7-AAD apoptosis kit was used to evaluate the apoptosis of cells treated with different samples. 4T1 cells were inoculated in a 6-well plate at a density of 3×10^5 cells per well and incubated overnight at 37 °C. The medium was replaced with fresh medium containing MSNs (100 μg/mL), CMSNs (100 μg/mL), free Dox (21 μg/mL) and CMSNs@Dox (121 μg/mL), respectively. After co-incubation for 48 h, cells were digested by trypsin and washed three times. According to the manufacture's instruction, the collected cells were stained with Annexin V-FITC for 5 min, then 400 μL PBS containing 10 μL PI was added to each group and analyzed by FCM.

3.12. Immunogenic Cell Death

Adenosine triphosphate (ATP) and high mobility group box 1 (HMGB1) secretion and calreticulin (CRT) eversion were regarded as the main indicators of immunogenic cell death (ICD). Extracellular ATP content was evaluated by an ATP Assay kit. In detail, 4T1 cells were inoculated in a 12-well plate with 1×10^5 cells per well and incubated overnight at 37 °C. Culture medium containing MSNs (100 μg/mL), CMSNs (100 μg/mL), free Dox (21 μg/mL) and CMSNs@Dox (121 μg/mL) was added. After incubation for 12 h, the culture medium supernatant in each group was collected and the ATP level was measured according to the manufacturer's guidelines. Extracellular HMGB1 level was detected using an enzyme linked immunosorbent assay (ELISA) kit. Cells were inoculated in a 24-well plate at a density of 7×10^4 per well. After 12 h incubation, the medium was replaced with fresh medium containing different samples (the same types and concentration as the ATP assay). The medium supernatant was collected and analyzed using an ELISA kit according to the manufacturer's protocol. The fluorescence intensity of the FITC anti-Calreticulin antibody could reflect the expression level of CRT on the cell surface. 4T cells were seeded in 12-well plate with a density of 1×10^5 cells per well and cultured overnight. After removing the cultured medium, cells were exposed to the fresh medium containing different samples (samples were identical to ATP assay) and incubated for 12 h continuously. Afterward, the cells were washed with PBS and digested by trypsin. After washing with PBS, the collected cells were co-incubated with FITC anti-Calreticulin antibody at room temperature for 30 min. Further, the stained cells were washed with PBS twice, then resuspended in 500 μL PBS and evaluated by FCM.

3.13. Dendritic Cells Maturation

To determine the maturation of dendritic cells, mouse bone marrow derived dendritic cell line 2.4 (DC2.4) was selected. The mouse bone marrow derived dendritic cell line 2.4 (DC2.4) was obtained from the Type Culture Collection of Chinese Academy of Science (Shanghai, China). DC2.4 cells were cultured in Dulbecco's Modified Eagle Medium containing 10% FBS and 1% penicillin-streptomycin under a humidified atmosphere with 5% CO_2 at 37 °C. 4T1 cells were inoculated (1.5×10^5) in a 12-well plate overnight. The medium was replaced with fresh medium containing different samples (MSNs 100 μg/mL, CMSNs 100 μg/mL, Dox 21 μg/mL, CMSNs@Dox 121 μg/mL), respectively. After 12 h post treatment, the medium of each treatment group was added to immature DC2.4 cells and co-incubated for another 12 h. The collected DC2.4 cells were stained with BV610 anti-mouse CD80 antibody and APC anti-mouse CD86 antibody in the dark for 30 min. Subsequently, the stained DC2.4 cells were washed with PBS twice, and the proportion of mature DCs was measured by FCM.

3.14. Statistical Analysis

All experimental data were expressed as mean ± standard deviation (SD), and statistical analysis was performed using GraphPad Prism 8 Software (GraphPad Software,

Inc., San Diego, CA, USA). Multiple sets of data were analyzed using one-way analysis of variance (ANOVA) following Tukey's test, with p value less than 0.05 considered statistically significant.

4. Conclusions

In summary, CMSNs with excellent specific biodegradability were prepared using the sol–gel method. Due to the multiple functions of the Ca^{2+} exerted in CMSNs, the obtained CMSNs nanoplatform not only possessed excellent specific biodegradability, but also easily realized the synergy of calcicoptosis and chemotherapy, serving a double purpose and significantly magnifying the efficacy of traditional chemotherapy. More importantly, the doping of Ca^{2+} also enabled the CMSNs@Dox DDSs to elicit more significant ICD and DCs maturation, which could effectively activate the anti-tumor immune response. This newly prepared Ca^{2+} doped nanoplatform provided a practical and feasible strategy for improving chemotherapy efficacy, and will provide more alternatives for tumor treatment by loading different therapeutic agents in the cancer therapeutic field.

Supplementary Materials: The following supporting information can be downloaded at: https://www.mdpi.com/article/10.3390/inorganics12060152/s1, Figure S1: (a) Wide scan XPS of MSN and (b) high solution XPS of MSNs. Figure S2: In vitro degradation of MSNs in PBS at pH (a) 5.6 and (b) 7.4. Figure S3: TEM images of CMSNs. Figure S4: TEM mapping of CMSNs. Figure S5: TEM images of CMSNs under PBS buffer at pH 5.6. Figure S6: TEM images of CMSNs under PBS buffer at pH 7.4. Figure S7: Cytotoxicity of 3T3 cells treatment with CMSNs and CMSNs@Dox for 48 h.

Author Contributions: Conceptualization, C.L. and X.T.; methodology, C.L.; validation, C.L. and X.T.; investigation, C.L., X.T. and G.H.; resources, C.L.; writing—original draft preparation, C.L., X.T. and G.H.; supervision, C.L.; project administration, C.L.; funding acquisition, C.L. All authors have read and agreed to the published version of the manuscript.

Funding: This research was funded by the Natural Science Foundation of Xinjiang, grant number 2021D01C279.

Data Availability Statement: The data presented in the study are included in the article and Supplementary Materials, further inquiries can be directed to the corresponding author.

Acknowledgments: The authors thank Kun Yang for his kind assistance with the preparation of materials.

Conflicts of Interest: The authors declare no conflicts of interest.

References

1. Anand, U.; Dey, A.; Chandel, A.K.S.; Sanyal, R.; Mishra, A.; Pandey, D.K.; De Falco, V.; Upadhyay, A.; Kandimalla, R.; Chaudhary, A.; et al. Cancer chemotherapy and beyond: Current status, drug candidates, associated risks and progress in targeted therapeutics. *Genes. Dis.* **2023**, *10*, 1367–1401. [CrossRef] [PubMed]
2. Li, D.; Hu, D.; Xu, H.; Patra, H.K.; Liu, X.; Zhou, Z.; Tang, J.; Slater, N.; Shen, Y. Progress and perspective of microneedle system for anti-cancer drug delivery. *Biomaterials* **2021**, *264*, 120410–120421. [CrossRef] [PubMed]
3. Johnson, Z.L.; Chen, J. ATP Binding Enables Substrate Release from Multidrug Resistance Protein 1. *Cell* **2018**, *172*, 81–89.e10. [CrossRef] [PubMed]
4. Bianchini, G.; Balko, J.M.; Mayer, I.A.; Sanders, M.E.; Gianni, L. Triple-negative breast cancer: Challenges and opportunities of a heterogeneous disease. *Nat. Rev. Clin. Oncol.* **2016**, *13*, 674–690. [CrossRef] [PubMed]
5. Ashrafizadeh, M.; Zarrabi, A.; Bigham, A.; Taheriazam, A.; Saghari, Y.; Mirzaei, S.; Hashemi, M.; Hushmandi, K.; Karimi-Maleh, H.; Nazarzadeh Zare, E.; et al. (Nano)platforms in breast cancer therapy: Drug/gene delivery, advanced nanocarriers and immunotherapy. *Med. Res. Rev.* **2023**, *43*, 2115–2176. [CrossRef] [PubMed]
6. Li, Z.; Zou, J.; Chen, X. In Response to Precision Medicine: Current Subcellular Targeting Strategies for Cancer Therapy. *Adv. Mater.* **2023**, *35*, e2209529. [CrossRef]
7. Palazzolo, S.; Bayda, S.; Hadla, M.; Caligiuri, I.; Corona, G.; Toffoli, G.; Rizzolio, F. The Clinical Translation of Organic Nanomaterials for Cancer Therapy: A Focus on Polymeric Nanoparticles, Micelles, Liposomes and Exosomes. *Curr. Med. Chem.* **2018**, *25*, 4224–4268. [CrossRef] [PubMed]
8. Gao, Y.; Gao, D.; Shen, J.; Wang, Q. A Review of Mesoporous Silica Nanoparticle Delivery Systems in Chemo-Based Combination Cancer Therapies. *Front. Chem.* **2020**, *8*, 598722–598738. [CrossRef] [PubMed]

9. Castillo, R.R.; Colilla, M.; Vallet-Regi, M. Advances in mesoporous silica-based nanocarriers for co-delivery and combination therapy against cancer. *Expert. Opin. Drug Deliv.* **2017**, *14*, 229–243. [CrossRef]
10. Hadipour Moghaddam, S.P.; Yazdimamaghani, M.; Ghandehari, H. Glutathione-sensitive hollow mesoporous silica nanoparticles for controlled drug delivery. *J. Control. Release* **2018**, *282*, 62–75. [CrossRef]
11. Theivendran, S.; Gu, Z.; Tang, J.; Yang, Y.; Song, H.; Yang, Y.; Zhang, M.; Cheng, D.; Yu, C. Nanostructured Organosilica Nitric Oxide Donors Intrinsically Regulate Macrophage Polarization with Antitumor Effect. *ACS Nano* **2022**, *16*, 10943–10957. [CrossRef] [PubMed]
12. Yang, Y.; Chen, F.; Xu, N.; Yao, Q.; Wang, R.; Xie, X.; Zhang, F.; He, Y.; Shao, D.; Dong, W.F.; et al. Red-light-triggered self-destructive mesoporous silica nanoparticles for cascade-amplifying chemo-photodynamic therapy favoring antitumor immune responses. *Biomaterials* **2022**, *281*, 121368–121377. [CrossRef] [PubMed]
13. Liu, B.; Feng, L.; Bian, Y.; Yuan, M.; Zhu, Y.; Yang, P.; Cheng, Z.; Lin, J. $Mn^{2+}/Fe^{3+}/Co^{2+}$ and Tetrasulfide Bond Co-Incorporated Dendritic Mesoporous Organosilica as Multifunctional Nanocarriers: One-Step Synthesis and Applications for Cancer Therapy. *Adv. Healthc. Mater.* **2022**, *11*, e2200665. [CrossRef] [PubMed]
14. Wang, X.; Li, X.; Ito, A.; Sogo, Y.; Watanabe, Y.; Tsuji, N.M.; Ohno, T. Biodegradable Metal Ion-Doped Mesoporous Silica Nanospheres Stimulate Anticancer Th1 Immune Response in Vivo. *ACS Appl. Mater. Inter.* **2017**, *9*, 43538–43544. [CrossRef] [PubMed]
15. Vallet-Regi, M.; Schuth, F.; Lozano, D.; Colilla, M.; Manzano, M. Engineering mesoporous silica nanoparticles for drug delivery: Where are we after two decades? *Chem. Soc. Rev.* **2022**, *51*, 5365–5451. [CrossRef] [PubMed]
16. Liu, C.-G.; Han, Y.-H.; Zhang, J.-T.; Kankala, R.K.; Wang, S.-B.; Chen, A.-Z. Rerouting engineered metal-dependent shapes of mesoporous silica nanocontainers to biodegradable Janus-type (sphero-ellipsoid) nanoreactors for chemodynamic therapy. *Chem. Eng. J.* **2019**, *370*, 1188–1199. [CrossRef]
17. Yu, X.; Wang, X.; Yamazaki, A.; Li, X. Tumor microenvironment-regulated nanoplatforms for the inhibition of tumor growth and metastasis in chemo-immunotherapy. *J. Mater. Chem. B* **2022**, *10*, 3637–3647. [CrossRef] [PubMed]
18. Shao, L.; Hu, T.; Fan, X.; Wu, X.; Zhou, F.; Chen, B.; Tan, S.; Xu, H.; Pan, A.; Liang, S.; et al. Intelligent Nanoplatform with Multi Therapeutic Modalities for Synergistic Cancer Therapy. *ACS Appl. Mater. Inter.* **2022**, *14*, 13122–13135. [CrossRef]
19. Qin, G.; Xie, W.; Luo, X.; Zou, G.; Mo, Q.; Zhong, W. Manganese-doped stellate mesoporous silica nanoparticles: A bifunctional nanoplatform for enhanced chemodynamic therapy and tumor imaging. *Microporous Mesoporous Mater.* **2024**, *370*, 113012. [CrossRef]
20. Lei, G.; Zhuang, L.; Gan, B. Targeting ferroptosis as a vulnerability in cancer. *Nat. Rev. Cancer* **2022**, *22*, 381–396. [CrossRef]
21. Liu, T.; Zhou, Z.; Zhang, M.; Lang, P.; Li, J.; Liu, Z.; Zhang, Z.; Li, L.; Zhang, L. Cuproptosis-immunotherapy using PD-1 overexpressing T cell membrane-coated nanosheets efficiently treats tumor. *J. Control. Release* **2023**, *362*, 502–512. [CrossRef]
22. Tsvetkov, P.; Coy, S.; Petrova, B.; Dreishpoon, M.; Verma, A.; Abdusamad, M.; Rossen, J.; Joesch-Cohen, L.; Humeidi, R.; Spangler, R.D.; et al. Copper induces cell death by targeting lipoylated TCA cycle proteins. *Science* **2022**, *375*, 1254–1261. [CrossRef]
23. Lv, M.; Chen, M.; Zhang, R.; Zhang, W.; Wang, C.; Zhang, Y.; Wei, X.; Guan, Y.; Liu, J.; Feng, K.; et al. Manganese is critical for antitumor immune responses via cGAS-STING and improves the efficacy of clinical immunotherapy. *Cell Res.* **2020**, *30*, 966–979. [CrossRef] [PubMed]
24. Yang, Y.; Fan, H.; Xu, X.; Yao, S.; Yu, W.; Guo, Z. Zinc Ion-Induced Immune Responses in Antitumor Immunotherapy. *CCS Chem.* **2024**, 1–20. [CrossRef]
25. Wang, X.; Li, X.; Onuma, K.; Sogo, Y.; Ohno, T.; Ito, A. Zn- and Mg-containing tricalcium phosphates-based adjuvants for cancer immunotherapy. *Sci. Rep.* **2013**, *3*, 2203. [CrossRef]
26. An, J.; Zhang, K.; Wang, B.; Wu, S.; Wang, Y.; Zhang, H.; Zhang, Z.; Liu, J.; Shi, J. Nanoenabled Disruption of Multiple Barriers in Antigen Cross-Presentation of Dendritic Cells via Calcium Interference for Enhanced Chemo-Immunotherapy. *ACS Nano* **2020**, *14*, 7639–7650. [CrossRef]
27. Awad, H.H.; El-Derany, M.O.; Mantawy, E.M.; Michel, H.E.; El-Naa, M.M.; Salah El-Din, R.A.; El-Brairy, A.I.; El-Demerdash, E. Comparative study on beneficial effects of vitamins B and D in attenuating doxorubicin induced cardiotoxicity in rats: Emphasis on calcium homeostasis. *Biomed. Pharmacother.* **2021**, *140*, 111679–111693. [CrossRef] [PubMed]
28. He, Y.; Zeng, B.; Liang, S.; Long, M.; Xu, H. Synthesis of pH-Responsive Biodegradable Mesoporous Silica-Calcium Phosphate Hybrid Nanoparticles as a High Potential Drug Carrier. *ACS Appl. Mater. Inter.* **2017**, *9*, 44402–44409. [CrossRef]
29. Choi, E.; Lim, D.-K.; Kim, S. Calcium-doped mesoporous silica nanoparticles as a lysosomolytic nanocarrier for amine-free loading and cytosolic delivery of siRNA. *J. Ind. Eng. Chem.* **2020**, *81*, 71–80. [CrossRef]
30. Chen, M.; Hu, J.; Bian, C.; Zhu, C.; Chen, C.; Guo, Z.; Zhang, Z.; Agyekum, G.A.; Zhang, Z.; Cao, X. pH-Responsive and Biodegradable ZnO-Capped Mesoporous Silica Composite Nanoparticles for Drug Delivery. *Materials* **2020**, *13*, 3950. [CrossRef]
31. Chang, L.; Yan, H.; Chang, J.; Gautrot, J.E. Cationic polymer brush-coated bioglass nanoparticles for the design of bioresorbable RNA delivery vectors. *Eur. Polym. J.* **2021**, *156*, 110593–110601. [CrossRef]
32. Gao, S.; Liu, Y.; Liu, M.; Yang, D.; Zhang, M.; Shi, K. Biodegradable mesoporous nanocomposites with dual-targeting function for enhanced anti-tumor therapy. *J. Control. Release* **2022**, *341*, 383–398. [CrossRef] [PubMed]
33. Guo, S.; Li, Z.; Feng, J.; Xiong, W.; Yang, J.; Lu, X.; Yang, S.; Xu, Y.; Wu, A.; Shen, Z. Cycloacceleration of ferroptosis and calcicoptosis for magnetic resonance imaging-guided colorectal cancer therapy. *Nano Today* **2022**, *47*, 101663–101677. [CrossRef]

34. Gong, F.; Xu, J.; Liu, B.; Yang, N.; Cheng, L.; Huang, P.; Wang, C.; Chen, Q.; Ni, C.; Liu, Z. Nanoscale CaH2 materials for synergistic hydrogen-immune cancer therapy. *Chem* **2022**, *8*, 268–286. [CrossRef]
35. de Barros, M.R.; Bittencourt, O.R.; Crocomo, P.Z.; Mafra, G.; Carasek, E.; Magosso, H.A.; Jost, C.L.; Winiarski, J.P. Adsorption of hazardous and noxious 4-nitrophenol by a silsesquioxane organic-inorganic hybrid material. *J. Sol-Gel Sci. Technol.* **2021**, *99*, 402–412. [CrossRef]
36. Li, Z.; Hu, M.; Shen, K.; Zhou, F.; Chen, Z.; Cheng, X.; Liu, Q.; Wu, X. Enhancing thermal safety of hydrophobic silica aerogels by incorporating sodium dodecyl sulfate intercalated layered double hydroxides. *J. Sol-Gel Sci. Technol.* **2024**. [CrossRef]
37. Narayan, R.; Nayak, U.Y.; Raichur, A.M.; Garg, S. Mesoporous Silica Nanoparticles: A Comprehensive Review on Synthesis and Recent Advances. *Pharmaceutics* **2018**, *10*, 118. [CrossRef] [PubMed]
38. Zhang, Y.; Huang, C.; Chang, J. Ca-Doped mesoporous SiO_2/dental resin composites with enhanced mechanical properties, bioactivity and antibacterial properties. *J. Mater. Chem. B* **2018**, *6*, 477–486. [CrossRef] [PubMed]
39. Li, X.; Zhang, L.; Dong, X.; Liang, J.; Shi, J. Preparation of mesoporous calcium doped silica spheres with narrow size dispersion and their drug loading and degradation behavior. *Microporous Mesoporous Mater.* **2007**, *102*, 151–158. [CrossRef]
40. dos Santos, G.A.; Abreu e Lima, R.S.; Pestana, C.R.; Lima, A.S.; Scheucher, P.S.; Thome, C.H.; Gimenes-Teixeira, H.L.; Santana-Lemos, B.A.; Lucena-Araujo, A.R.; Rodrigues, F.P.; et al. (+)alpha-Tocopheryl succinate inhibits the mitochondrial respiratory chain complex I and is as effective as arsenic trioxide or ATRA against acute promyelocytic leukemia in vivo. *Leukemia* **2012**, *26*, 451–460. [CrossRef]
41. Bock, F.J.; Tait, S.W.G. Mitochondria as multifaceted regulators of cell death. *Nat. Rev. Mol. Cell Biol.* **2020**, *21*, 85–100. [CrossRef] [PubMed]
42. Kroemer, G.; Galassi, C.; Zitvogel, L.; Galluzzi, L. Immunogenic cell stress and death. *Nat. Immunol.* **2022**, *23*, 487–500. [CrossRef]
43. Galluzzi, L.; Humeau, J.; Buque, A.; Zitvogel, L.; Kroemer, G. Immunostimulation with chemotherapy in the era of immune checkpoint inhibitors. *Nat. Rev. Clin. Oncol.* **2020**, *17*, 725–741. [CrossRef] [PubMed]
44. Zhang, L.; Montesdeoca, N.; Karges, J.; Xiao, H. Immunogenic Cell Death Inducing Metal Complexes for Cancer Therapy. *Angew. Chem. Int. Ed. Engl.* **2023**, *62*, e202300662. [CrossRef] [PubMed]
45. Li, X.; Wu, M.; Pan, L.; Shi, J. Tumor vascular-targeted co-delivery of anti-angiogenesis and chemotherapeutic agents by mesoporous silica nanoparticle-based drug delivery system for synergetic therapy of tumor. *Int. J. Nanomed.* **2016**, *11*, 93–105. [CrossRef]

inorganics

MDPI

Article

Luminescence Efficiency and Spectral Compatibility of Cerium Fluoride (CeF_3) Inorganic Scintillator with Various Optical Sensors in the Diagnostic Radiology X-ray Energy Range

Vasileios Ntoupis [1], Christos Michail [1,*], Nektarios Kalyvas [1], Athanasios Bakas [2], Ioannis Kandarakis [1], George Fountos [1] and Ioannis Valais [1]

1 Radiation Physics, Materials Technology and Biomedical Imaging Laboratory, Department of Biomedical Engineering, University of West Attica, Ag. Spyridonos, 12210 Athens, Greece; vdoupis@uniwa.gr (V.N.); nkalyvas@uniwa.gr (N.K.); kandarakis@uniwa.gr (I.K.); gfoun@uniwa.gr (G.F.); valais@uniwa.gr (I.V.)

2 Department of Biomedical Sciences, University of West Attica, Ag. Spyridonos, 12210 Athens, Greece; abakas@uniwa.gr

* Correspondence: cmichail@uniwa.gr

Abstract: The aim of this study was to experimentally assess the luminescence efficiency of a cerium fluoride (CeF_3) inorganic scintillator in crystal form as a possible alternative to high-luminescence but hygroscopic cerium bromide ($CeBr_3$). The experiments were performed under typical diagnostic radiology X-rays (50–140 kVp). Parameters such as the crystal's absolute luminescence efficiency (AE) and the spectral matching with a series of optical detectors were examined. The replacement of bromine with fluorine appeared to drastically reduce the AE of CeF_3 compared to $CeBr_3$ and other commercially available inorganic scintillators such as bismuth germanate ($Bi_4Ge_3O_{12}$-BGO). CeF_3 reaches a maximum luminescence efficiency value of only 0.8334 efficiency units (EUs) at 140 kVp, whereas the corresponding values for $CeBr_3$ and BGO were 29.49 and 3.41, respectively. Furthermore, the emission maximum (at around 313 nm) moved towards the lower part of the visible spectrum, making CeF_3 suitable for spectral coupling with various photocathodes and photomultipliers applied in nuclear medicine detectors, but completely unsuitable for spectral matching with CCDs and CMOS. The obtained luminescence efficiency results denote that CeF_3 cannot be applied in medical imaging applications covering the range 50–140 kVp; however, examination of its luminescence output in the nuclear medicine energy range (~70 to 511 keV) could reveal possible applicability in these modalities.

Keywords: inorganic scintillators; crystals; X-ray radiation; medical detectors; cerium fluoride; medical imaging

1. Introduction

Research on novel inorganic scintillators for a variety of applications, including medical applications, is an ongoing task for research groups worldwide [1–11]. An efficient single-crystal scintillator with a light yield of the order of 6×10^4 photons/MeV is cerium bromide ($CeBr_3$). It also has a decay time of only 19 ns [12] and a spectrum lying in the lower part of the visible range, making it suitable for coupling with various optical detectors used in nuclear medicine [13–21].

However, $CeBr_3$ is hygroscopic and should be encapsulated in order to maintain its structural integrity, leading to a reduction in the luminescence signal [22]. A material with a similar structure with the bromine (Br) atom replaced by fluorine (F) is cerium fluoride (CeF_3) [13,23–28].

CeF_3 has been in the spotlight in the last 30 years due to experiments including the Large Hadron Collider (LHC) [29–38]. Since CeF_3 single crystals have attractive optical, mechanical and luminescence characteristics, they are suitable for a variety of applications [25,29,39–42]. CeF_3 crystals are rare-earth trifluorides with a tysonite structure [43].

The crystalline structure can be hexagonal, as per the crystal sample used in our study, at high temperatures and trigonal at low ones [43]. CeF_3's hexagonal-shape crystalline structure includes layers of cerium ions situated very close together. At the same time, fluorine ions occupy three distinct positions inside the lattice [43–45]. Previous X-ray diffraction (XRD) analysis exhibits characteristic peaks that agree with the hexagonal structure of the crystal [43–45]. Optical and structural properties, such as a large Verdet constant (247 rad/Tm at 450 nm), high transmittance of the crystal, about 92% compared to terbium gallium garnet (TGG) and a great cutoff absorption edge (at 270 nm compared to 400 nm for the TGG crystal), render it a good scintillator candidate for magneto-optical materials and mostly for various optical applications in the region of high-energy physics [46,47].

Furthermore, a CeF_3 crystal's mechanical properties reveal an important anisotropy. Young's modulus anisotropy (E_{max}/E_{min} = 195 GPa/100 GPa = 1.95), shear modulus anisotropy (G_{max}/G_{min} = 65.7 GPa/34.2 GPa = 1.92), Poisson's ratio (0.16–0.48) and Vickers microhardness (2.3–2.9 GPa) make CeF_3 crystals very useful for technologies that need flexible and durable materials [45].

It has been discovered in earlier studies that CeF_3 responds non-proportionally to gamma rays up to 5.1 MeV [48,49]. Nonetheless, the analysis of CeF_3 monocrystals demonstrates that they are appropriate for high-rate calorimetry. Tests utilizing CeF_3-based prototype calorimeters (such as the high-luminosity Large Hadron Collider, HL-LHC) demonstrated encouraging results in terms of energy resolution [36–38].

According to recent studies, a CeF_3 crystal is an excellent Faraday rotator for broadband high-precision spectroscopy, particularly suitable for high-power laser applications [40,50–52]. This material appears with better characteristics than the commercial TGG crystal in terms of transparency in the UV-NIR region [46]. Additionally, a CeF_3 mono-crystal as an optical isolator is an advantageous alternative for protecting laser devices from back reflection, particularly in the UV-NIR wavelength ranges [40,46,50].

An essential disadvantage of a CeF_3 inorganic scintillator is that it tends strongly to overlap the luminescence band with the absorption cutoff, which is due to the fact that the Stokes shift is small and the material is a self-activated scintillator [39–41,43,44,46–49,53,54].

The aim of this work was to experimentally assess the general performance (emission efficiency and spectral compatibility with optical sensors) of a CeF_3 crystal under conditions found in medical laboratories as well as in various other applications such as industrial, geological, archeological, in security, etc. Most of these applications use mainly continuous spectrum X-rays produced by suitable X-ray tubes.

The findings of this work were compared with a $CeBr_3$ crystal that was recently examined by our group as well as to a crystal that has been incorporated in commercially available nuclear medicine scanners, i.e., bismuth germanate ($Bi_4Ge_3O_{12}$-BGO) [1,55]. The properties of the three materials are shown in Table 1.

Table 1. Properties of CeF_3, $CeBr_3$ and BGO single crystals [1,18,19,26,39,40,56–75].

	Units	CeF_3	$CeBr_3$	BGO
Wavelength of max emission	Nm	340 (slow), 300–310 (fast) [40,68]	380 [1,18]	480 [19,74]
Emission wavelength range	Nm	250–425 [68]	325–475 [1]	375–650 [1]
Decay times	Ns	30 (slow), 8 (fast) [40]	19 [1,18]	300 [19,56]
Light yield	photons/MeV	4.4×10^3 [40]	6×10^4 [1,18,26]	8.9×10^3 [19]
Photoelectron yield	% of NaI:Tl	8 [68]	122 [1]	15–20 [74]
Radiation length	Cm	1.654 [68]	1.96 [1]	1.12 [76]
Refractive index @ max nm		1.62 @ 440 nm [40]	2.09 @ 340 nm [1]	2.15 @ 480 nm [77]
Density	g/cm^3	6.16 [40]	5.1 [1]	7.13 [19]

Table 1. *Cont.*

	Units	CeF_3	$CeBr_3$	BGO
Melting point	°C	1443–1450 [39,40,71]	722 [1]	1050 [74]
Mechanical hardness	Mohs	4.5–5 [39]	5–6 [1]	5 [78]
Radiation hardness	Rad	$>10^6$ [73]	2×10^3 [1]	10^7 [79]
Hygroscopic		No [40]	Yes [1,18]	No [19]

2. Results

Figure 1 shows the signal reading produced by the CeF_3 crystal, after X-ray irradiation, compared to the X-ray exposure rate. In the same Figure can be seen the output signal of the $CeBr_3$ and BGO crystals, under similar irradiation conditions. All the crystal samples had a purity of >99.99% (personal communication with Advatech (London, UK). From Figure 1, it can be depicted that the output signal and the exposure rate of the medical X-ray tube can be considered linear in the examined energy range (0 to 372 mR/s), except for $CeBr_3$ where linearity is lost for less than 50 mR/s. However, the replacement of bromine with fluorine in $CeBr_3$ results in a significant reduction in the signal output curve (from 29.54 in $CeBr_3$ to 0.834 in CeF_3). The output signal of BGO, which is a commercially available crystal in medical imaging applications, is also greater than that of CeF_3, denoting that CeF_3 has a low light yield for medical applications in the energy range 50–140 kVp.

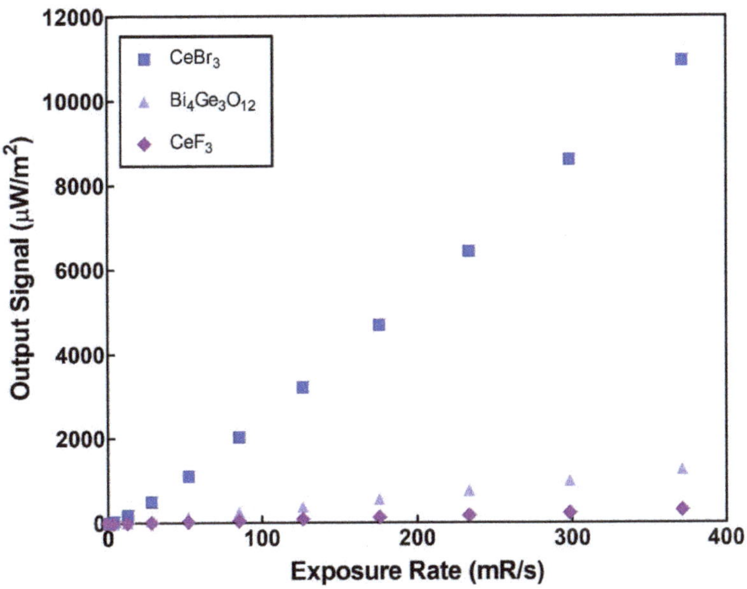

Figure 1. CeF_3's output signal compared to $CeBr_3$ and BGO.

Figure 2 shows the experimental results regarding the absolute luminescence efficiency of CeF_3 compared to the $CeBr_3$ and $Bi_4Ge_3O_{12}$ crystals. The results are shown for the available kVp of the X-ray tube, with steps of 10 kVp. A common finding, for the three examined crystals, is that in the examined X-ray range, the luminescence efficiency increases when the tube voltage increases. The increase in the luminescence efficiency is more evident in the range 50–80 kVp, due to the fact that the K-edges of all three materials are in this energy range. CeF_3 reaches a maximum value of only 0.8334 EU (EU is the S.I. equivalent $\mu Wm^{-2}/(mGy/s)$ at 140 kVp), whereas the corresponding values for $CeBr_3$ and BGO are

29.49 and 3.41, respectively, despite the higher density (6.16 compared to 5.1 g/cm^3) of CeF_3 compared with $CeBr_3$ (Table 1) [1]. Luminescence efficiency measurements in higher available X-ray energies could reveal a more efficient performance that possibly fulfils the requirements for nuclear medicine applications.

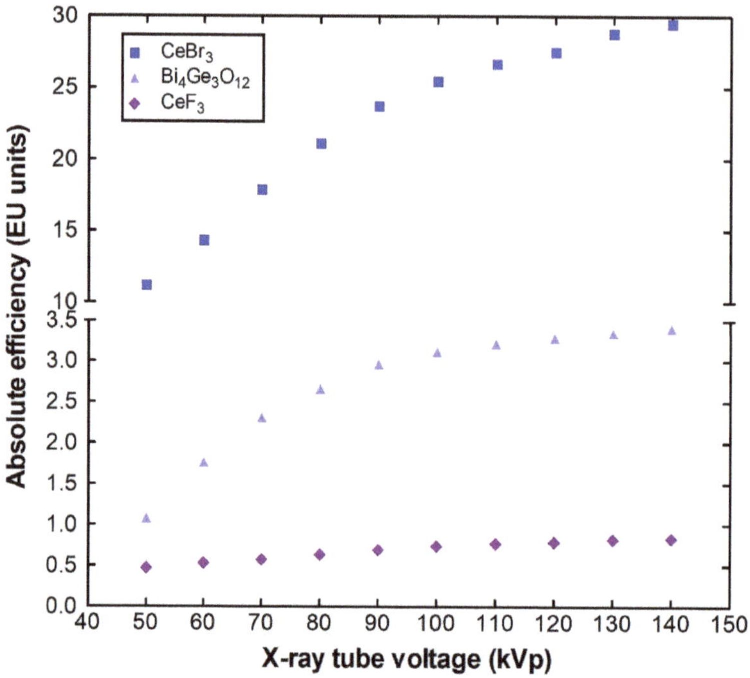

Figure 2. CeF_3's absolute luminescence efficiency compared to $CeBr_3$ and BGO.

Figure 3 shows the spectral matching values of CeF_3 with a broad variety of optical sensors. The emission spectrum of CeF_3 in the range of 260–400 nm and with an emission maximum at around 313 nm, which shifted to the shortwave region by 67 nm compared to $CeBr_3$, limits the available sensors commonly used in medical imaging applications that could be coupled with, such as charge coupled devices (CCDs) or complementary metal oxide semiconductors (CMOS) [80].

The main luminescence mechanism, observed in the CeF_3 crystal, is mainly due to 5d-4f transitions of the cerium (Ce^{3+}) ions [25]. Electron transitions from the 4f to the 5d energy level occur when these ions (CeF_3) are excited. The return of electrons to the 4f state produces photons. Due to this mechanism, the two primary luminescence bands of the crystal are created. These bands are located at around 290 nm and 340 nm, respectively [25,81,82]. Previous studies have demonstrated an association between the emission at 290 nm and the 5d-4f transitions in cerium (Ce^{3+}) ions. This band is produced by the electrons that radiatively decay from the excited 5d ground state to the 4f ground state [25,81,82]. Frenkel self-trapped excitons are produced as a result of the presence of sub bands in the conduction band, which are obtained from the 5d states of the ions and provide the procedure of the luminescence [25,43–45]. The electronic structure and the luminescence mechanism of CeF_3 crystal are valuable in evolving scintillators for radiological applications and for other applications in high-energy physics [40,46,47,50–52].

As can be depicted from Figure 3, the spectral matching factors between CeF_3 and most of the CCDs and CMOS are close to zero. However, CeF_3 can be matched with a variety of optical sensors, such as silicon (SiPMs), position-sensitive photomultipliers, etc., which are used in nuclear medicine applications [83,84]. For example, the emission spectrum of CeF_3 has a 97% match with the spectral sensitivity of multialkali photocathodes and flat-panel position-sensitive (PS) photomultipliers (H8500D-03) (for both, the spectral matching factor is 0.97). Lower compatibility values can be found with gallium arsenide (GaAs) photocathodes (α_s = 0.8) and the extended (E-S20) photocathode used to measure the luminescence efficiency signal in this work (α_s = 0.76). This value is also used for the calculation of the AE.

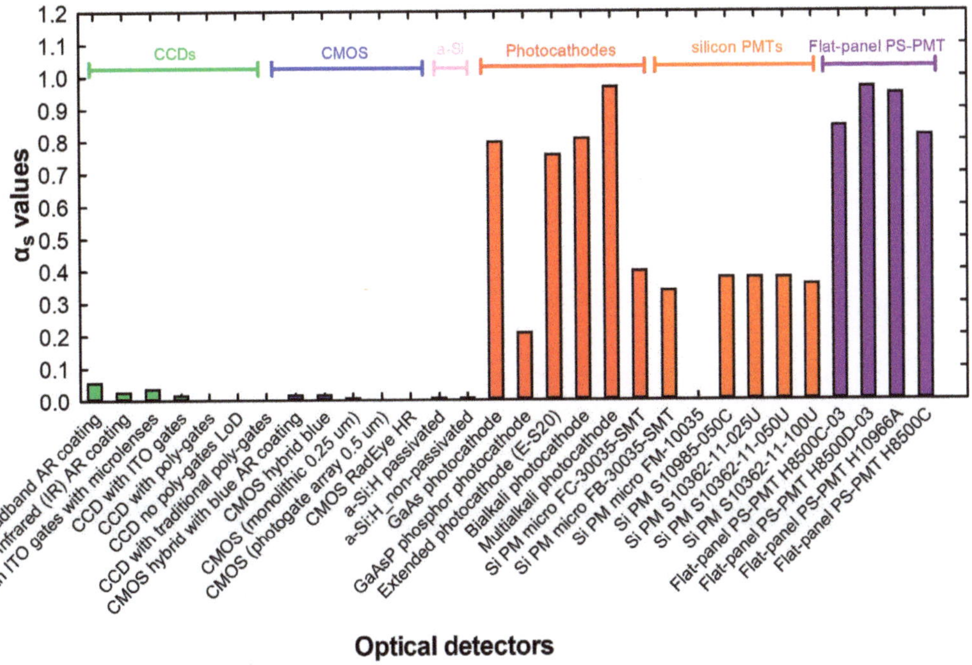

Figure 3. CeF_3 inorganic crystal's spectral matching factors with a series of digital optical detectors available for medical imaging applications.

The spectral responses that were used for the calculation of the α_s between CeF_3 with a series of light sensors are shown in Figures 4 and 5, along with the emitted spectrum of CeF_3 [68]. Figure 4 shows the grouped spectral responses of the photocathodes (Figure 4a) and silicon photomultipliers (Figure 4b), whereas Figure 5 shows the grouped spectral responses of the CCDs (Figure 5a) and CMOS (Figure 5b).

The CeF_3 spectral matching results, presented in the spectral matching factors data of Figure 3, were used to calculate the EE that could be produced upon coupling the specific crystal with the examined optical sensors. Figures 6 and 7 show the calculated effective luminescence efficiency values with the X-ray tube voltage for the above-mentioned optical sensors. Figure 6 shows the grouped spectral responses of the photocathodes (Figure 6a) and silicon photomultipliers (Figure 6b), whereas Figure 7 shows the grouped spectral responses of the CCDs (Figure 7a) and CMOS (Figure 7b).

Following the spectral matching factor results, the effective efficiency is maximized for multialkali photocathodes and flat-panel position-sensitive photomultipliers, since in these cases, most of the emitted light from the CeF_3 scintillator can be adequately collected by the detectors.

The combination of the multialkali photocathode with CeF_3 results in the effective efficiency values being reduced by 2.58% compared to the absolute efficiency values at

140 kVp. In the case of the combination of CeF_3 with the gallium arsenide photocathode, the corresponding absolute efficiency reduction is 24.26% at 140 kVp.

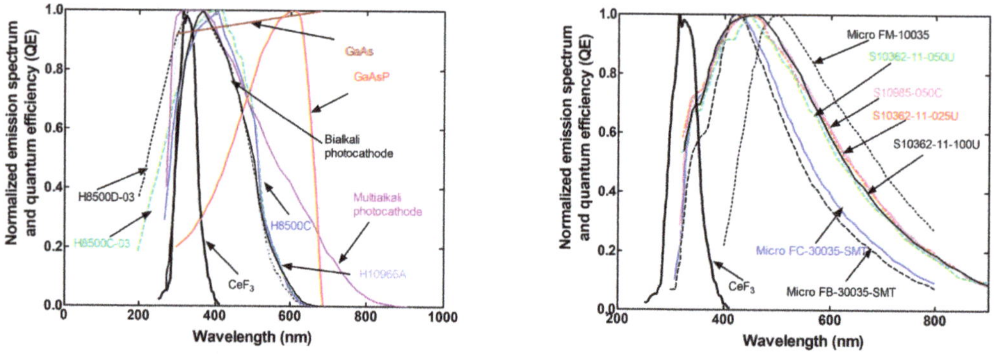

(**a**) (**b**)

Figure 4. Spectral responses of (**a**) photocathodes and (**b**) silicon PMs, along with the emission spectrum of CeF_3.

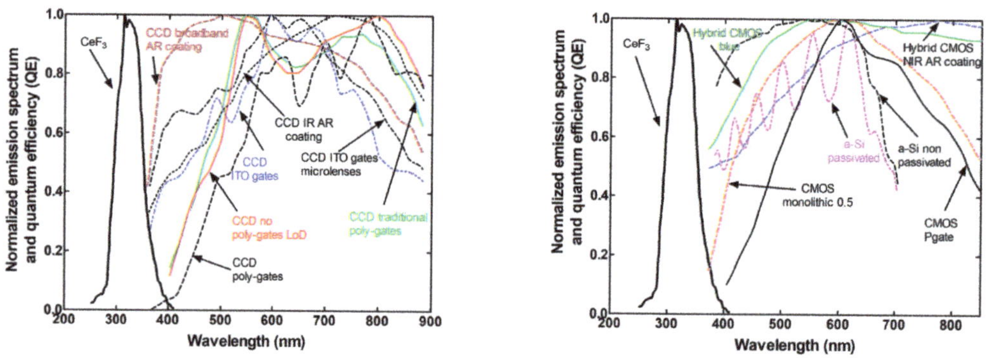

(**a**) (**b**)

Figure 5. Spectral responses of (**a**) CCDs and (**b**) CMOS, along with the emission spectrum of CeF_3.

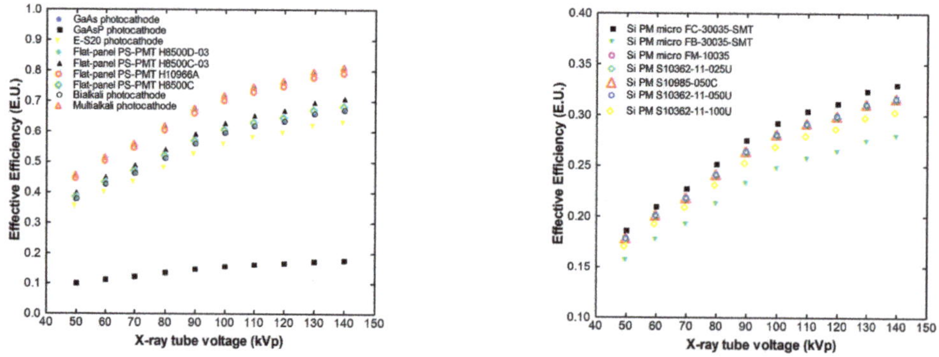

(**a**) (**b**)

Figure 6. Effective luminescence efficiency of CeF_3 with photocathodes (**a**) and silicon PMTs (**b**).

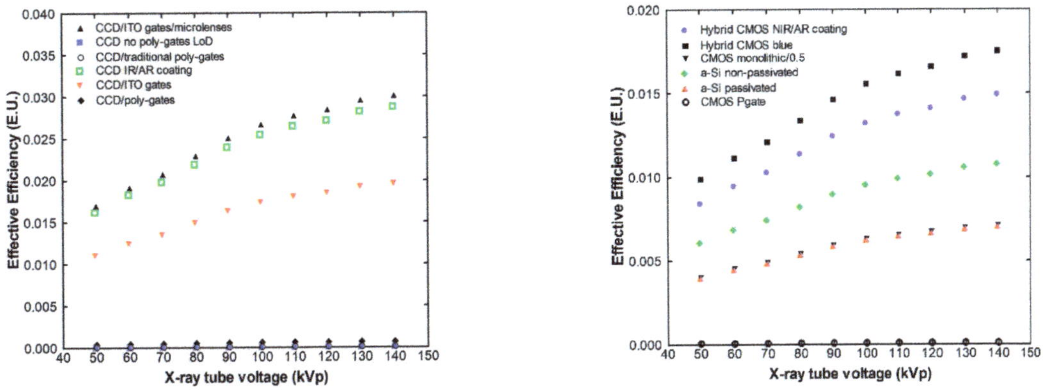

(a) (b)

Figure 7. Effective luminescence efficiency of CeF_3 with CCDs (**a**) and CMOS (**b**).

Typical EE values for single-crystal scintillators, already integrated in medical imaging systems, range from 3 EE for BGO (used, for example, in the GE Discovery IQ scanner, GE Healthcare, Milwaukee, WI, USA) [85], 5 EE for GSO:Ce (Philips Gemini GXL PET/CT, Philips Medical Systems, Eindhoven, The Netherlands) [86], 8 for LYSO:Ce (Philips Gemini TF PET/CT, Philips Medical Systems, Eindhoven, The Netherlands) [87] to 12 for LSO:Ce (Siemens Biograph TruePoint PET/CT, Siemens Healthineers, Forchheim, Germany) [88,89].

Figure 8 shows the X-ray luminescence efficiency of the CeF_3 crystal compared to the data for $CeBr_3$ and BGO, in the examined X-ray tube voltage range. The XLE is an important parameter for detectors that are operated as energy integrators. This type of sensor is used in medical imaging systems, in which the signal output is related to the amount of energy that is captured by the crystal. The CeF_3 XLE values are below the corresponding ones for $CeBr_3$ and BGO as was expected from the previous findings (Figures 1 and 2). They start from 3.2×10^{-4} at 50 kVp and drop down to 2.3×10^{-4} at 140 kVp. The corresponding values for $CeBr_3$ and BGO were 7.6×10^{-3} and 7.3×10^{-4} at 50 kVp, reaching a maximum of 8.9×10^{-3} ($CeBr_3$ at 90 kVp) and 1.1×10^{-3} (BGO at 80 kVp) and afterwards, both drop down to 8.2×10^{-3} and 9.4×10^{-4} at 140 kVp, respectively. In the case of CeF_3, even the characteristic K-edge (at ~40 keV as shown in Figure 9) does not contribute to an XLE increase, like in the case of $CeBr_3$, with a K-edge at almost the same energy with CeF_3 (~40 keV) or BGO (~90.5 keV).

Figure 9 shows the calculated attenuation coefficients (μ_{att}) and the energy absorption (μ_{en}) coefficients of CeF_3, $CeBr_3$ and BGO, in the energy range 8 to 150 keV. The elemental coefficients (cerium, fluorine, bromine, bismuth (Bi), germanium (Ge) and oxygen (O)) that were used to calculate the CeF_3, $CeBr_3$ and BGO mixtures were obtained from the NIST XmuDat database [90,91]. The decrease in the coefficients, except from the sudden increase due to the characteristic K-absorption, for every mixture point, where the probability of photoelectric interaction maximizes, can also be seen in Figure 9.

Figure 10 shows the EAE values of CeF_3 compared to $CeBr_3$ and BGO. The energy absorption efficiency strongly depends on the μ_{en} and μ_{att} coefficients, shown in Figure 9, as well as on their ratio (μ_{en}/μ_{att}, Equation (4)), shown as an inset in Figure 10. Initially, for energies up to 39 keV, CeF_3 maintained a higher attenuation coefficient ratio than both $CeBr_3$ and BGO, leading to a higher energy absorption efficiency value at 40 kVp. However, when the energy increases, the energy absorption of BGO showed superior energy absorption efficiency values across the examined energy range, due to the combined effects of the density, higher μ_{en}/μ_{att} ratio (from 39 to 90 keV) and the increase in absorption due to the K-edge. Both $CeBr_3$ and CeF_3 have close attenuation coefficient values due to the presence of cerium (gram atomic mass = 140.116 g/mol) in both crystals; however, the influence of the three atoms of bromine (gram atomic mass = 79.904 g/mol) is stronger than the corresponding three atoms of fluorine (gram atomic mass = 18.998 g/mol), leading to a gram molecular mass for CeF_3 equal to 197.111 g/mol compared to 172.549 g/mol for $CeBr_3$. After the K-edge absorption, CeF_3 exhibits a lower μ_{en}/μ_{att} ratio and consequently lower EAE values up to 129 keV across the examined energy range.

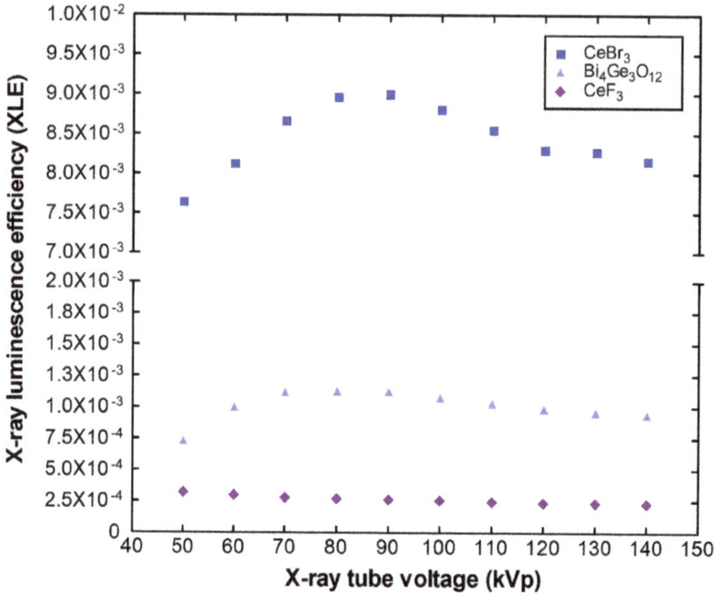

Figure 8. XLE of CeF_3 compared to $CeBr_3$ and BGO.

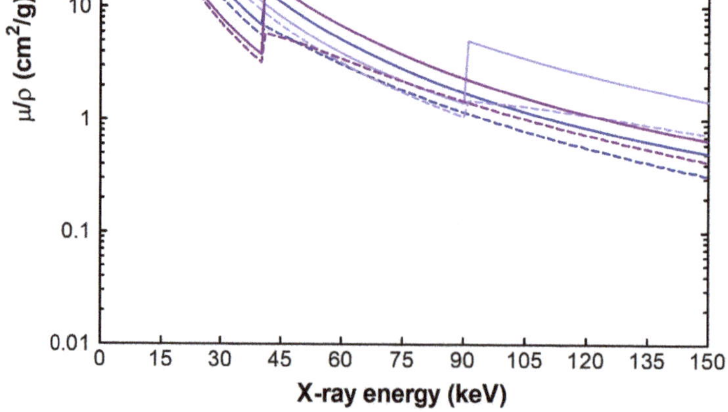

Figure 9. Mass attenuation coefficients of CeF_3 compared to $CeBr_3$ and BGO.

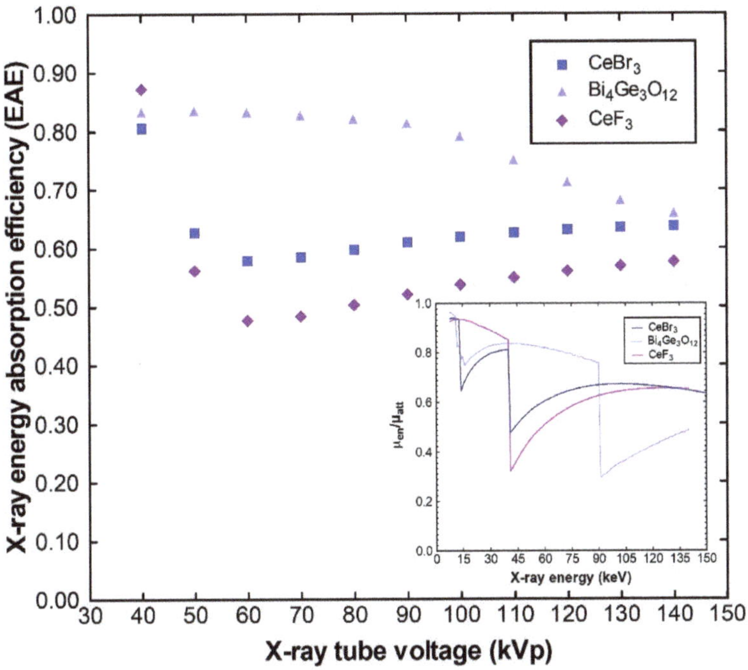

Figure 10. Energy absorption efficiency and μ_{en}/μ_{att} ratios of CeF_3 compared to $CeBr_3$ and BGO.

3. Materials and Methods

The cubic-shaped (edge of 10 mm) and polished CeF_3 crystal was obtained from Advatech (London, UK) [68]. For the irradiation of the sample, a typical radiological X-ray tube (CPI Inc., Georgetown, ON, Canada, model CMP 200 DR high-frequency X-ray generator and an X-ray tube (IAE SpA, Milano, Italy, model SpA-RTM90HS) was used, with voltage settings of 50–140 kVp, tube current fixed at 63 mAs, large focus (1.2 mm), a source to detector distance of 72.5 cm and an additional aluminum (Al) filter with a thickness of 20 mm.

The emitted light energy produced by the crystals per unit area (Ψ_Λ) was determined as the output signal in the energy range that diagnostic medical X-ray tubes allow (50–140 kVp). The methodology is described in detail in [1].

The CeF_3 sample was wrapped with Teflon tape (only one face was free) and inserted in the input port of an integrating sphere ((Newport Corp., Irvine, CA, USA, Oriel model 70451). The light that was produced by the sample, after X-ray excitation, was collected by a photomultiplier tube (PMT, (EMI, London, UK, EMI model 9798). The PMT's photocathode was the extended sensitivity (S20) type. The dose was monitored using an RTI Piranha P100B dosimeter (Mölndal, Sweden). In our experiments, the PMT works with a 22 volts difference between the initial dynode (all dynodes shorted) and the photocathode. The produced current is recorded by a sub-femtoamp remote source meter (Keithley, Cleveland, OH, USA, model 6430).

The ratio of the light energy flux emitted by the examined sample, normalized by the X-ray exposure rate, can be expressed as the absolute luminescence efficiency:

$$AE = \eta_A = \frac{\Psi_\Lambda}{\dot{X}} = \left(\frac{i_{elec}}{S\eta_p\alpha_s c_g}\right)\dot{X}^{-1}, \tag{1}$$

In Equation (1), Ψ_Λ is the light energy flux (output signal) in units of $\mu W\ m^{-2}$. $\dot{X}$ is the exposure rate ($mR\ s^{-1}$). i_{elec} is the current produced by the electrometer in pA and S denotes the surface of the crystal, excited by X-rays (mm^2). The peak sensitivity of the photocathode (η_p) is expressed in units of pA/W. α_s is the spectral matching between the light source (in this case, crystal) to the spectral response of the optical sensor (in this case, the photocathode). Finally, the percentage of the sample's emitted light escaping the integrating sphere and reacting with the PMT is the geometric light collection efficiency (c_g) with a measured value of 15.6. The units of the luminescence efficiency are EU = ($\mu W\ m^{-2}$)/(mR s^{-1}). The experimental error is within ±3.4% due to the combined effects of the reproducibility of the X-ray tube output and optical measurement setups.

If the AE is reduced by the spectral matching factor (α_s) (denoting the percent of the light produced by the scintillator, which is within the same wavelength range as the optical response of a light sensor), then the effective efficiency is derived [1]:

$$EE = \eta_{eff} = \eta_A \alpha_s \tag{2}$$

The effective efficiency was calculated for various optical detectors, often used in medical applications, while the spectrum of CeF_3 was obtained from manufacturer data [68].

The X-ray luminescence efficiency (XLE) is the ratio of the crystal's emitted light energy flux over the X-ray energy flux ($\eta_\Psi = \Psi_\Lambda/\Psi_0$) [1].

Calculation of the XLE can be performed by the conversion of the X-ray exposure (in Equation (1)) to the X-ray energy flux (Ψ_o), using $\Psi_o = X\hat{\Psi}$ where $\hat{\Psi}$ is the X-ray energy flux per exposure rate:

$$\hat{\Psi} = \frac{\int \Psi_0(E)dE}{\int \Psi_0(E)\left[\frac{X}{\Psi_0(E)}\right]dE} = \frac{\int \Psi_0(E)dE}{\int \Psi_0(E)\left[\left(\frac{\mu_{en}(E)}{\rho}\right)_{air}\cdot\left(\frac{W_A}{e}\right)^{-1}\right]dE} \tag{3}$$

$(\mu_{en}(E)/\rho)_{air}$ is the X-ray mass energy absorption coefficient of the air and W_A/e is the average energy/unit of charge, for the production of an electron/ion pair in the air.

The energy absorption efficiency is used to account for the energy locally absorbed at the points of X-ray interaction, within the crystal [1]:

$$EAE(E) = \frac{\int_0^{E_0} \Phi_0(E)E\left(\frac{\mu_{en}(E)/\rho}{\mu_{att}(E)/\rho}\right)\left(1 - e^{-(\mu_{att}(E)/\rho)\rho W}\right)dE}{\int_0^{E_0} \Phi_0(E)EdE} \tag{4}$$

In Equation (4), the X-ray photon fluence is depicted by $\Phi_0(E)$ in units of photons/unit of area and is multiplied by the energy of the photons (E), in order to reproduce the X-ray energy fluence. The X-ray mass attenuation ($\mu_{att}(E)/\rho$) and the energy absorption coefficients ($\mu_{en}(E)/\rho$) were calculated from the NIST XmuDat database [90,91]. The crystal thickness is denoted by W and the density by ρ, in units of g/cm^3 [92].

4. Conclusions

This study aimed to investigate the luminescence properties of a CeF_3 single-crystal inorganic scintillator as a possible alternative to high-luminous but hygroscopic $CeBr_3$. The experiments were performed under the exposure of a typical X-ray unit, with X-rays spanning from 50 to 140 kVp. The fluorine in CeF_3 appeared to drastically reduce the luminescence efficiency, compared to the $CeBr_3$ crystal and other commercially used crystals in medical applications such as BGO. Furthermore, the emission maximum moved even lower than $CeBr_3$, towards the lower part of the visible spectrum and beyond, making CeF_3 suitable for spectral coupling with some multialkali photocathodes and flat-panel position-sensitive photomultipliers applied in nuclear medicine detectors, but completely unsuitable for spectral matching with CCDs and CMOS, used in a variety of medical imaging sensors. The obtained luminescence efficiency results denote that CeF_3 cannot be applied in medical imaging applications, covering the range 50–140 kVp; however, examination of its luminescence efficiency at higher energies could reveal a possible applicability for nuclear medicine detectors.

Author Contributions: Conceptualization, C.M. and V.N.; methodology, C.M., V.N., I.V., N.K., G.F. and I.K.; software, V.N.; validation, N.K. and I.K.; formal analysis, C.M., N.K. and V.N.; investigation, C.M., N.K., G.F., V.N., A.B. and I.V.; resources, A.B.; data curation, C.M., N.K., V.N., G.F. and I.K.; writing—original draft preparation, V.N. and C.M.; writing—review and editing, C.M., I.K., N.K., A.B. and I.V.; visualization, I.V.; supervision, C.M.; project administration, C.M. and I.V. All authors have read and agreed to the published version of the manuscript.

Funding: This research received no external funding.

Data Availability Statement: Data are contained within the article.

Conflicts of Interest: The authors declare no conflicts of interest.

References

1. Linardatos, D.; Michail, C.; Kalyvas, N.; Ninos, K.; Bakas, A.; Valais, I.; Fountos, G.; Kandarakis, I. Luminescence Efficiency of Cerium Bromide Single Crystal under X-ray Radiation. *Crystals* **2022**, *12*, 909. [CrossRef]
2. Zou, Z.; Weng, J.; Liu, C.; Lin, Y.; Zhu, J.; Sun, Y.; Huang, J.; Gong, G.; Wen, H. Crystal Growth, Photoluminescence and Radioluminescence Properties of Ce^{3+}-Doped $Ba_3Y(PO_4)_3$ Crystal. *Crystals* **2024**, *14*, 431. [CrossRef]
3. Ciomaga Hatnean, V.C.; Pui, A.; Simonov, A.; Ciomaga Hatnean, M. Crystal Growth of the R_2SiO_5 Compounds (R = Dy, Ho, and Er) by the Floating Zone Method Using a Laser-Diode-Heated Furnace. *Crystals* **2023**, *13*, 1687. [CrossRef]
4. Sahani, R.M.; Pandya, A. Novel Epoxy-bPBD-BisMSB Composite Plastic Scintillator for Alpha, Beta and Gamma Radiation Detection. *Sci. Rep.* **2024**, *14*, 6531. [CrossRef] [PubMed]
5. Pagano, F.; Král, J.; Děcká, K.; Pizzichemi, M.; Mihóková, E.; Čuba, V.; Auffray, E. Nanocrystalline Lead Halide Perovskites to Boost Time-of-Flight Performance of Medical Imaging Detectors. *Adv. Mater. Interfaces* **2024**, *11*, 2300659. [CrossRef]
6. Lai, Z.; Wang, W.; Liu, F.; Liu, Z.; Wang, S.; Lai, J.; Chen, J.; Zhang, H.; Xu, R.; Wang, L. High-Quality CsI(Tl) Single-Crystal Flake Scintillators Grown by the Space-Confined Solution Method. *Opt. Mater.* **2024**, *151*, 115333. [CrossRef]
7. Mahato, S.; Makowski, M.; Bose, S.; Kowal, D.; Kuddus Sheikh, M.A.; Braueninger-Wemer, P.; Witkowski, M.E.; Ray, S.K.; Drozdowski, W.; Birowosuto, M.D. Improvement of Light Output of $MAPbBr_3$ Single Crystal for Ultrafast and Bright Cryogenic Scintillator. *J. Phys. Chem. Lett.* **2024**, *15*, 3713–3720. [CrossRef]
8. Glodo, J.; van Loef, E.; Wang, Y.; Bhattacharya, P.; Pandian, L.S.; Shirwadkar, U.; Hubble, I.; Schott, J.; Muller, M. Novel High-Stopping Power Scintillators for Medical Applications. In Proceedings of the Medical Imaging 2024: Physics of Medical Imaging, SPIE, San Diego, CA, USA, 1 April 2024; Volume 12925, pp. 585–595.
9. Nazarov, M.; Tsukerblat, B. Optical Lines in Europium and Terbium-Activated Yttrium Tantalate Phosphor: Combined Experimental and Group-Theoretical Analysis. *Optics* **2023**, *4*, 510–524. [CrossRef]
10. Komendo, I.; Mechinsky, V.; Fedorov, A.; Dosovitskiy, G.; Schukin, V.; Kuznetsova, D.; Zykova, M.; Velikodny, Y.; Korjik, M. Effect of the Synthesis Conditions on the Morphology, Luminescence and Scintillation Properties of a New Light Scintillation Material Li_2CaSiO_4:Eu^{2+} for Neutron and Charged Particle Detection. *Inorganics* **2022**, *10*, 127. [CrossRef]
11. Cheng, Y.; Huang, Y.; Yu, G. N-S-Co-Doped Carbon Dot Blue Fluorescence Preparation and Baicalein Detection. *Inorganics* **2024**, *12*, 154. [CrossRef]
12. Higgins, W.M.; Churilov, A.; van Loef, E.; Glodo, J.; Squillante, M.; Shah, K. Crystal Growth of Large Diameter $LaBr_3$:Ce and $CeBr_3$. *J. Cryst. Growth* **2008**, *310*, 2085–2089. [CrossRef]
13. Wei, H.; Martin, V.; Lindsey, A.; Zhuravleva, M.; Melcher, C.L. The Scintillation Properties of $CeBr_{3-x}Cl_x$ Single Crystals. *J. Lumin.* **2014**, *156*, 175–179. [CrossRef]
14. Loyd, M.; Stand, L.; Rutstrom, D.; Wu, Y.; Glodo, J.; Shah, K.; Koschan, M.; Melcher, C.L.; Zhuravleva, M. Investigation of $CeBr_{3-x}I_x$ Scintillators. *J. Cryst. Growth* **2020**, *531*, 125365. [CrossRef]
15. Kandarakis, I. Luminescence in Medical Image Science. *J. Lumin.* **2016**, *169*, 553–558. [CrossRef]
16. Madej, A.; Witkowski, M.E.; Wojtowicz, A.J.; Zych, E. Photo- and Radioluminescent Properties of Undoped and Bi-Doped Lu_2WO_6 Powders at 10–300K. *J. Lumin.* **2015**, *160*, 50–56. [CrossRef]
17. Xie, S.; Zhang, X.; Zhang, Y.; Ying, G.; Huang, Q.; Xu, J.; Peng, Q. Evaluation of Various Scintillator Materials in Radiation Detector Design for Positron Emission Tomography (PET). *Crystals* **2020**, *10*, 869. [CrossRef]
18. Koppert, W.J.C.; Dietze, M.M.A.; Velden, S.; Steenbergen, J.H.L.; Jong, H.W.A.M. A Comparative Study of NaI(Tl), $CeBr_3$ and CZT for Use in a Real-Time Simultaneous Nuclear and Fluoroscopic Dual-Layer Detector. *Phys. Med. Biol.* **2019**, *64*, 135012. [CrossRef] [PubMed]
19. Lecoq, P. Development of New Scintillators for Medical Applications. *Nucl. Instrum. Methods Phys. Res. Sect. Accel. Spectrom. Detect. Assoc. Equip.* **2016**, *809*, 130–139. [CrossRef]
20. Kim, K.J.; Furuya, Y.; Kamada, K.; Murakami, R.; Kochurikhin, V.V.; Yoshino, M.; Chiba, H.; Kurosawa, S.; Yamaji, A.; Shoji, Y.; et al. Growth and Scintillation Properties of Directionally Solidified Ce:$LaBr_3$/$AEBr_2$ (AE = Mg, Ca, Sr, Ba) Eutectic System. *Crystals* **2020**, *10*, 584. [CrossRef]

21. Carone, D.; Klepov, V.V.; Misture, S.T.; Schaeperkoetter, J.C.; Jacobsohn, L.G.; Aziziha, M.; Schorne-Pinto, J.; Thomson, S.A.J.; Hines, A.T.; Besmann, T.M.; et al. Luminescence and Scintillation in the Niobium Doped Oxyfluoride $Rb_4Ge_5O_9F_6$:Nb. *Inorganics* **2022**, *10*, 83. [CrossRef]
22. Linardatos, D.; Ntoupis, V.; Tseremoglou, S.; Valais, I.; Ninos, K.; Bakas, A.; Lavdas, E.; Kandarakis, I.; Kalyvas, N.; Fountos, G.; et al. Light Output Dependence of $CeBr_3$ Hygroscopic Scintillator upon Temperature. *Procedia Struct. Integr.* **2023**, *47*, 80–86. [CrossRef]
23. Li, P.; Gridin, S.; Ucer, K.B.; Williams, R.T.; Menge, P.R. Picosecond Absorption Spectroscopy of Self-Trapped Excitons and Ce Excited States in $CeBr_3$ and $La_{1-x}Ce_xBr_3$. *Phys. Rev. B* **2019**, *99*, 104301. [CrossRef]
24. Lecoq, P.; Gektin, A.; Korzhik, M. Scintillation and Inorganic Scintillators. In *Inorganic Scintillators for Detector Systems: Physical Principles and Crystal Engineering*; Lecoq, P., Gektin, A., Korzhik, M., Eds.; Springer International Publishing: Cham, Switzerland, 2017; pp. 1–41, ISBN 978-3-319-45522-8.
25. Kochan, O.; Chornodolskyy, Y.; Selech, J.; Karnaushenko, V.; Przystupa, K.; Kotlov, A.; Demkiv, T.; Vistovskyy, V.; Stryhanyuk, H.; Rodnyi, P.; et al. Energy Structure and Luminescence of CeF_3 Crystals. *Materials* **2021**, *14*, 4243. [CrossRef]
26. Przystupa, K.; Chornodolskyy, Y.M.; Selech, J.; Karnaushenko, V.O.; Demkiv, T.M.; Kochan, O.; Syrotyuk, S.V.; Voloshinovskii, A.S. The Influence of Halide Ion Substitution on Energy Structure and Luminescence Efficiency in $CeBr_2I$ and $CeBrI_2$ Crystals. *Materials* **2023**, *16*, 5085. [CrossRef]
27. Chornodolskyy, Y.M.; Karnaushenko, V.O.; Vistovskyy, V.V.; Syrotyuk, S.V.; Gektin, A.V.; Voloshinovskii, A.S. Energy Band Structure Peculiarities and Luminescent Parameters of CeX_3 (X = Cl, Br, I) Crystals. *J. Lumin.* **2021**, *237*, 118147. [CrossRef]
28. Russell-Webster, B.; Abboud, K.A.; Christou, G. Molecular Nanoparticles of Cerium Dioxide: Structure-Directing Effect of Halide Ions. *Chem. Commun.* **2020**, *56*, 5382–5385. [CrossRef]
29. Chipaux, R.; Faure, J.-L.; Rebourgeard, P.; Dauphin, G.; Safieh, J. Behaviour of CeF_3 Scintillator in an LHC-like Environment. *Nucl. Instrum. Methods Phys. Res. Sect. Accel. Spectrom. Detect. Assoc. Equip.* **1994**, *345*, 440–444. [CrossRef]
30. Schneegans, M.A. Cerium Fluoride Crystals for Calorimetry at LHC. *Nucl. Instrum. Methods Phys. Res. Sect. Accel. Spectrom. Detect. Assoc. Equip.* **1994**, *344*, 47–56. [CrossRef]
31. Ferrere, D.; Lebeau, M.; Schneegans, M.; Vivargent, M.; Lecoq, P. High Resolution Crystal Calorimetry at LHC. *Nucl. Instrum. Methods Phys. Res. Sect. Accel. Spectrom. Detect. Assoc. Equip.* **1992**, *315*, 332–336. [CrossRef]
32. Auffray, E.; Beckers, T.; Bourotte, J.; Chipaux, R.; Commichau, V.; Dafinei, I.; Depasse, P.; Djambazov, L.; Dydak, U.; El Mamouni, H.; et al. Performance of a Cerium Fluoride Crystal Matrix Measured in High-Energy Particle Beams. *Nucl. Instrum. Methods Phys. Res. Sect. Accel. Spectrom. Detect. Assoc. Equip.* **1996**, *378*, 171–178. [CrossRef]
33. Nikl, M.; Mares, J.A.; Dusek, M.; Lecoq, P.; Dafinei, I.; Auffray, E.; Pazzi, G.P.; Fabeni, P.; Jindra, J.; Skoda, Z. Decay Kinetics of Ce^{3+} Ions under Gamma and KrF Excimer Laser Excitation in CeF_3 Single Crystals. *J. Phys. Condens. Matter* **1995**, *7*, 6355. [CrossRef]
34. Auffray, E.; Beckers, T.; Chipaux, R.; Dafinei, I.; Depasse, P.; El Mamouni, H.; Faure, J.L.; Fay, J.; Hillemanns, H.; Ille, B.; et al. First Results on Large Cerium Fluoride Crystals in A Test Beam. *MRS Online Proc. Libr.* **1994**, *348*, 117–122. [CrossRef]
35. Pedrini, C.; Moine, B.; Bouttet, D.; Belsky, A.N.; Mikhailin, V.V.; Vasil'ev, A.N.; Zinin, E.I. Time-Resolved Luminescence of CeF_3 Crystals Excited by X-ray Synchrotron Radiation. *Chem. Phys. Lett.* **1993**, *206*, 470–474. [CrossRef]
36. Auffray, E.; Baccaro, S.; Beckers, T.; Benhammou, Y.; Belsky, A.N.; Borgia, B.; Boutet, D.; Chipaux, R.; Dafinei, I.; de Notaristefani, F.; et al. Extensive Studies on CeF_3 Crystals, a Good Candidate for Electromagnetic Calorimetry at Future Accelerators. *Nucl. Instrum. Methods Phys. Res. Sect. Accel. Spectrom. Detect. Assoc. Equip.* **1996**, *383*, 367–390. [CrossRef]
37. Inagaki, T.; Yoshimura, Y.; Kanda, Y.; Matsumoto, Y.; Minami, K. Development of CeF_3 Crystal for High-Energy Electromagnetic Calorimetry. *Nucl. Instrum. Methods Phys. Res. Sect. Accel. Spectrom. Detect. Assoc. Equip.* **2000**, *443*, 126–135. [CrossRef]
38. Bianchini, L.; Dissertori, G.; Donegà, M.; Lustermann, W.; Marini, A.; Micheli, F.; Nessi-Tedaldi, F.; Pandolfi, F.; Peruzzi, M.; Schonenberger, M.; et al. High-Energy Electron Test Results of a Calorimeter Prototype Based on CeF_3 for HL-LHC Applications. In Proceedings of the 2015 IEEE Nuclear Science Symposium and Medical Imaging Conference (NSS/MIC), San Diego, CA, USA, 31 October–7 November 2015; pp. 1–2.
39. Karimov, D.N.; Lisovenko, D.S.; Ivanova, A.G.; Grebenev, V.V.; Popov, P.A.; Sizova, N.L. Bridgman Growth and Physical Properties Anisotropy of CeF_3 Single Crystals. *Crystals* **2021**, *11*, 793. [CrossRef]
40. Víllora, E.G.; Yuan, D.; Shimamura, K. CeF_3 Single Crystals for UV-VIS-IR Optical Isolators. *Int. J. Appl. Ceram. Technol.* **2023**, *20*, 2047–2054. [CrossRef]
41. Yang, W.; Kong, X.; Fu, B.; Yang, Y.; Chen, R.; Zuo, C.; Liu, H.; Yu, Y.; Zeng, F.; Li, C. Optical Properties of CeF_3 Crystal at High Temperature or Pressure by First Principles and Its Application in Isolators. *Opt. Mater.* **2024**, *154*, 115758. [CrossRef]
42. Yuan, D.; Víllora, E.G.; Shimamura, K. Regular Hexagonal-Prism Microvoids in CeF_3 Single Crystals. *J. Cryst. Growth* **2021**, *558*, 126024. [CrossRef]
43. Chylii, M.; Loghina, L.; Kaderavkova, A.; Slang, S.; Rodriguez-Pereira, J.; Frumarova, B.; Vlcek, M. Morphology and Optical Properties of CeF_3 and CeF_3:Tb Nanocrystals: The Dominant Role of the Reaction Thermal Mode. *Mater. Chem. Phys.* **2021**, *260*, 124161. [CrossRef]
44. Zhu, L.; Li, Q.; Liu, X.; Li, J.; Zhang, Y.; Meng, J.; Cao, X. Morphological Control and Luminescent Properties of CeF_3 Nanocrystals. *J. Phys. Chem. C* **2007**, *111*, 5898–5903. [CrossRef]

45. Sizova, N.L.; Karimov, D.N.; Kosova, T.B.; Lisovenko, D.S. Mechanical Properties of CeF_3 Single Crystals. *Crystallogr. Rep.* **2019**, *64*, 942–946. [CrossRef]
46. Li, H.; Wang, J.; Chen, J.; Dai, Y.; Su, L.; Li, X.; Kalashnikova, A.M.; Wu, A. Bridgman Growth and Magneto-Optical Properties of CeF_3 Crystal as Faraday Rotator. *Opt. Mater.* **2020**, *100*, 109675. [CrossRef]
47. Kamenskikh, I.; Tishchenko, E.; Kirm, M.; Omelkov, S.; Belsky, A.; Vasil'ev, A. Decay Kinetics of CeF_3 under VUV and X-ray Synchrotron Radiation. *Symmetry* **2020**, *12*, 914. [CrossRef]
48. Klamra, W.; Sibczynski, P.; Moszynski, M.; Czarnacki, W.; Kozlov, V. Extensive Studies on Light Yield Non-Proportional Response of Undoped CeF_3 at Room and Liquid Nitrogen Temperatures. *J. Instrum.* **2013**, *8*, P06003. [CrossRef]
49. Thongpool, V.; Phunpueok, A.; Jaiyen, S.; Choosakul, N.; Aphairaj, D.; Thongchai, P.; Singhaseree, C. Preparation of CeF_3 Nanoparticles Loaded PPO/PVT Composites for Radiation Detection. *Dig. J. Nanomater. Biostructures* **2022**, *16*, 621–625. [CrossRef]
50. Lacy, J.H.; Patenotte, G.E.; Kinney, A.C.; Majumder, P.K. Broadband High-Precision Faraday Rotation Spectroscopy with Uniaxial Single Crystal CeF_3 Modulator. *Photonics* **2024**, *11*, 304. [CrossRef]
51. Starobor, A.; Mironov, E.; Palashov, O. High-Power Faraday Isolator on a Uniaxial CeF_3 Crystal. *Opt. Lett.* **2019**, *44*, 1297–1299. [CrossRef]
52. Vojna, D.; Slezák, O.; Yasuhara, R.; Furuse, H.; Lucianetti, A.; Mocek, T. Faraday Rotation of Dy_2O_3, CeF_3 and $Y_3Fe_5O_{12}$ at the Mid-Infrared Wavelengths. *Materials* **2020**, *13*, 5324. [CrossRef]
53. Lecoq, P. How High Energy Physics Is Driving the Development of New Scintillators. In Proceedings of the Fifth International Conference on Inorganic Scintillators and Their Applications, Moscow, Russia, 16–20 August 1999; CERN: Moscow, Russia, 1999.
54. Grynyov, B.; Ryzhikov, V.; Kim, J.K.; Jae, M. *Scintillator Crystals, Radiation Detectors & Instruments on Their Base*; V. Ryzhikov: Kharkiv, Ukraine, 2004; ISBN 966-02-3314-0.
55. Michail, C.; Liaparinos, P.; Kalyvas, N.; Kandarakis, I.; Fountos, G.; Valais, I. Phosphors and Scintillators in Biomedical Imaging. *Crystals* **2024**, *14*, 169. [CrossRef]
56. Valais, I.; Michail, C.; David, S.; Nomicos, C.D.; Panayiotakis, G.S.; Kandarakis, I. A Comparative Study of the Luminescence Properties of LYSO:Ce, LSO:Ce, GSO:Ce and BGO Single Crystal Scintillators for Use in Medical X-ray Imaging. *Phys. Med.* **2008**, *24*, 122–125. [CrossRef] [PubMed]
57. van Eijk, C.W.E. Inorganic Scintillators in Medical Imaging. *Phys. Med. Biol.* **2002**, *47*, R85–R106. [CrossRef]
58. Balcerzyk, M.; Moszynski, M.; Kapusta, M.; Wolski, D.; Pawelke, J.; Melcher, C.L. YSO, LSO, GSO and LGSO. A Study of Energy Resolution and Nonproportionality. *IEEE Trans. Nucl. Sci.* **2000**, *47*, 1319–1323. [CrossRef]
59. Kozma, P.; Kozma, P. Radiation Sensitivity of GSO and LSO Scintillation Detectors. *Nucl. Instrum. Methods Phys. Res. Sect. Accel. Spectrom. Detect. Assoc. Equip.* **2005**, *539*, 132–136. [CrossRef]
60. Holl, I.; Lorenz, E.; Mageras, G. A Measurement of the Light Yield of Common Inorganic Scintillators. *IEEE Trans. Nucl. Sci.* **1988**, *35*, 105–109. [CrossRef]
61. Passeri, A.; Formiconi, A.R. SPECT and Planar Imaging in Nuclear Medicine. In *Ionizing Radiation Detectors for Medical Imaging*; World Scientific: Singapore, 2004; pp. 235–285, ISBN 978-981-238-674-8.
62. Christian, J. Advances in CMOS SSPM Detectors. In *Biological and Medical Sensor Technologies*; Iniewski, K., Ed.; CRC Press: Boca Raton, FL, USA, 2017; p. 327, ISBN 978-1-138-07321-0.
63. Stolberg-Rohr, T.; Hawkins, G.J. Spectral Design of Temperature-Invariant Narrow Bandpass Filters for the Mid-Infrared. *Opt. Express* **2015**, *23*, 580. [CrossRef]
64. Rodnyj, P.A. Interaction of Ionizing Radiation with Scintillators. In *Physical Processes in Inorganic Scintillators*; The CRC Press Laser and Optical Science and Technology Series; CRC Press: Boca Raton, FL, USA, 1997; p. 17, ISBN 978-0-8493-3788-8.
65. Kawaguchi, N.; Kimura, H.; Nakauchi, D.; Kato, T.; Yanagida, T. Inorganic Fluoride Scintillators. In *Phosphors for Radiation Detectors*; John Wiley & Sons, Ltd.: Hoboken, NJ, USA, 2022; pp. 91–120, ISBN 978-1-119-58336-3.
66. van Loef, E.V.; Shah, K.S. Advances in Scintillators for Medical Imaging Applications. In Proceedings of the SPIE, San Diego, CA, USA, 12 September 2014; Barber, H.B., Furenlid, L.R., Roehrig, H.N., Eds.; p. 92140A.
67. Drozdowski, W.; Dorenbos, P.; Bos, A.J.J.; Owens, A.; Richaud, D. Gamma Radiation Hardness of Ø1″ × 1″ $LaBr_3$:Ce, $LaCl_3$:Ce, and $CeBr_3$ Scintillators. In Proceedings of the 2008 IEEE Nuclear Science Symposium Conference Record, Dresden, Germany, 19–25 October 2008; pp. 2856–2858.
68. CeF3—Cerium Fluoride Scintillator Crystal | Advatech UK. Available online: https://www.advatech-uk.co.uk/cef3.html (accessed on 2 May 2024).
69. CeBr3—Cerium Bromide Scintillator Crystal | Advatech UK. Available online: https://www.advatech-uk.co.uk/cebr3.html (accessed on 2 May 2024).
70. Shen, H.; Xu, J.Y.; Ping, W.J.; He, Q.B.; Zhang, Y.; Jin, M.; Jiang, G.J. Growth, Mechanical and Thermal Properties of $Bi_4Si_3O_{12}$ Single Crystals. *Chin. Phys. Lett.* **2012**, *29*, 076501. [CrossRef]
71. Cerium Fluoride Crystal CeF3 Scintillation Crystal–MSE Supplies LLC. Available online: https://www.msesupplies.com/products/cerium-fluoride-cef3-crystal?variant=40204171444282 (accessed on 7 June 2024).
72. Li, H.; Ge, W.; Li, J.; Dai, Y.; Chen, J.; Su, L.; Jin, Z.; Ma, G.; Li, X.; Wu, A. Temperature-Dependent Terahertz Dielectric Modulation of a High-Performance Magneto-Optic CeF_3 Crystal. *Phys. B Condens. Matter* **2020**, *599*, 412468. [CrossRef]

73. Kobayashi, M.; Ishii, M.; Krivandina, E.A.; Litvinov, M.M.; Peresypkin, A.I.; Prokoshkin, Y.D.; Rykalin, V.I.; Sobolev, B.P.; Takamatsu, K.; Vasil'chenko, V.G. Cerium Fluoride, a Highly Radiation-Resistive Scintillator. *Nucl. Instrum. Methods Phys. Res. Sect. Accel. Spectrom. Detect. Assoc. Equip.* **1991**, *302*, 443–446. [CrossRef]
74. Kozma, P.; Kozma, P. Radiation Resistivity of BGO Crystals Due to Low-Energy Gamma-Rays. *Nucl. Instrum. Methods Phys. Res. Sect. Accel. Spectrom. Detect. Assoc. Equip.* **2003**, *501*, 499–504. [CrossRef]
75. Snigireva, O.A.; Solomonov, V.I. Role of the Ce^{2+} Ions in Cerium Fluoride Luminescence. *Phys. Solid State* **2005**, *47*, 1443–1445. [CrossRef]
76. Yang, F.; Mao, R.; Zhang, L.; Zhu, R.-Y. A Study on Radiation Damage in BGO and PWO-II Crystals. *J. Phys. Conf. Ser.* **2012**, *404*, 012025. [CrossRef]
77. Sasano, M.; Nishioka, H.; Okuyama, S.; Nakazawa, K.; Makishima, K.; Yamada, S.; Yuasa, T.; Okumura, A.; Kataoka, J.; Fukazawa, Y.; et al. Geometry Dependence of the Light Collection Efficiency of BGO Crystal Scintillators Read out by Avalanche Photo Diodes. *Nucl. Instrum. Methods Phys. Res. Sect. Accel. Spectrom. Detect. Assoc. Equip.* **2013**, *715*, 105–111. [CrossRef]
78. Chaiphaksa, W.; Limkitjaroenporn, P.; Kim, H.J.; Kaewkhao, J. Moh's Hardness Scale and Micro Vicker's Hardness Study of Bgo and Lyso Inorganic Scintillators. *J. Phys. Conf. Ser.* **2018**, *970*, 012005. [CrossRef]
79. Kozma, P.; Bajgar, R.; Kozma, P. Radiation Resistivity of PbF_2 Crystals. *Nucl. Instrum. Methods Phys. Res. Sect. Accel. Spectrom. Detect. Assoc. Equip.* **2002**, *484*, 149–152. [CrossRef]
80. Magnan, P. Detection of Visible Photons in CCD and CMOS: A Comparative View. *Nucl. Instrum. Methods Phys. Res. Sect. Accel. Spectrom. Detect. Assoc. Equip.* **2003**, *504*, 199–212. [CrossRef]
81. Yang, Y.; Zhang, L.; Quan, C.; Li, S.; Zhang, S.; Zhang, Y.; Fang, Q.; He, M.; Xu, M.; Hang, Y. Growth, Thermal, and Polarized Spectroscopic Properties of Nd:CeF_3 Crystal for Dual-Wavelength Lasers. *J. Lumin.* **2020**, *227*, 117558. [CrossRef]
82. Shimamura, K.; Víllora, E.G.; Nakakita, S.; Nikl, M.; Ichinose, N. Growth and Scintillation Characteristics of CeF_3, PrF_3 and NdF_3 Single Crystals. *J. Cryst. Growth* **2004**, *264*, 208–215. [CrossRef]
83. Schaber, J.; Xiang, R.; Gaponik, N. Review of Photocathodes for Electron Beam Sources in Particle Accelerators. *J. Mater. Chem. C* **2023**, *11*, 3162–3179. [CrossRef]
84. MA, L.; Chen, L.; Huang, G.; Hu, J.; Han, X.; Hua, Z.; Huang, X.; Jin, M.; Jiang, X.; Jin, Z.; et al. The Time Resolution Improvement of Cherenkov-Radiator-Window Photomultiplier Tube. *J. Instrum.* **2023**, *18*, C12020. [CrossRef]
85. Ponti, E.D.; Crivellaro, C.; Morzenti, S.; Monaco, L.; Todde, S.; Landoni, C.; Elisei, F.; Musarra, M.; Guerra, L. Clinical Application of a High Sensitivity BGO PET/CT Scanner: Effects of Acquisition Protocols and Reconstruction Parameters on Lesions Quantification. *Curr. Radiopharm.* **2022**, *15*, 218–227. [CrossRef] [PubMed]
86. Sathiakumar, C.; Som, S.; Eberl, S.; Lin, P. NEMA NU 2-2001 Performance Testing of a Philips Gemini GXL PET/CT Scanner. *Australas. Phys. Eng. Sci. Med.* **2010**, *33*, 199–209. [CrossRef] [PubMed]
87. Surti, S.; Kuhn, A.; Werner, M.E.; Perkins, A.E.; Kolthammer, J.; Karp, J.S. Performance of Philips Gemini TF PET/CT Scanner with Special Consideration for Its Time-of-Flight Imaging Capabilities. *J. Nucl. Med.* **2007**, *48*, 471–480.
88. Jakoby, B.W.; Bercier, Y.; Watson, C.C.; Bendriem, B.; Townsend, D.W. Performance Characteristics of a New LSO PET/CT Scanner with Extended Axial Field-of-View and PSF Reconstruction. *IEEE Trans. Nucl. Sci.* **2009**, *56*, 633–639. [CrossRef]
89. Valais, I.; David, S.; Michail, C.; Konstantinidis, A.; Kandarakis, I.; Panayiotakis, G.S. Investigation of Luminescent Properties of LSO:Ce, LYSO:Ce and GSO:Ce Crystal Scintillators under Low-Energy γ-Ray Excitation Used in Nuclear Imaging. *Nucl. Instrum. Methods Phys. Res. Sect. Accel. Spectrom. Detect. Assoc. Equip.* **2007**, *581*, 99–102. [CrossRef]
90. Storm, L.; Israel, H.I. Photon Cross Sections from 1 keV to 100 MeV for Elements Z = 1 to Z = 100. *At. Data Nucl. Data Tables* **1970**, *7*, 565–681. [CrossRef]
91. Hubbell, J.H.; Seltzer, S.M. *Tables of X-ray Mass Attenuation Coefficients and Mass Energy-Absorption Coefficients*; NIST Standard Reference Database 126; National Institute of Standards and Technology-PL: Gaithersburg, MD, USA, 1995.
92. Boone, J. X-ray Production, Interaction, and Detection in Diagnostic Imaging. In *Handbook of Medical Imaging. Volume 1: Physics and Psychophysics*; Beutel, J., Kundel, H.L., Van Metter, R.L., Eds.; SPIE Press: Bellingham, WA, USA, 2000; Volume 1, pp. 36–57, ISBN 978-0-8194-7772-9.

inorganics

Article

Antimicrobial, Optical, and Mechanical Properties of Saliva-Contaminated Silver–Zeolite Nanoparticle-Incorporated Dental Acrylic Resins

Çisel Kısa Yaman [1], Necla Demir [1,*], Uğur Arslan [2] and Nurullah Çiftçi [3]

1 Department of Prosthetic Dentistry, School of Dentistry, University of Selcuk, Konya 42079, Turkey; cisellkisa@gmail.com
2 Department of Medical Microbiology, School of Medicine, University of Kafkas, Kars 36100, Turkey; drarslanugur@gmail.com
3 Department of Medical Microbiology, School of Medicine, University of Selcuk, Konya 421331, Turkey; ciftcinurullah72@gmail.com
* Correspondence: necladt@gmail.com

Abstract: Background and Purpose: This study aimed to evaluate the flexural strength, color change and antimicrobial effect of silver–zeolite nanoparticles (NPs) in acrylic resin materials. **Methods:** Fifty-six disc-shaped acrylic resin samples were divided into four groups (n = 7) according to concentrations of silver–zeolite NPs (0%, 2%, 4%, 5%). Discs were contaminated with *C. albicans* and *S. mutans*. The antimicrobial effect was tested by inoculating contaminated discs on Tryptic soy agar (TSA), Sabouraud Dextrose Agar (SDA), Tryptic soy broth (TSB), and Sabouraud dextrose broth (SDB). Forty rectangular 65 × 10 × 2.5 mm acrylic resin specimens were also classified into four groups (n = 10) according to concentrations of silver–zeolite NPs. For the color change, L, a, and b values of rectangular specimens were examined with a spectrophotometer. A three-point bending test was also performed using a Devotrans device to determine the flexural bond strength of rectangular specimens. Scanning electron microscope analysis (SEM/EDX analysis) was also performed. **Results:** In this study, the antimicrobial effect increased with the concentration of silver–zeolite NPs added to acrylic resin discs. In our study, adding 2% silver–zeolite NPs was more effective against *C. albicans*. The antimicrobial effect against *S. mutans* increased with concentration of silver–zeolite NPs (<0.001). The colonization of *C. albicans* was significantly reduced by silver–zeolite NPs. A significant increase was observed in the color change as the nanoparticle percentage ratio increased ($p < 0.001$). The flexural strength values of the groups containing 2% and 4% nanoparticles were found to be clinically acceptable. **Conclusions:** The study showed that bacterial and fungal colonization is significantly reduced by adding silver–zeolite nanoparticles to acrylic resin discs. Based on its antimicrobial, physical, and mechanical properties, we recommend adding 2% silver–zeolite nanoparticles to the acrylic resin material for optimal results.

Keywords: acrylic resin; antimicrobial activity; nanoparticle; silver; zeolite; color change; flexural strength

1. Introduction

Polymethyl methacrylate (PMMA) is commonly used in prosthetic dentistry for edentulous patients due to its aesthetic appeal, easy handling, affordability, and stability in the oral cavity. Denture base materials are also widely used in the manufacture of orthodontic removable appliances such as temporary dentures, placeholders, and functional appliances [1]. One of the most common problems is that the space maintainer, while using these appliances, has the potential for plaque accumulation due to their surface porosity and food retention configuration, which increases the bacterial activity of the cariogenic oral flora for patients and dentists [2]. PMMA is considered to be the material of choice for prosthetic base fabrication due to its low cost, water solubility/absorption, and adequate durability. However, some of the disadvantages of this material are its dimensional instability, residual monomer content, and poor mechanical properties, particularly its poor impact resistance. Poor mechanical properties often lead to internal and external fractures of the denture base. Different approaches have been adapted to improve the mechanical and microbiological properties of PMMA denture base material [3]. The insertion of prosthetic devices into the oral cavity affects the oral microflora and results in accumulation of microorganisms on the prosthetic surface. Such fouling of denture surfaces can lead to a number of oral diseases and consequently to the replacement of the denture [4].

The oral cavity is a complex environment where microorganisms compete for space and nutrients. Especially in elderly and immunocompromised patients, these microorganisms can distribute and trigger systemic diseases such as infective endocarditis and aspiration pneumonia, which are the main causes of morbidity and mortality [5]. Since microorganisms accumulate on removable prostheses' surfaces, they can damage soft tissue and oral health [6], impairing the effectiveness and quality of prosthetic treatment [7]. Although systemic or topical antifungal therapies have been recommended to reduce fungal infection, many barriers limit their use, such as antimicrobial resistance, poor oral hygiene, cost, and the routine of applying proper denture cleaning and strict local treatment [8]. Studies have shown that *Streptococcus mutans* and *Lactobacillus casei* may have cariogenic potential due to their acidogenic and aciduric abilities [9]. Oral candidiasis is the most common infection of the oral mucosa in patients (approximately 72%) using full dentures [10,11].

Geriatric patients are unable to provide effective hygiene and care due to poor motor coordination, cognitive impairment, or memory loss [12]. For this reason, clinical trials have been made to prevent or reduce the adhesion and accumulation of microorganisms on the prosthesis surface [13]. The oral microflora is a complex ecosystem that includes a wide variety of microbial species [14]. More than 700 different types of bacteria have been identified in the human oral cavity so far, and most of them are associated with plaque [15]. Bacteria help Candida adhesion to the denture base and mucosa by producing extracellular polymers and increasing acidity, creating optimal conditions for fungal growth. The oral cavity consists of many surfaces covered with a large number of bacteria, and some of them may form biofilm [16]. The aesthetic properties of dental materials have become very popular and essential today. These materials should have a shiny and smooth surface and maintain their smoothness in the oral environment for a long time. Smooth surfaces provide an aesthetic appearance and prevent the formation of the biofilm layer that causes plaque retention. Among the bacterial species in the oral flora, *S. mutans* plays the most important role in plaque formation since it is the first microorganism to adhere to the surface [17]. Candida species are members of oral flora, and they are commonly responsible for opportunistic infections in the oral cavity. Using glycoprotein adhesins, *Candida albicans* attaches to the epithelial cells in the mouth, including the buccal mucosa, tongue, teeth, and oral prostheses. Although the conditions for growth in the oral cavity are unfavorable, *C. albicans* can still proliferate in saliva because it contains glucose [18]. Opportunistic oral pathogens can colonize on acrylic materials and cause dental infections such as denture stomatitis. Currently, it is a common practice to add antimicrobial agents to dental materials. Silver nanoparticles can be added to acrylic resin materials to prevent bacterial colonization by *S. mutans*, *Escherichia coli*, and *Staphylococcus aureus*. In addition to their antibacterial effect, nanoparticle-added acrylic resins have shown antifungal properties against the adhesion of *C. albicans* [19]. In the design of a removable prosthesis, durability is one of the most important mechanical properties to ensure that the prosthesis is effective, safe, and serves for a reasonable period of time. Flexibility must be high to protect the underlying soft tissues by absorbing masticatory forces. It must have a high elastic limit to prevent permanent deformation when stretched [20]. It must have sufficient mechanical strength to resist fracture under repeated chewing forces. It should be dimensionally stable. It must have adequate wear resistance. In the last two decades, experiments have been carried out involving particles, fibers, and fillers of different shapes, sizes, forms, and orientations [21,22].

Recent advances in nanotechnology, such as nanoparticles, nanofibers, and nanotubes, have been used for PMMA reinforcement [23]. Among inorganic antimicrobial agents, silver nanoparticles have been most widely used to fight infections since ancient times [24,25]. Nano-sized (nm) inorganic particles of silver have rapid and broad-spectrum efficacy due to their sustained release of silver cations (Ag^+) [26,27]. This form appears to be more effective compared to micro-sized silver powder (μm), which shows lower antimicrobial activity due to its limited surface area [28].

Silver nanoparticles are one of the most commonly used nanoparticles because of their ductility, electrical and thermal conductivity, and antimicrobial activity [29]. Zeolites are crystalline made of aluminum silicate with void spaces ranging from 3 to 10 angstroms. Antimicrobial cations, such as silver and zinc, can get trapped in zeolite pores and be replaced with other cations from the environment over time [30]. When ions are available, they contact microorganisms and suppress their growth by inactivating enzymes, interrupting RNA replication, and blocking respiration through oxidative stress [30,31]. In dentistry, dental caries and candidiasis are common illnesses in the oral cavity, causing health problems for many people worldwide. Bacteria and yeast in the mouth are responsible for the reason of these conditions. *Candida albicans* is the most common cause of fungal infections in the oral mucosa, particularly in elderly patients who wear full dentures. Studies have shown that antimicrobial zeolites are effective against aerobic and anaerobic bacteria and Candida species [32,33]. This study aimed to investigate the antimicrobial effect of acrylic resins containing different amounts of silver–zeolite nanoparticles. Drawbacks associated with the use of silver nanoparticles and other silver-based materials include nanoparticle aggregation and deterioration of composite mechanical properties, aesthetic incompatibility due to color changes resulting from the reduction of Ag^+ to Ag°, and short-term antimicrobial activity [4]. While improving the mechanical properties of a denture base acrylic resin, it is also of great importance not to compromise its physical properties [34].

This in vitro study aims to investigate the effect of silver zeolite, an antimicrobial nanoparticle, on the microbiological, optical, and mechanical properties of acrylic resin when added to heat-polymerized acrylic resin at different ratios. The null hypothesis of the study is that silver–zeolite nanoparticles added to dental acrylic resins will not have an antimicrobial effect on saliva-contaminated dental acrylic and will not change its mechanical and physical properties.

2. Results

2.1. Surface Topography

The surface topography roughness of the silver–zeolite nanoparticle (NP) samples incorporated into heat-cured acrylic resin was examined using Scanning Electron Microscopy (SEM) (Figure 1). SEM and EDX analysis confirmed the presence of silver–zeolite nanoparticles within the dental acrylic. Surface roughness was assessed through the SEM images, which revealed that roughness increased with higher nanoparticle concentrations compared to the control group (Figure 1). Higher nanoparticle concentrations led to the formation of hollow areas and noticeable indented and protruding surfaces. In the 4% and 5% nanoparticle groups, the pit surface areas significantly expanded (Figure 1). This increased surface roughness, characterized by visible porous structures, could facilitate the colonization of rapid biofilm-forming microorganisms such as *Candida*. Therefore, it is crucial to carefully adjust the nanoparticle concentration in acrylic resin materials to minimize surface roughness and mitigate potential microbial colonization.

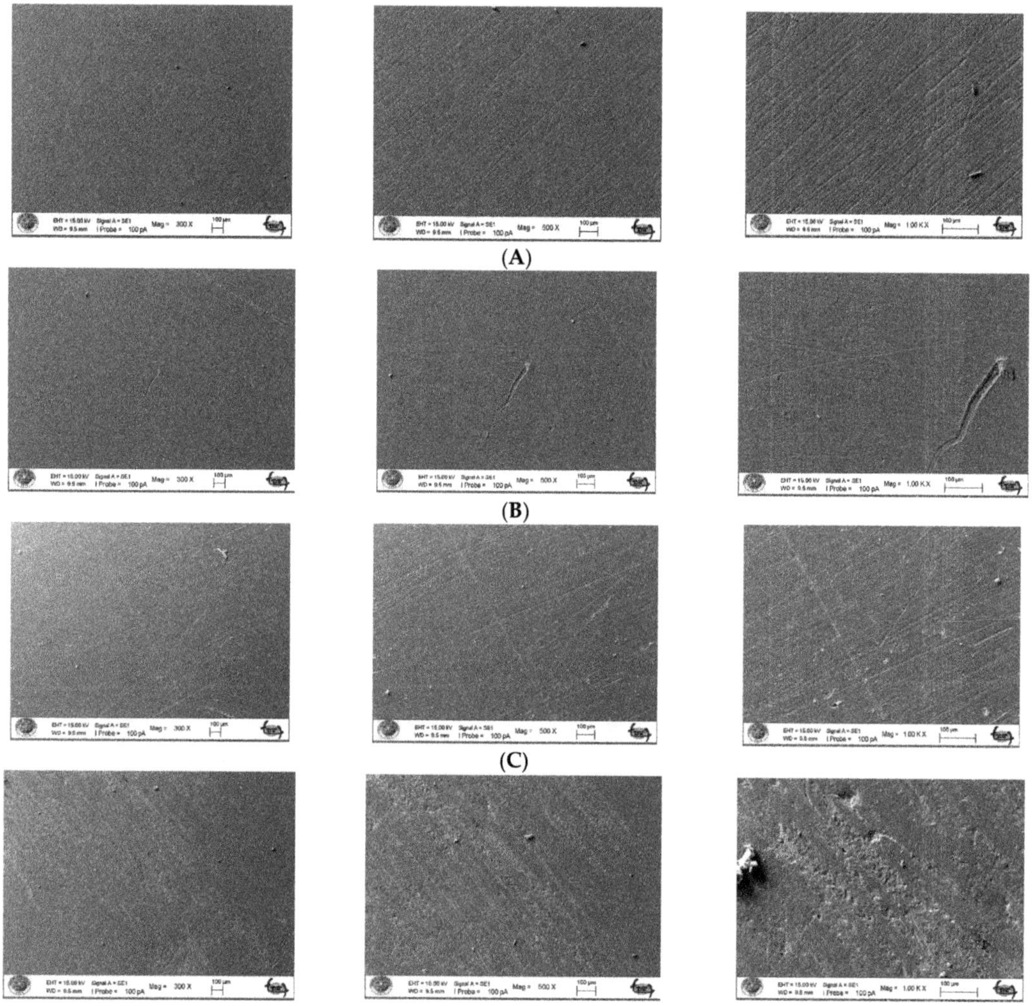

Figure 1. Surface topography of the discs by SEM images. (**A**) Control group surface topography (from left to right 300×/500×/1000× magnification); (**B**) 2% NPs added discs surface topography (from left to right 300×/500×/1000× magnification); (**C**) 4% NPs added discs surface topography (from left to right 300×/500×/1000× magnification); (**D**) 5% NPs added discs surface topography (from left to right 300×/500×/1000× magnification).

2.2. Energy-Dispersive X-ray Analysis

Scanning electron microscope and energy-dispersive X-ray (SEM/EDX) analysis (Hitachi S-4100 FE-SEM/EDS, Tokyo, Japan) of Ag-denture acrylic was carried out in a field emission electron microscope attached to an SEM at an accelerated voltage of 20 keV. Energy-dispersive X-ray (EDX) analysis was carried out to determine the presence of elements in the samples from each group (2%, 4%, 5%) containing NPs (Figure 2). As a result of EDX analysis, the calculated atomic percentage of silver was found to be 1.01 in the 2% group. These results confirmed the inclusion of silver in the 2% group in dental acrylic samples. EDX analysis showed 7.14 as the calculated atomic percentage of silver in the 4% group. EDX results also highlighted the presence of silver in the 4% group in dental acrylic samples. As another result of EDX analysis, the calculated atomic percentage of silver was found to be 11.25 in the 5% group (Figure 2). These results also confirmed the inclusion of silver in the 5% group of dental acrylic specimens. As seen in (4%) and (5%) SEM/EDX graphics and images, as the percentage of nanoparticles increased, the displayed amount of silver nanoparticle increased and the presence of silver was proven (Figure 2).

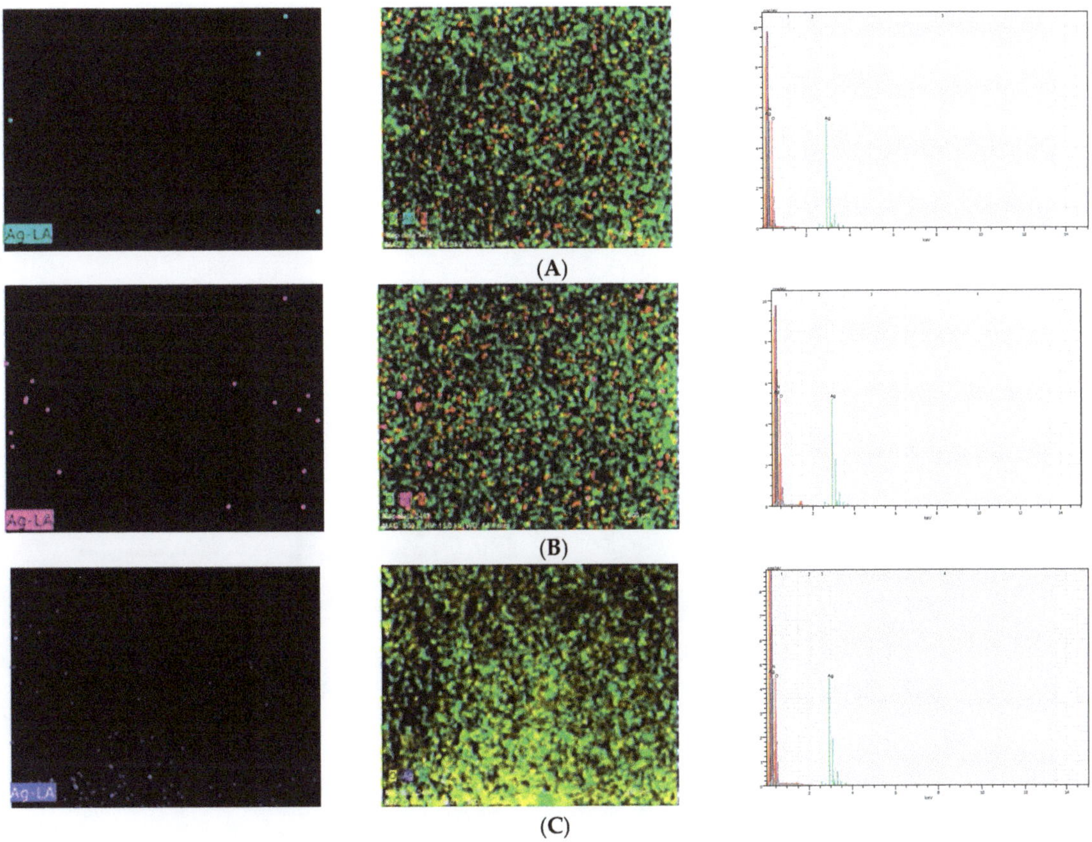

Figure 2. Ag Images with SEM/EDX in the (**A**) 2%, (**B**) 4%, and (**C**) 5% (respectively, up to down) groups.

2.3. *Optical Test Results*

In line with the findings of our study, there was a significant difference between the groups in the ΔE2000 color analysis of all groups with each other. A significant increase was observed in the 2%, 4%, and 5% groups compared to the control group ($p < 0.001$). Compared to the 0% group, the highest color change was observed in the 5% group (Table 1). It was determined that there was a color change with the increase in nanoparticle ratio (Figure 3).

Table 1. Comparison of ΔE2000 values of heat-polymerized acrylic resin samples according to groups.

Group	n	Min	Max	Mean	SDPMA	*p*
0–2%	10	2.48	3.48	2.87	0.27 [a]	<0.001
0–4%	10	4.91	5.88	5.41	0.36 [b]	
0–5%	10	5.72	6.63	6.17	0.27 [c]	
2–4%	10	2.78	4.06	3.23	0.44 [d]	
2–5%	10	3.46	5.09	4.22	0.44 [e]	
4–5%	10	1.25	1.84	1.52	0.21 [f]	

N: Number of samples, Min: Minimum, Max: Maximum, Mean: Mean SD: Standard deviation. One-way analysis of variance, a–f: There is no difference between groups with the same letter.

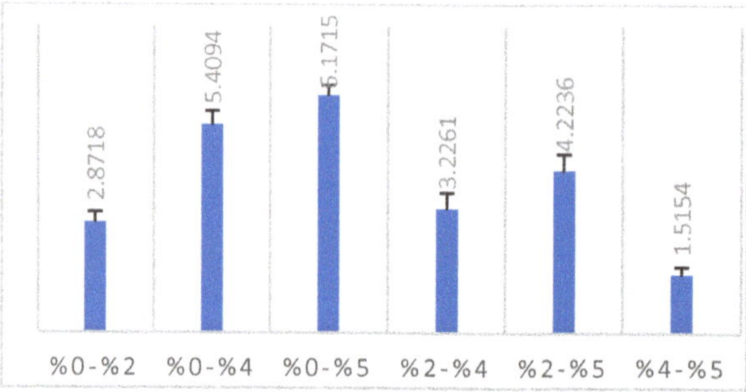

Figure 3. ΔE2000 color change of the groups.

2.4. Flexural Strength Test Results

The Kruskal–Wallis test was used statistically to test whether there was a difference between the average flexural strength values of the groups to which silver–zeolite nanoparticles were added at different rates. The results of this analysis are given in Table 2.

Table 2. Flexural strength test results by groups.

Groups n	Min	Max	Mean	SD	*p*
0% Ag/Z 10	76.80	84.00	80.28 [a]	2.49	<0.001
2% Ag/Z 10	56.40	75.60	65.40 [b]	5.46	
4% Ag/Z 10	48.00	79.20	65.40 [b]	10.99	
5% Ag/Z 10	37.20	69.60	60.36 [c]	10.25	

Kruskal–Wallis test, a–c: There is no difference between groups with the same letter. N: Number of samples, Min: Minimum, Max: Maximum, Mean: Mean, SD: Standard deviation.

No significant difference was observed in the mean values between the 2% and 4% groups. There was a significant difference between the groups ($p < 0.001$). When Figure 4 is examined, it is determined that the average values of the groups containing 2% and 4% nanoparticles were the same, and the strength value was lower than the 0% group. The group containing 5% nanoparticles was found to have lower flexural strength than all groups. In our results, a decrease was observed from 0% (80.28) to 2% (65.40), and no flexural strength change was observed between 2% (65.40) and 4% (65.40). However, the flexural strength decreased in the group in which we increased the nanoparticle ratio from 4% to 5% (Figure 4).

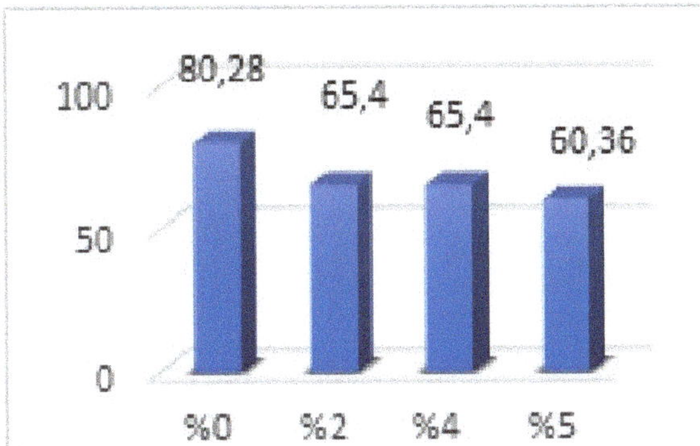

Figure 4. Flexural strength values of three-point bending strength test.

2.5. SEM Analysis

As a result of the three-point bending strength test, it was observed that the fracture line in acrylic samples increased as the ratio of silver–zeolite nanoparticles increased in the SEM analysis. In general, mirror, mist, and hackle fracture regions are formed in the images on the fracture surface of polymers. The mirror region is the initial stage of fracture, and crack growth is very slow. This region was observed to be very large in the 0% control group (Figure 5a). In Figure 5c, the crack line in the 0% control group is very few and small. In the 0% control group, the fracture line is less pronounced, and the number of cracks is less. A thin mist zone surrounds the mirror zone. When the energy is concentrated in the bifurcation form where the crack line reaches rapid propagation, the hackle zone surrounds the mist zone (Figure 5b/Hackle zone). The number of nano-fractures increased in the 2% group (Figure 6c). When the SEM images of the 4% group were examined, it was observed that the nano-cracks became larger as the nanoparticle percentage increased (Figure 7b). It was observed that there was an increase in nanoparticle agglomeration compared to the 2% group. In the group containing 5% nanoparticles, the hackle zone was enlarged, and the cracks (Figure 8c) were indicated by arrows, and the lines appeared longer and wider.

(a) (b) (c)

Figure 5. SEM Images of Fracture Surface of Control Group 300×/500×/1.00K× magnification: (**a**) Mist zone, (**b**) Hackle zone, (**c**) Nano-crack.

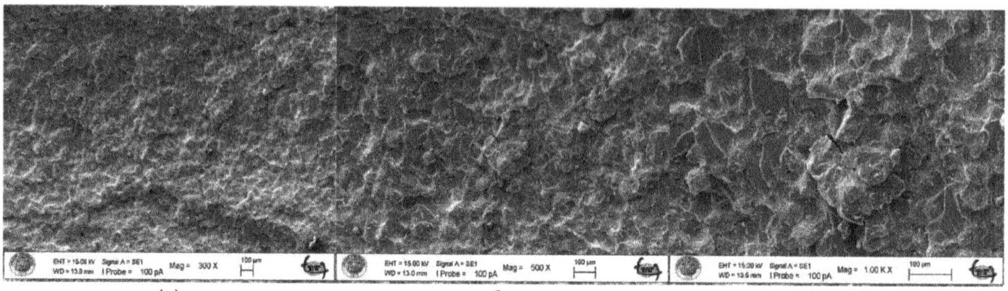

(a) (b) (c)

Figure 6. SEM Images of Fracture Surface of 2% Group 300× /500× /1.00K× Magnification: (**a**) 2% Ag/Z group agglomeration, (**b**) 2% Ag/Z group agglomeration, (**c**) Increase in the number of nano-cracks.

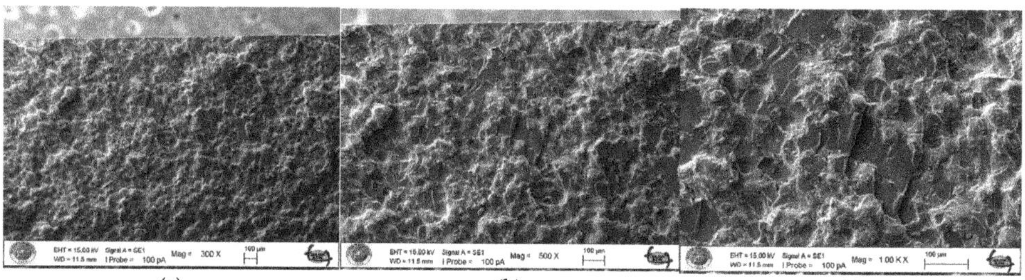

(a) (b) (c)

Figure 7. 4% Group Fracture Surface SEM Images 300× /500× /1.00K× Magnification: (**a**) Increase in the number of nano-fractures, (**b**) 4% Ag/Z group agglomeration, (**c**) Increase in the size of nano-fractures.

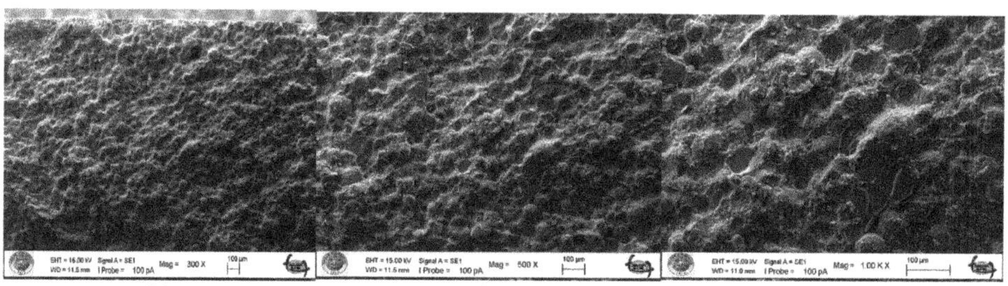

(a) (b) (c)

Figure 8. 5% Group Fracture Surface SEM Images 300× /500× /1.00K× Magnification: (**a**,**b**) 5% Ag/Z group agglomeration, (**c**) Hackle zone expansion, nano-crack length increase.

2.6. Antimicrobial Test Results

Bacterial and fungal colonization on the discs was evaluated using three different methods. First, XTT and PMS solutions were added to the discs, and color changes, which correlated with microorganism concentration, were measured using a spectrophotometer (Figure 9). Second, contaminated discs were inoculated onto TSA and SDA plates, and the results were evaluated after 24–48 h of incubation. Third, discs were incubated in TSB and SDB broth for 6 h, then plated onto TSA and SDA media, with colony counts assessed after 24–48 h of incubation. None of the control groups exhibited antimicrobial activity. For all acrylic resins, the addition of 2% silver zeolite was sufficient to impart antimicrobial activity against the test strains.

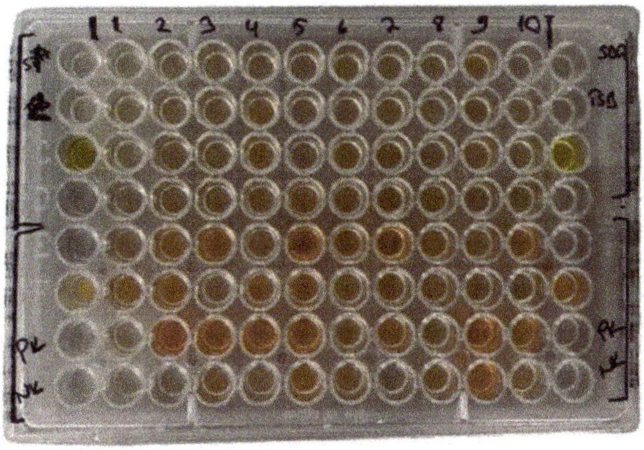

Figure 9. The color differences after XTT and PMS added.

In each well, concentrations were measured using a spectrophotometer. The minimum, maximum, mean, and standard deviation values were obtained by measuring the cell density of *C. albicans* at a wavelength of 490 nm (Table 3). One-way analysis of variance (ANOVA) revealed a statistically significant difference among the *C. albicans* groups ($p = 0.001$).

Table 3. Measurement of microorganism concentration with spectrophotometer.

	Group	Mean ± SD	Min–Max	*p*
S. mutans (490 nm)	0% Group	0.272 ± 0.079^{a}	0.128–0.361	**<0.001**
	2% Group	0.120 ± 0.014^{bc}	0.103–0.140	
	4% Group	0.128 ± 0.015^{b}	0.117–0.158	
	5% Group	0.097 ± 0.018^{c}	0.080–0.131	
C. albicans (490 nm)	%0 Group	0.154 ± 0.010^{b}	0.138–0.170	**<0.001**
	2% Group	0.119 ± 0.017^{a}	0.100–0.143	
	4% Group	0.155 ± 0.023^{b}	0.132–0.199	
	5% Group	0.145 ± 0.012^{b}	0.128–0.159	

One-way variance analysis, a–c: There is no difference between groups with the same letter. Min: Minimum, Max: Maximum, SD: Standard deviation.

The mean fungal concentration in the control group was 0.154, which decreased to 0.119 in the 2% silver–zeolite group. In contrast, the concentrations in the 4% and 5% groups increased compared to the 2% group. Statistically, no significant difference was observed between the 4% and 5% groups and the control group. The acrylic resins with 2% silver–zeolite nanoparticles demonstrated the optimal antimicrobial effect against *C. albicans* colonization (Figure 10). Additionally, the minimum, maximum, mean, and standard deviation values for bacterial concentration were obtained by measuring *S. mutans* with a spectrophotometer at 490 nm (Table 3). One-way analysis of variance (ANOVA) revealed a statistically significant difference between the mean values of the *S. mutans* groups ($p < 0.001$). Concentration levels decreased with increasing silver–zeolite concentration among the groups (Figure 10).

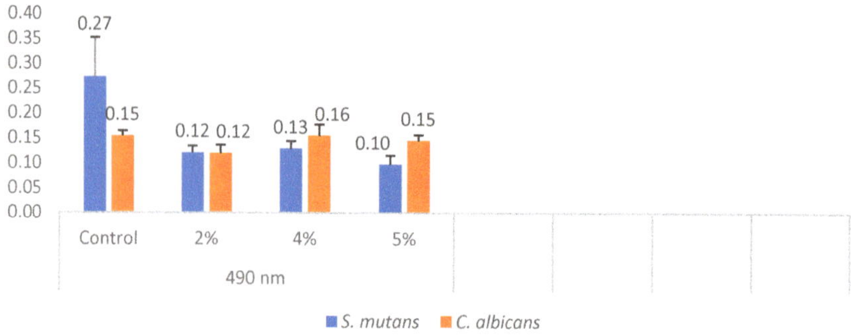

Figure 10. Differences in *C. albicans and S. mutans* concentrations.

The optimal antimicrobial effect against *S. mutans* was achieved with acrylic resins incorporating 5% silver–zeolite nanoparticles. Contaminated discs were inoculated onto TSA (for *S. mutans*) and SDA (for *C. albicans*) media. Discs contaminated with *C. albicans* were incubated at 37 °C for 24 h, whereas discs contaminated with *S. mutans* were incubated at the same temperature for 48 h. The control group exhibited no antimicrobial activity. A reduction in the colony number of *C. albicans* and *S. mutans* was observed in accordance with the concentration of silver–zeolite nanoparticles incorporated into the acrylic resin discs (Figure 11). Additionally, contaminated discs were inoculated into TSB and SDB media, incubated for 4–6 h, and then 10 μL samples were plated onto SDA and TSA. After 24–48 h of incubation, a substantial decrease in colony numbers was noted, correlated with increasing concentrations of silver–zeolite nanoparticles (Figures 12 and 13). The mean colony number (CFU/mL) of *C. albicans* is detailed in Table 4.

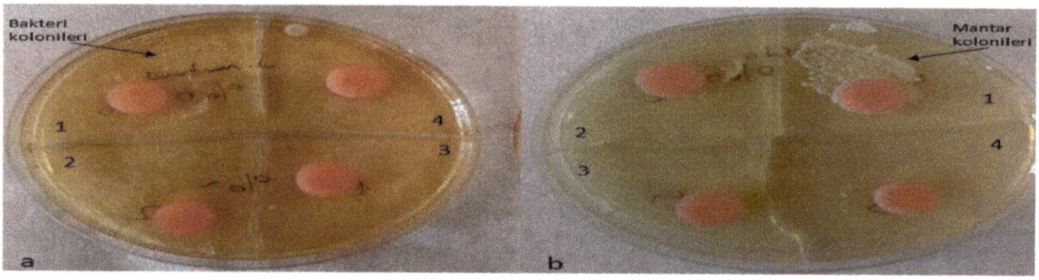

Figure 11. Inoculation of contaminated discs with *S. mutans* and *C. albicans*. (**a**) *S. mutans*, a^1 no silver–zeolite NPs added, a^2 discs with 2% silver–zeolite NPs, a^3 discs with 4% silver–zeolite NPs, a^4 discs with 5% silver–zeolite NPs; (**b**) *C. albicans*, b^1 no silver–zeolite NPs added, b^2 discs with 2% silver–zeolite NPs, b^3 discs with 4% silver–zeolite NPs, b^4 discs with 5% silver–zeolite NPs.

Figure 12. Colony count of *C. albicans* strains. 1: no silver–zeolite NPs added (Control group), 2: discs with 2% silver–zeolite NPs, 3: discs with 4% silver–zeolite NPs, 4: discs with 5% silver–zeolite NPs.

Figure 13. Colony count of *S. mutans* strains; 1: no silver–zeolite NPs added (Control group), 2: discs with 2% silver–zeolite NPs, 3: discs with 4% silver–zeolite NPs, 4: discs with 5% silver–zeolite NPs.

Table 4. Colony number of different groups (CFU/mL).

	C. albicans	*S. mutans*
0%	100,000 (100,000–100,000) ^{c}	100,000 (50,000–100,000) ^{b}
2%	80,000 (50,000–90,000) ^{bc}	0 (0–0) ^{a}
4%	10,000 (5000–50,000) ^{ab}	0 (0–0) ^{a}
5%	5000 (5000–20,000) ^{a}	0 (0–0) ^{a}
p	**<0.001**	**<0.001**

Kruskal–Wallis test; a–c There is no difference between groups with the same letter for each column.

According to the Kruskal–Wallis test analysis, there was a statistically significant difference on the *C. albicans* colony number between the groups with silver–zeolite NPs and the control group ($p < 0.001$). As seen in Table 4, there was a significant decrease in the amount of *C. albicans*, but the bacterial growth did not stop totally. Therefore, *C. albicans* was found in each group even in a small number. While it was 100,000 CFU/mL in the 0% control group, it was 80,000 CFU/mL in the 2% group; It was calculated as 10,000 CFU/mL in the 4% group and 5000 CFU/mL in the 5% group. The mean values of *S. mutans* colonies (CFU/mL) are shown in Table 2. According to the results of Kruskal–Wallis test analysis, the groups with silver–zeolite NPs had a statistically significant difference compared with the control group ($p < 0.001$). According to our results, on average, 100,000 CFU/mL bacteria was found in the control group. There was no bacterial growth in the other groups. Therefore, our results show that silver–zeolite NPs have effect on bacterial growth. As seen in the graphics, the number of *S. mutans* colonies decreased significantly, and statistically significant difference was found between the control group and the others ($p < 0.001$).

3. Discussion

Acrylic resins are commonly used for making denture bases due to their easy manufacturing process, low cost, light weight, good appearance, and suitable properties for the oral environment [35]. Acrylic resins used for removable denture bases require improved mechanical properties to withstand repetitive chewing forces in the oral cavity [36]. Studies have shown that the strength properties of acrylic resins can be enhanced by integrating them with fibers, nanoparticles, and nanotubes [4,8,36]. Researchers are still searching for biocompatible nanomaterials to enhance the mechanical and biological properties of acrylic resins [37]. The accumulation of microorganisms that cause denture-associated stomatitis can be caused by surface roughness exceeding the acceptable limit of acrylic resins [38]. It is well established that denture bases made from heat-cured acrylic resin can serve as a breeding ground for microorganisms, leading to recurring infections in denture wearers [39]. Currently, there is a trend of incorporating antimicrobial agents into dental materials. The process has benefits such as the ability to prevent or reduce bacterial and fungal colonization by releasing agents [19]. The null hypothesis of the study was that silver–zeolite nanoparticles added to dental acrylic resins would not have an antimicrobial effect on saliva-contaminated dental acrylic and would not change its mechanical and physical properties. Based on the study's analysis results, the null hypothesis of the study was rejected.

One of the most popular antimicrobial investigation method is colorimetric measurement with XTT. It converts live bacterial cells' mitochondrial activities into tetrazolium salts that turn into a water-soluble orange formazan dye after the reaction [40,41]. The formazan products formed as a result of the rapid reaction are measured in the colorimetric spectrophotometer [30]. We used XTT colorimetric substrate in our study due to its numerous beneficial properties. In accordance with the scientific literature, we used a spectrophotometer to measure the optical density at 490 nm [6,42].

In Li et al.'s study, dental acrylic samples with 1%, 2%, 3%, and 5% silver were examined using a confocal scanning laser microscope. No significant effect was observed in the groups added at rates of 1%, 2%, and 3%, except for the 5% group which showed a significant decrease [43]. In a study, different concentrations (0–5%) of zeolite nanoparticles were added to heat-polymerized PMMA resin. Results showed that 2% zeolite addition rate was optimal for both mechanical and antimicrobial properties against *C. albicans* [44]. The results of this study are fully consistent with our findings. Our study concludes that the ideal nanoparticle concentration for optimum antimicrobial effect against *C. albicans* was 2%. In Casemiro et al.'s study, they added silver chino zeolite nanoparticles in varying concentrations (2.5%, 5%, 7.5%, 10%) to dental acrylic, and they reported that it had an antimicrobial effect against *C. albicans* and *S. mutans* strains [36]. Kuroki et al. found out that adding 1% silver–zeolite nanoparticles to dental acrylic reduced colonization of *S. mutans* compared to the control group [45]. We have found that adding silver nanoparticles to dental resin materials reduces bacterial and fungal colonization, similar to previous studies [46,47].

To minimize the risk of microbial attachment and plaque formation, it is crucial to properly finish and polish the external surfaces of prostheses. This is an essential step in the clinical protocol for fabricating the prostheses. After dentures are used in the oral environment, surface changes can occur, leading to the formation of rough surfaces that are more favorable for plaque accumulation. Therefore, regular maintenance and professional polishing of prosthetic surfaces are necessary to maintain oral hygiene and minimize the risk of oral infections. In our study, antimicrobial effect against *C. albicans* at 4% and 5% of nanoparticles concentration were lower than 2%. This is as a result of the surface roughness having increased with the nanoparticle concentration, and *C. albicans* can grow fast and adhere easily to rough surfaces [48,49]. However, in these groups, the number of *Candida* colonies decreased according to silver NPs' concentration. Silver nanoparticle treatment has previously been reported to induce alterations in the cellular morphology of *C. albicans*, with inhibition of Candida hyphae formation [50]. The reason for the increase in absorbance may be due to the fact that Candida species develop hyphae by progressing towards 'mycelial' development during the incubation period [51]. Hyphae may have reduced the optical transmittance [51]. It is thought that it may have developed hyphae later in the experiment, and therefore, the absorbance may have increased.

Spectrophotometer values include the sum of viable and non-viable cells of the measured solution. However, the optical density measurement of cells is only an approximation of the colony count (CFU). Therefore, confirming with agar plate is an important control. The number of viable pathogens present in the suspension is determined by counting colony (CFU/mL) on agar plate [52,53]. As a result of our statistical analysis, a significant difference was observed between the groups in *C. albicans*. The number of *C. albicans* colonies had a significant decrease as nanoparticle percentages increased, but the fungal growth did not stop totally. A statistically significant difference was found among the groups tested for *S. mutans* ($p < 0.001$). Based on our analysis, we found that the addition of nanoparticles led to an increase in the antimicrobial effect, which was directly proportional to their concentration. There was a significant difference among the groups in the *S. mutans* bacteria. As a result of *S. mutans* colony count, it was seen that the growth stopped totally with silver NPs. Optical density measurements and medium count values with a spectrophotometer have resulted in supporting each other. In the study of Malic et al., 0.2, 0.4, 0.7, and 2% silver zeolite was added to dental acrylic, and they reported that it had an antimicrobial effect against *C. albicans* and *S. mutans* as a result of CFU count [4]. In Abe et al.'s study, silver zeolite was added to the tissue, resulting in a significant decrease in the number of *C. albicans* colonies [19]. Kuromi et al. counted the colony number to evaluate the antimicrobial effect of dental acrylic added with zeolite against *S. mutans* bacteria. It was reported that 1% added zeolite resulted in a significant decrease in the number of *S. mutans* colonies compared to the control group [45].

Micrometer and nanometer images of the samples are observed by SEM. In our study, we observed that the roughness of disc surfaces increased with silver NPs' concentration as shown by SEM images of surface topography. We can state that the rough structure on the surface supports the colonization of yeast on the discs. Ranjitha et al. also studied the effects of 1% and 2% silver nanoparticle addition on surface roughness of heat-polymerized acrylic resin. They found higher surface roughness values in the 2% group compared with the 1% and control group as we reported in our study [29].

In order to confirm the presence of nanoparticles in PMMA, Aljafery and his colleagues reported the presence of silver and zinc synthesis forms of silver–zinc–zeolite nanoparticle added to dental acrylic by EDX analysis [47]. Nam et al. also incorporated silver, gold, and platinum to dental acrylic and visualized the presence of nanoparticles by SEM/EDX [54]. In the study conducted by Kasraei et al., silver nanoparticles were added to polymethylmethacrylate, and the presence of nanoparticles was examined by SEM/EDX [55]. In our study, we performed SEM/EDX analysis on one sample from each group before the test. Similarly, in our study we performed SEM/EDX analysis on one sample from each group before the test. The presence of silver was detected in all groups with the EDX images.

This study was conducted under in vitro conditions, which means that the results may differ from the in vivo activity. Therefore, it is important to carefully control the in vivo activities of silver NPs in clinical applications through animal models or human volunteers. The effect of thermal aging in the mouth was not evaluated in our study as a limitation. In order to support the clinical use of silver-zeolite NPs in dental acrylics, its biocompatibility can be investigated in further studies. Further studies are still required if the cations may precipitate within the oral epithelial cell for a long period of time, possibly causing argyria and disruption of normal microflora [56,57]. The results of Nam et al.'s study implicate that Ag-denture acrylic might act as a low-releasing antifungal device and therefore could help geriatric denture wearers who have restricted manual dexterity or cognitive disturbances to improve their oral hygiene status [12].

It is important to evaluate the mechanical properties of acrylic resins containing zeolites because removable and complete dentures are subjected to repeated forces [8]. Nevertheless, the addition of small percentages of zeolite to poly (methylmethacrylate) may be effective against microorganisms; therefore, its effect on mechanical properties may be less significant than the potential benefits, especially for patients who do not follow an adequate denture cleaning protocol [9,36,58].

Mechanical properties of the rectangular resin samples with 0, 2, 4, and 5% silver–zeolite nanoparticles were evaluated, and color change was also examined after the added nanoparticles. Researchers have been curiously looking for biocompatible nanomaterials to improve acrylic resins' mechanical and biological properties [37]. The flexural strength of acrylic resin, also called the modulus of rupture, is the property that occurs when the prosthesis is subjected to cyclic deformation. It represents the highest stress experienced in this material at the moment of rupture. The low ability to withstand such deformation under force is the factor most responsible for the clinical incidence of early fracture of the prosthesis [59]. The frequent fracture of this material in the clinic may be due to low resistance to impact or bending forces [34]. Kamonkhantikul et al. evaluated the antimycotic, optical, and mechanical properties of different ratios of zinc oxide nanoparticles in thermosetting PMMA material [60]. Sodagar et al. found that SiO_2 and TiO_2 nanoparticles added at 0.5%

and 0.1% decreased the flexural strength of acrylic resins [61]. In light of these studies, many methods have been tried to strengthen acrylic resin-based materials. In recent years, inorganic fillers containing various silver and silver ions have been used in different forms, such as silver–zirconium phosphate, silver–titanium dioxide, and silver–zeolite [62]. However, despite the antimicrobial advantages of the added modifying antimicrobial agents, their effect on the mechanical properties of the resins should also be evaluated [62]. Craig reported that the flexural strength test is the test that best simulates prostheses subjected to intraoral forces [63]. The flexural strength of materials is measured by uniaxial or biaxial testing with static and dynamic force loading. Today, the uniaxial three-point flexural test has been accepted as the standard test by ISO for acrylic resin materials due to its ease of application and reproducibility. This standard states that the minimum flexural strength values should be 65 MPa [64]. Azeez et al. [8] added 0.5% silver–zinc–zeolite nanoparticles to the heat-polymerized acrylic base material and reported a non-significant difference in the flexural strength test. They stated that the low ratio had little effect on the material. Casemiro et al. added silver–zinc–zeolite nanoparticles to heat-polymerized acrylic base material at 0, 2.5, 5, 7.5, and 10% ratios and applied a flexural strength test. They reported that the results of the mechanical tests indicated that the addition of zeolite may decrease the flexural and impact strength of acrylic resins in accordance with the findings of previous studies [36,65,66]. In the study by Murat et al., silver–zeolite nanoparticles were added to heat-polymerized acrylic at 0, 0.5, 2, and 4%, and flexural strength was evaluated. It was reported that the mechanical properties of microbial agents added to methacrylate were in line with similar studies and weakened the mechanical properties [65,66]. Yadav et al. reported that silver–zinc–zeolite added to acrylic resin at a rate of 5% caused lower flexural strength than 65 MPa specified in the ISO standard [67]. In this study, similar results were found, and the average value of silver–zeolite nanoparticles added at 2% was 65.40, while it decreased to 60.36 in the 5% group. The result in this study was partially similar to the result of Shibate et al. and Sodagar et al., who found that the flexural strength decreased as the percentage of nanoparticles increased [60,68].

It is thought that there is a decrease in flexural strength due to the effect of nano-sized oxides on the internal structure of the polymerized acrylic resin. In addition, the dispersion of the added nanoparticles in the PMMA matrix affects the conversion of monomers to polymer, resulting in an increase in unreacted residual monomers that act as plasticizers [60]. The addition of nanoparticles to acrylic resins causes the particles to aggregate and cluster together in the matrix, forming stress concentration zones and weakening the mechanical properties of the materials [66,69]. It should be considered that the decrease in flexural strength may also be due to the porous structure of the zeolite nanoparticle [46]. Unlike these studies, Kassae et al. reported that adding 0.5% silver nanoparticles did not cause a significant effect on the flexural strength of PMMA [70]. This result is thought to be due to the low nanoparticle ratio.

When an antibacterial agent was added, a color change in the denture base resin was reported due to the silver ion in the antibacterial agent [45,71]. Wakasa et al. reported that the addition of 1% and 2% silver–zeolite nanoparticles to autopolymerized acrylic resin did not affect the polymerization of the resin and discoloration of the dental material due to the silver in it [71]. Such a discoloration was reported to occur due to silver ions in the antimicrobial agent. It is believed that such discoloration is due to the oxidation of silver ions on the surface of the material as well as the formation of metal oxides from antibacterial metal ions during an oxidation–reduction reaction that occurs during polymerization [44]. In the study of Kuroki et al., 1% silver–zeolite nanoparticles were added to PMMA, and three color parameters (L*, a*, and b*) were measured using the CIE LAB color system color analysis method to examine the color change [45]. In line with these studies, the color differences between the groups in our study were quantitatively evaluated by averaging the measurements made from two points in a spectrophotometer device (Konica Minolta, Tokyo, Japan). Ren et al. used the CIE ΔE 2000 analysis for the evaluation of two different colors in dental acrylic [72]. In the present study, color analysis was performed with the CIE

ΔE 2000 LAB color system, which is a more up-to-date approach. Wakasa et al. reported that the addition of 1% and 2% silver–zeolite nanoparticles to autopolymerized acrylic resin did not affect the polymerization of the resin and that the dental material changed color due to the silver content [30]. The color analysis findings of this study were similar to those of Wakasa et al. In another zeolite nanoparticle study, Kuroki et al. added 1% nanoparticles and reported a color change [32]. Azeez et al. added 0% and 0.5% silver–zinc–zeolite and concluded that both groups had a statistically insignificant difference in color change. They stated that this result was due to the fact that the percentage of zeolite added was less than 2% [8].

In another study by Malic and et al. [4], they showed that silver–zeolite nanoparticles added with increasing percentages caused intense brown coloration. They stated that this dark coloration is unacceptable for dental applications. They reported that silver zeolite treated with silver nitrate did not show such an intense coloration [4]. Therefore, silver nitrate-treated zeolite was preferred for this study. In this regard, it can be stated that coloration occurs, but nitrate treatment reduces the color change. Casemiro et al. added silver–zinc–zeolite nanoparticles to dental acrylic at 0, 2.5, 5, 7.5, and 10% and observed that the opacity of polymethylmethacrylate increased as the nanoparticle ratio increased [36]. In this study, similar to the results of the studies in the literature, it was observed that ΔE values changed as the percentage of nanoparticles added increased as a result of color analysis.

Paravina et al. reported 1.8 CIEΔE2000 units, which corresponds to the acceptability threshold of color differences for dentists [73]. Ren et al. reported that the 50% acceptability threshold for acrylic resin materials for denture bases was 4.08 CIEΔE2000 units [74]. Similar to this study, Alp et al. accepted the 50% acceptability threshold as 4.08 CIEΔE2000 units [75]. Color differences are graded with numbers from 1 to 5. ΔE00 and ΔEab based on perceptibility and acceptability thresholds are derived from value analysis research findings [73]. Grades 5 and 4 correspond to excellent ($\Delta E00 < 0.8$) and acceptable color stability (>0.8 and <1.8), respectively. Grade 3 corresponds to the type of mismatch considered as moderate color mismatch (>1.8 and <3.6). Grade 2 corresponds to a clearly unacceptable color mismatch (>3.6 and <5.4), while grade 1 is considered a marked unacceptable mismatch (>5.4) [73]. According to the clinical acceptability of the ΔE2000 value [76,77], it can be stated that the 2% group has clinically acceptable results. The color changes of the 4% and 5% groups are not clinically acceptable values compared to the control group [73].

While the in vitro methodology used in this study offers precise control over experimental variables, it does not fully reflect the complex intraoral environment, especially due to the exclusion of thermal aging effects commonly encountered in the mouth. Such conditions, including temperature variations from food and beverages, significantly affect the longevity and performance of dental materials, a factor not accounted for in laboratory set-up. The findings of the current study need to be confirmed by in vivo studies. These subsequent investigations involving the multifaceted dynamics of the human oral cavity will prove the results' applicability in a real-world clinical context, thus bridging the gap between laboratory research and practical dental application.

4. Materials and Methods

A mixture of pure PMMA powder and liquid monomer without silver–zeolite (Ag-Z) nanoparticles served as the control group. Silver–zeolite (Ag-Z) nanoparticles (Nanokar, Istanbul, Turkey) with a purity of 99.99% and a particle size of less than one μm were added in varying amounts: 0% (control), 2%, 4%, and 5% weighed with an analytical balance (Imıcryl, Konya, Turkey) and mixed with the acrylic in a Speed Mixer device (Imıcryl, Konya, Turkey) for 2 min at 1000 rpm by the same operator. The materials were then proportioned, mixed, packed and polymerized following the manufacturers' instructions. Fifty-six disc-shaped specimens (12 × 2 mm) were prepared for the antimicrobial assay, and 40 bars (65 × 10 × 2.5 mm) were produced for testing optical and mechanical properties. These bar samples were fabricated into rectangular prisms, measuring 65 × 10 × 2.5 mm, in alignment with American Dental Association (ADA) No. 12 standards, and their dimensions were verified using calipers. A special wax (Kışlık, Istanbul, Türkiye) was melted into the molds to prepare rectangular-shaped patterns. The stainless steel disc and rectangular-shaped patterns were invested in flasks with dental stone. After the dental stone set, the flasks were opened and the patterns were removed, leaving rectangular-shaped and disc-shaped cavities that were used as matrixes for the fabrication of heat-polymerized acrylic resin specimens. The PMMA powder and liquid monomer (Imicryl, Konya, Türkiye) were mixed at a powder/liquid ratio of 23.4 g:10 mL and left until the mixture reached the dough stage. The mixture was then placed into the flask and hydraulic-pressed using a conventional method. The flasks were placed in a water bath that was heated up to 100 °C and allowed to boil for 45 min to achieve polymerization according to the recommendations. After cooling to room temperature, the specimens were deflasked, the excess material was removed using a carbide bur in a low-speed rotary instrument, and wet-polished using 800-, 1000-, 1200-, and 2000-grit silicon carbide paper with a polishing machine (Brightway, Bursa, Turkiye) at 300 rpm for 30 s. The thickness of all specimens was checked by measuring with an electronic caliper (Absolute Digimatic Caliper, Mitutoyo, Kawasaki, Japan). The specimens were ultrasonically cleaned for 5 min to remove any debris and dried with compressed air.

4.1. *Antimicrobial Susceptibility Tests*

Before microbial assay, samples were incubated in distilled water for 24 h to leach excess residual monomer. The disc specimens were autoclaved at 121 °C and 1 atm pressure in 20 min and then immersed in sterile artificial saliva. Artificial saliva was prepared at neutral pH (7–7.4), and chemicals were used as follows: 0.4 g of NaCl (Sigma-Aldrich, Darmstadt, Germany), 0.4 g of KCl (Sigma Aldrich, Germany), 0.795 g of $CaCl_2 \cdot 2(H_2O)$, 0.78 g of $NaH_2PO_4 \cdot 2(H_2O)$, 0.0005 g of $Na_2S \cdot 9(H_2O)$, 1 g of urea, and 1000 mL of distilled water. After the artificial saliva was prepared, it was sterilized in the autoclave and stored in the dark at room temperature

In this study, the standard strains ATCC 97012 *C. albicans* and RSHM 676 *S. mutans* were used to assess bacterial and fungal colonization on the surfaces of acrylic discs incorporating silver–zeolite nanoparticles. Tryptic soy broth (TSB) and Tryptic soy agar (TSA) (Biomerieux, Marcy l'Etoile, France) were used for *S. mutans*, while Sabouraud dextrose agar (SDA) and Sabouraud dextrose broth (SDB) (Sigma Aldrich, St. Louis, MO, USA) were used for *C. albicans*. All the cultural media were prepared according to the manufacturer's recommendations and sterilized in an autoclave at 121 °C at 1 atm pressure for 20 min and transferred to the Petri dishes and tubes after cooling.

4.1.1. Contamination of Discs with Microorganisms

The discs were placed into sterile 6 × 4 cell culture dishes. For bacterial studies, 1 mL of Tryptic soy broth (TSB) was added to each well, and for fungal studies, 1 mL of Sabouraud dextrose broth (SDB) was used. To simulate the human oral environment, 1 mL of artificial saliva was also added to each well. Suspensions of bacterial and fungal colonies, adjusted to a 0.5 McFarland standard, were prepared in their respective culture media, and 1 mL of each suspension was added to the wells to contaminate the discs. Each study included one well as a sterility control and one as a growth control. The plates were then covered with parafilm and incubated for 24 h at 37 °C. After incubation, the liquid from each well was removed, and the discs were washed twice with 600 µL of PBS to eliminate non-adherent microorganisms. The discs were then re-contaminated and incubated for an additional 24 h.

4.1.2. Preparation of XTT and PMS Solutions

XTT (SERVA, Heidelberg, Germany) and Phenazine metasulfate (PMS) (Sigma Aldrich, USA) were dissolved in sterile Phosphate Buffer Solution (PBS) according to the manufacturer's recommendations. All studies with XTT and PMS were performed in dark room conditions. The solutions were prepared fresh on the day of the study, covered with aluminum foil to avoid light, and stored in the dark. Then, 790 μL of PBS, 200 μL of XTT, and 10 μL of phenazine methosulfate PMS solution were added to each well and incubated at 37 °C for 3 h. After incubation, 100 μL of the solution from each well was transferred to a new sterile 12 × 8 cell culture plate using a sterile pipette tip.

4.1.3. Bacterial Colonization

Contamination of microorganisms on the discs were studied in several stages after all discs were contaminated with microorganisms. At first, the discs were taken from the wells and inoculated by moving them over the appropriate media. Secondly, the discs were transferred in broth media and incubated for 6 h at 37 °C. Following this, 10 μL of the sample was taken in each well and inoculated onto an agar plate for colony counting. Thirdly, after adding XTT and PMS solutions to the discs, bacterial concentrations were measured at 492 nm wavelengths using a spectrophotometer.

4.2. *Optical Properties*

Before optical and mechanical tests, forty bar specimens according to ADA specification No. 12 were incubated in distilled water for 48 h to leach excess residual monomer. To mimic the oral environment in vitro, all samples were kept in artificial saliva for 1 week in an oven (Nüve 1020, Ankara, Turkiye). The containers with artificial saliva were rinsed daily. The tested samples were exposed to light within a wavelength range of 380–780 nm via a measuring head positioned at a right angle. For each specimen, three repeated measurements were taken to determine the colorimetric values, i.e., L* (brightness), a* (red–green proportion), and b* (yellow–blue proportion). For color change, L, a, and b values were measured from two points on the smoothest surface with a spectrophotometer (Konica Minolta, Osaka, Japan) in the control group (0%) and samples containing silver–zeolite nanoparticles at 2, 4, and 5% ratios, and the values were averaged (Figure 11). The color change was calculated with the formula ΔE2000:

$$\Delta E_{00} = [(\frac{\Delta L'}{K_L S_L})^2 + (\frac{\Delta C'}{K_C S_C})^2 + (\frac{\Delta H'}{K_H S_H})^2 + R_T(\frac{\Delta C'}{K_C S_C})(\frac{\Delta H'}{K_H S_H})]^{1/2}$$

The parametric factors K_L, K_C, and K_H were all set to 1. The mean of three readings was recorded, and the value of each specimen was calculated using the CIE-L*a*b*-color system with a colorimeter.

4.3. *Mechanical Properties*

After spectrophotometric evaluation, flexural strength was evaluated by a three-point bending using a Devotrans Device (Imicryl, Konya, Turkiye) to determine the flexural strength of the same forty bar-shaped specimens according to ADA specification No. 12 at 5 mm/min speed (Figure 12). The samples were placed on the apparatus, which was placed on supports parallel to each other and 50 mm apart. A load was applied until the specimen fractured. Flexural strength was calculated as follows:

σ = 3Fl\2bd2

where σ = flexural strength (N/mm^2 = MPa), F = the maximum load during fracture (N), l = the distance between the supports (50 mm), b = the specimen width (mm), and d = specimen thickness (mm). Scanning electron microscope (SEM) analysis was performed via the SEM device FEI (Quanta 250 FEG; FEI Ltd., Brno, Czech Republic) at the Selcuk University Advanced Technology Research and Application Center.

Scanning Electron Microscope (SEM)

Fractured surfaces of the bar-shaped specimens of the control and other groups after the three-point bending test were randomly selected, gold-sputter-coated, and observed using SEM (Quanta 250; FEI Company, Eindhoven, The Netherlands) under acceleration of 20 kV and magnification of 35,000×.

4.4. Statistical Analysis

One-way Analysis of Variance (ANOVA) test was used to determine the significant difference between groups. When there was a significant difference, Tukey HSD and Tamhane's T2 tests were used for multiple comparisons to identify specific differences between groups. These tests offer a significant determination of the relations among various groups in the field of microbiological research. The data analysis was performed using IBM SPSS V23 software. The Shapiro–Wilk test was used to assess normal distribution conformity with a significance level of $p < 0.05$.

5. Conclusions

Three different rates of nanoparticles (0%, 2%, 4%, 5%) were added to the heat-polymerized acrylic resin used as the denture base material. In vitro antimicrobial effects of nanoparticles were investigated against *C. albicans* and *S. mutans* strains. The surfaces of the samples were analyzed using topographic imaging to determine their level of surface roughness. The presence of silver nanoparticles was evaluated using SEM/EDX imaging.

According to the spectrophotometer analysis, discs with 2% silver–zeolite nanoparticles were more effective against *C. albicans* than those with 4% and 5%. Discs with 2% silver–zeolite nanoparticles showed a significant antimicrobial effect compared to the 0% group. The reason for the increase in colonization of *C. albicans* in the 4% and 5% groups can be attributed to the increased surface roughness. The concentration of nanoparticles had a significant impact on the decrease of *S. mutans* colonization. Based on its antimicrobial and mechanical properties, we recommend adding 2% silver–zeolite nanoparticles to the acrylic resin material used for denture base material for optimal results. While adding silver–zeolite nanoparticles provides antimicrobial properties, it also reduces the material's structural integrity and alters its aesthetic appearance, highlighting the necessity of balancing antimicrobial efficacy with mechanical and esthetic properties in dental materials. In the current study, it was found that the addition of 2% silver nitrate-treated zeolite nanoparticles to acrylic material used in dental prostheses was satisfactory in terms of both discoloration and mechanical strength, i.e., the acrylic material was discolored at an acceptable level, and its mechanical strength remained adequate. However, when the amount of nanoparticles was increased to 5%, the strength of the material decreased significantly, i.e., the material became weaker. Specimens prepared with nanoparticles added at 2% and 4% were found to be clinically suitable with a strength value of 65.40 MPa. Moreover, as the proportion of nanoparticles added to increase the antimicrobial properties increased, an increase in the color of the material was also observed. However, the silver–zeolite nanoparticles added at 2% were still within clinically acceptable limits regarding color change. As a clinical significance; it is advisable to incorporate 2% silver–zeolite nanoparticles into the material of acrylic dental prostheses to impart antimicrobial features while preserving a balance between mechanical and aesthetic properties. Therefore, this will make the acrylic dentures more durable and reduce microbial growth, providing both an aesthetic and functional solution.

Author Contributions: Conceptualization, N.D. and U.A.; data curation, Ç.K.Y.; investigation, Ç.K.Y., N.D., U.A. and N.Ç.; methodology, Ç.K.Y., U.A. and N.Ç.; project administration, N.D.; supervision, N.D. and U.A.; writing—original draft preparation, Ç.K.Y. and N.Ç.; writing—review and editing, N.D. and N.Ç. All authors have read and agreed to the published version of the manuscript.

Funding: The research leading to these results received funding from Selcuk University Science Research Thesis Projects (under Grant Agreement No: 21212060).

Institutional Review Board Statement: The authors state that they have obtained appropriate institutional review board approval or have followed the principles outlined in the Declaration of Helsinki for all human or animal experimental investigations. This study was conducted with Selcuk University Medical Faculty Ethical Committee approval with decision No: 2022/18. The manuscript does not report on or involve any animals, humans, human data, human tissue or plants for consent to participate.

Data Availability Statement: The original contributions presented in the study are included in the article, further inquiries can be directed to the corresponding author.

Acknowledgments: The researchers appreciate Murat Aydin (who has a doctorate degree from the microbiology department) for his contributions to this research study.

Conflicts of Interest: The authors have no other relevant affiliations or financial involvement with any organization or entity with a financial interest in or financial conflict with the subject matter or materials discussed in the manuscript apart from those disclosed. This includes employment, consultancies, honoraria, stock ownership or options, expert testimony, grants or patents received or pending, or royalties.

References

1. Price, C. A history of dental polymers. *Aust. Prosthodont. J.* **1994**, *8*, 47–54. [PubMed]
2. Lessa, F.C.R.; Enoki, C.; Ito, I.Y.; Faria, G.; Matsumoto, M.A.N.; Nelson-Filho, P. In-vivo evaluation of the bacterial contamination and disinfection of acrylic baseplates of removable orthodontic appliances. *Am. J. Orthod. Dentofac. Orthop.* **2007**, *131*, 705.e11–705.e17. [CrossRef] [PubMed]
3. Rickman, L.J.; Padipatvuthikul, P.; Satterthwaite, J.D. Contemporary denture base resins: Part 1. *Dent. Update* **2012**, *39*, 25–28. [CrossRef] [PubMed]
4. Malic, S.; Rai, S.; Redfern, J.; Pritchett, J.; Liauw, C.M.; Verran, J.; Tosheva, L. Zeolite-embedded silver extends antimicrobial activity of dental acrylics. *Colloids Surf. B Biointerfaces* **2019**, *173*, 52–57. [CrossRef] [PubMed]
5. Sumi, Y.; Miura, H.; Michiwaki, Y.; Nagaosa, S.; Nagaya, M. Colonization of dental plaque by respiratory pathogens in dependent elderly. *Arch. Gerontol. Geriatr.* **2007**, *44*, 119–124. [CrossRef]
6. Wady, A.; Machado, A.L.; Zucolotto, V.; Zamperini, C.; Berni, E.; Vergani, C.E. Evaluation of Candida albicans adhesion and biofilm formation on a denture base acrylic resin containing silver nanoparticles. *J. Appl. Microbiol.* **2012**, *112*, 1163–1172. [CrossRef]
7. Valentini, F.; Luz, M.S.; Boscato, N.; Pereira-Cenci, T. Biofilm formation on denture liners in a randomised controlled in situ trial. *J. Dent.* **2013**, *41*, 420–427.
8. Azeez, Z.A.; Fatah, N.A. The effect of incorporation of prepared Ag-Zn Zeolite on some properties of heat polymerized acrylic denture base materials. *J. Baghdad Coll. Dent.* **2015**, *27*, 63–69. [CrossRef]
9. Saengmee-Anupharb, S.; Srikhirin, T.; Thaweboon, B.; Thaweboon, S.; Amornsakchai, T.; Dechkunakorn, S.; Suddhasthira, T. Antimicrobial effects of silver zeolite, silver zirconium phosphate silicate and silver zirconium phosphate against oral microorganisms. *Asian Pac. J. Trop. Biomed.* **2013**, *3*, 47–52. [CrossRef]
10. Schneid, T.R. An in vitro analysis of a sustained release system for the treatment of denture stomatitis. *Spec. Care Dent.* **1992**, *12*, 245–250.
11. Budtz-Jørgensen, E.; Mojon, P.; Banon-Clément, J.M.; Baehni, P. Oral candidiasis in long-term hospital care: Comparison of edentulous and dentate subjects. *Oral Dis.* **1996**, *2*, 285–290. [CrossRef] [PubMed]
12. Nam, K.Y.; Lee, C.H.; Lee, C.J. Antifungal and physical characteristics of modified denture base acrylic incorporated with silver nanoparticles. *Gerodontology* **2012**, *29*, e413–e419.
13. Buergers, R.; Rosentritt, M.; Handel, G. Bacterial adhesion of Streptococcus mutans to provisional fixed prosthodontic material. *J. Prosthet. Dent.* **2007**, *98*, 461–469. [PubMed]
14. McCarthy, C.; Snyder, M.; Parker, R. The indigenous oral flora of man-I: The newborn to the 1-year-old infant. *Arch. Oral Biol.* **1965**, *10*, 61–70. [CrossRef] [PubMed]
15. Paster, B.J.; Olsen, I.; Aas, J.A.; Dewhirst, F.E. The breadth of bacterial diversity in the human periodontal pocket and other oral sites. *Periodontology* **2006**, *42*, 80–87.
16. Albandar, J.; Brunelle, J.; Kingman, A. Destructive periodontal disease in adults 30 years of age and older in the United States. *J. Periodontol.* **1999**, *70*, 13–29.
17. Uçar, Y.; Bakar, O.; Ekinci, M.; Kayar, B. Poliamid ile farklı polimetilmetakrilat protez kaide materyallerinin mikroorganizma tutulumunun karşılaştırılması. *SDÜ Tıp Fak. Derg.* **2013**, *20*, 8–13.
18. Li, L.; Redding, S.; Dongari-Bagtzoglou, A. Candida glabrata, an emerging oral opportunistic pathogen. *J. Dent. Res.* **2007**, *86*, 204–215.

19. Abe, Y.; Ishii, M.; Takeuchi, M.; Ueshige, M.; Tanaka, S.; Akagawa, Y. Effect of saliva on an antimicrobial tissue conditioner containing silver–zeolite. *J. Oral Rehabil.* **2004**, *31*, 568–573. [CrossRef]
20. Alla, R.; Raghavendra, K.; Vyas, R.; Konakanchi, A. Conventional and contemporary polymers for the fabrication of denture prosthesis: Part I–overview, composition and properties. *Int. J. Appl. Dent. Sci.* **2015**, *1*, 82–89.
21. Galav, A.; Deogade, S.C.; Mantri, S.; Sumathi, K.; Galav, S. Effect of water storage on the flexural strength of heat-cured denture base resin reinforced with stick(s) glass fibers. *Contemp. Clin. Dent.* **2017**, *8*, 264–271. [PubMed]
22. Khosravani, M.R. Mechanical behavior of restorative dental composites under various loading conditions. *J. Mech. Behav. Biomed. Mater.* **2019**, *93*, 151–157. [CrossRef] [PubMed]
23. Gad, M.M.; Fouda, S.M.; Al-Harbi, F.A.; Näpänkangas, R.; Raustia, A.N. PMMA denture base material enhancement: A review of fiber, filler, and nanofiller addition. *Int. J. Nanomed.* **2017**, *12*, 3801–3812. [CrossRef] [PubMed]
24. Oka, H.; Tomioka, T.; Tomita, K.; Nishino, A.; Ueda, S. Inactivation of enveloped viruses by a silver-thiosulfate complex. *Met. Drugs* **1994**, *1*, 511. [CrossRef] [PubMed]
25. Tokumaru, T.; Shimizu, Y.; Fox, J.R.C. Antiviral activities of silver sulfadiazine in ocular infection. *Res. Commun. Chem. Pathol. Pharmacol.* **1974**, *8*, 151–158.
26. Alt, V.; Bechert, T.; Steinrucke, P.; Wagener, M.; Seidel, P.; Dingeldein, E.; Domann, E.; Schnettler, R. An in vitro assessment of the antibacterial properties and cytotoxicity of the nanoparticles silver bone cement. *Biomaterials* **2004**, *25*, 4383–4391. [CrossRef]
27. Samuel, U.; Guggenbichler, J.P. Prevention of catheter- related infections: The potential of a new nanosilver impregnated catheter. *Int. J. Antimicrob. Agents* **2004**, *23*, 75–78. [CrossRef]
28. Wright, J.B.; Lam, K.; Hansen, D.; Burrell, R. Efficacy of topical silver against fungal burn wound pathogens. *Am. J. Infect. Control* **1999**, *27*, 344–350. [CrossRef]
29. Ranjitha, V.; Reddy, K.M.; Shastry, Y.M. Effects of silver nanoparticles addition on flexural strength and surface roughness of heat polymerized acrylic resin-An in-vitro study. *J. Adv. Med. Dent. Sci. Res.* **2022**, *10*, 87–91.
30. Nikawa, H.; Yamamoto, T.; Hamada, T.; Rahardjo, M.; Murata, H.; Nakanoda, S. Antifungal effect of zeolite-incorporated tissue conditioner against Candida albicans growth and/or acid production. *J. Oral Rehabil.* **1997**, *24*, 350–357.
31. Feng, Q.L.; Wu, J.; Chen, G.Q.; Cui, F.; Kim, T.; Kim, J.A. mechanistic study of the antibacterial effect of silver ions on Escherichia coli and Staphylococcus aureus. *Biomed. Mater. Res.* **2000**, *52*, 662–668. [CrossRef]
32. Cheun, K. Antibacterial activity of A-type zeolite supporting silver ions in deionized distilled water. *J. Antibact. Antifung. Agent.* **1996**, *24*, 269–274.
33. Inoue, Y.; Hoshino, M.; Takahashi, H.; Noguchi, T.; Murata, T.; Kanzaki, Y.; Hamashima, H.; Sasatsu, M. Bactericidal activity of Ag–zeolite mediated by reactive oxygen species under aerated conditions. *J. Inorgan. Biochem.* **2002**, *92*, 37–42.
34. Asar, N.V.; Albayrak, H.; Korkmaz, T.; Turkyilmaz, I. Influence of various metal oxides on mechanical and physical properties of heat-cured polymethyl methacrylate denture base resins. *J. Adv. Prosthodont.* **2013**, *5*, 241–247. [CrossRef]
35. Abdulridha, W.M.; Almusawi, R.M.; Al-Jubouri, O.M.; Wally, Z.J.; Zidan, S.; Haider, J.; Al-Quraine, N.T. Studying the effect of adding Titanium Dioxide (TiO_2) nanoparticles on the compressive strength of chemical and heat-activated acrylic denture base resins. *Adv. Mater. Process. Technol.* **2020**, *8*, 1058–1070. [CrossRef]
36. Casemiro, L.A.; Martins, C.H.G.; Pires-de-Souza, F.C.P.; Panzeri, H. Antimicrobial and mechanical properties of acrylic resins with incorporated silver-zinc zeolite part I. *Gerodontology* **2008**, *25*, 187–194. [CrossRef]
37. de Souza Leão, R.; de Moraes, S.L.D.; de Luna Gomes, J.M.; Lemos, C.A.A.; da Silva Casado, B.G.; do Egito Vasconcelos, B.C.; Pellizzer, E.P. Influence of addition of zirconia on PMMA: A systematic review. *Mater. Sci. Eng. C* **2020**, *106*, 110292.
38. Darwish, M.; Nassani, M.Z. Evaluation of the effect of denture adhesives on surface roughness of two chemically different denture base resins. *Eur. J. Dent.* **2016**, *10*, 321–326. [CrossRef]
39. Keng, S.B.; Lim, M. Denture plaque distribution and the effectiveness of a perborate-containing denture cleanser. *Quintessen Intern.* **1996**, *27*, 341–345.
40. Sen, B.H.; Chugal, N.M.; Liu, H.; Fleischmann, J.A. new method for studying the adhesion of Candida albicans to dentin in the presence or absence of smear layer. *Oral Surg. Oral Med. Oral Pathol. Oral Radiol. Endodontology* **2003**, *96*, 201–206.
41. Koban, I.; Matthes, R.; Hübner, N.O.; Welk, A.; Sietmann, R.; Lademann, J.; Kramer, A.; Kocher, T. XTT assay of ex vivo saliva biofilms to test antimicrobial influences. *GMS Krankenhaushygiene Interdiszip.* **2012**, *7*. [CrossRef]
42. Ozel, G.S.; Guneser, M.B.; Inan, O.; Eldeniz, A.U. Evaluation of C. albicans and S. mutans adherence on different provisional crown materials. *J. Adv. Prosthodont.* **2017**, *9*, 335–340. [CrossRef] [PubMed]
43. Li, Z.; Sun, J.; Lan, J.; Qi, Q. Effect of a denture base acrylic resin containing silver nanoparticles on Candida albicans adhesion and biofilm formation. *Gerodontology* **2016**, *3*, 209–216. [CrossRef] [PubMed]
44. Nakanoda, S.; Nikawa, H.; Hamada, T.; Yamamoto, T.; Nakamoto, K. The material and antifungal properties of antibiotic zeolite incorporated acrylic resin. *J. Jpn. Prosthodont. Soc.* **1995**, *39*, 919–926. [CrossRef]
45. Kuroki, K.; Hayashi, T.; Sato, K.; Asai, T.; Okano, M.; Kominami, Y.; Takahashi, Y.; Kawai, T. Effect of self-cured acrylic resin added with an inorganic antibacterial agent on Streptococcus mutans. *Dent. Mater. J.* **2010**, *29*, 277–285. [CrossRef]
46. Dao, P.H.; Nguyen, T.D.; Nguyen, T.C.; Nguyen, A.H.; Mac, V.P.; Tran, H.T.; Phung, T.L.; Vu, Q.T.; Vu, D.H.; Ngo, T.C.Q.; et al. Assessment of some characteristics, properties of a novel waterborne acrylic coating incorporated TiO_2 nanoparticles modified with silane coupling agent and Ag/Zn zeolite. *Prog. Org. Coat.* **2022**, *163*, 106641. [CrossRef]

47. Aljafery, A.M.; Al-Jubouri, O.M.; Wally, Z.J.; Almusawi, R.M.; Abdulrudha, N.H.; Haider, J. The Effects of Incorporating Ag-Zn Zeolite on the Surface Roughness and Hardness of Heat and Cold Cure Acrylic Resins. *J. Compos. Sci.* **2022**, *6*, 85. [CrossRef]
48. Karcı, M.; Demir, N.; Yazman, Ş. Farklı Nanopartiküllerle Güçlendirilen Mikrodalga Enerjisi İle Polimerize Olan Akrilik Rezinin Mekanik Ve Fiziksel Özelliklerinin Değerlendirilmesi. *Atatürk Üniversitesi Diş Hekim. Fakültesi Derg.* **2017**, *30*, 101–106.
49. Karamustafa, F. Akrilik Rezin Materyallerine Farklı Oranlarda çay Ağacı Yağı Eklenmesinin Candida Albicans İnhibisyonuna ve Materyallerin Yüzey Pürüzlülüğüne Etkisi. Ph.D. Thesis, Karadeniz Teknik Üniversitesi, Trabzon, Türkiye, 2021.
50. Khatoon, N.; Sharma, Y.; Sardar, M.; Manzoor, N. Mode of action and anti-Candida activity of Artemisia annuamediated-synthesized silver nanoparticles. *J. Mycol. Med.* **2019**, *29*, 201–209. [CrossRef]
51. Haque, F.; Alfatah, M.; Ganesan, K.; Bhattacharyya, M.S. Inhibitory effect of sophorolipid on Candida albicans biofilm formation and hyphal growth. *Sci. Rep.* **2016**, *6*, 235–275. [CrossRef]
52. Jimenez, L.; Ignar, R.; Smalls, S.; Grech, P.; Hamilton, J.; Bosko, Y.; English, D. Molecular detection of bacterial indicators in cosmetic/pharmaceuticals and raw materials. *J. Ind. Microbiol. Biotechnol.* **1999**, *22*, 93–95. [CrossRef]
53. Huang, J.; Hitchins, A.D.; Tran, T.T.; McCarron, J.E. Bacteriological Analytical Manual Chapter 23: Methods for Cosmetics. July 2024 Edition. pp. 1–15. Available online: https://www.fda.gov/media/177960/download (accessed on 22 September 2024).
54. Nam, K.Y. Characterization and antifungal activity of the modified PMMA denture base acrylic: Nanocomposites impregnated with gold, platinum, and silver nanoparticles. *Nanobiomater. Dent.* **2016**, *11*, 309–336.
55. Kasraei, S.; Azarsina, M. Addition of silver nanoparticles reduces the wettability of methacrylate and silorane-based composites. *Br. Dent. J.* **2012**, *26*, 505–510. [CrossRef] [PubMed]
56. Petering, H.G. Pharmacology and toxicology of heavy metals: Silver. *Pharmacol. Ther.* **1976**, *1*, 127–130. [CrossRef]
57. Ramage, G.; Vande Walle, K.; Wickes, B.L.; López-Ribot, J.L. Standardized method for in vitro antifungal susceptibility testing of Candida albicans biofilms. *Antimicrob. Agents Chemother.* **2001**, *45*, 2475–2479. [CrossRef]
58. Mutneja, P.; Raghavendraswamy, K.N.; Gujjari, A.K. Flexural strength of heat cure acrylic resin after incorporation of different percentage of Silver zinc zeolite- an in-vitro study. *IJCI* **2013**, *4*, 25–31.
59. Takahashi, T.; Gonda, T.; Maeda, Y. The influence of reinforcement on strain in maxillary complete dentures: A preliminary report. *Int. J. Prosthodont.* **2011**, *24*, 273–276.
60. Kamonkhantikul, K.; Arksornnukit, M.; Takahashi, H. Antifungal, optical, and mechanical properties of polymethylmethacrylate material incorporated with silanized zinc oxide nanoparticles. *Int. J. Nanomed.* **2017**, *12*, 2353–2360. [CrossRef]
61. Sodagar, A.; Khalil, S.; Kassaee, M.Z.; Shahroudi, A.S.; Pourakbari, B.; Bahador, A. Antimicrobial properties of poly (methyl methacrylate) acrylic resins incorporated with silicon dioxide and titanium dioxide nanoparticles on cariogenic bacteria. *J. Orthodont. Sci.* **2016**, *5*, 7.
62. Ferreira, L.; Guedes, J.F.; Almeida-Aguiar, C.; Fonseca, A.M.; Neves, I.C. Microbial growth inhibition caused by Zn/Ag-Y zeolite materials with different amounts of silver. *Colloids Surf. B Biointerfaces* **2016**, *142*, 141–147. [CrossRef]
63. Craig, R.G.P.J.; Wataha, J.C. Dental Polymers. In *Dent Mater, Properties and Manipulation*, 11th ed.; C. V. Mosby Company: St. Louis, MO, USA, 2001.
64. *Standardization of Dentistry: Denture Base Polymer*; International Organization for Standardization: Geneva, Switzerland, 1999.
65. Addy, M.; Handley, R. The effects of the incorporation of chlorhexidine acetate on some physical properties of polymerized and plasticized acrylics. *J. Oral Rehabil.* **1981**, *8*, 155–163. [CrossRef] [PubMed]
66. Murat, S. Gümüş zeolit ilave edilen akrilik rezinlerin eğilme dayanimlarinin in-vitro değerlendirilmesi. *Atatürk Ünv. Diş Hek. Fak. Derg.* **2018**, *4*, 539–544.
67. Yadav, N.S.; Saraf, S.; Mishra, S.K.; Hazari, P. Effects of fluconazole, chlorhexidine gluconate, and silver-zinc zeolite on flexural strength of heat-cured polymethyl methacrylate resin. *J. Nat. Sci. Biol. Med.* **2015**, *6*, 340.
68. Shibata, T.; Hamada, N.; Kimoto, K.; Sawada, T.; Kumada, H.; Umemoto, T.; Toyoda, M. Antifungal effect of acrylic resin containing apatite-coated TiO_2 photocatalyst. *Dent. Mater. J.* **2007**, *26*, 437–444. [CrossRef] [PubMed]
69. Shan, Y.; Bai, Y.; Yang, S.; Zhou, Q.; Wang, G.; Zhu, B.; Zhou, Y.; Fang, W.; Wen, N.; He, R.; et al. 3D-printed strontium-incorporated β-TCP bioceramic triply periodic minimal surface scaffolds with simultaneous high porosity, enhanced strength, and excellent bioactivity. *J. Adv. Ceram.* **2023**, *12*, 1671–1684. [CrossRef]
70. Kassaee, M.; Akhavan, A.; Sheikh, N.; Sodagar, A. Antibacterial effects of a new dental acrylic resin containing silver nanoparticles. *J. Appl. Polym. Sci.* **2008**, *110*, 1699–1703. [CrossRef]
71. Wakasa, K. A fundamental study on dental application of silver zeolite. *J. Hiroshima Univ. Dent. Soc.* **1997**, *29*, 87–98.
72. Ren, J.; Lin, H.; Huang, Q.; Liang, Q.; Zheng, G. Color difference threshold determination for acrylic denture base resins. *Bio-Med. Mater. Eng.* **2015**, *26*, S35–S43. [CrossRef]
73. Paravina, R.D.; Ghinea, R.; Herrera, L.J.; Bona, A.D.; Igiel, C.; Linninger, M.; Sakai, M.; Takahashi, H.; Tashkandi, E.; Perez, M.d.M. Color difference thresholds in dentistry. *J. Esthet. Restor. Dent.* **2015**, *27*, S1–S9. [CrossRef]
74. Ren, J.; Lin, H.; Huang, Q.; Zheng, G. Determining color difference thresholds in denture base acrylic resin. *J. Prosthet. Dent.* **2015**, *114*, 702–708. [CrossRef]
75. Alp, G.; Johnston, W.M.; Yilmaz, B. Optical properties and surface roughness of prepolymerized poly (methyl methacrylate) denture base materials. *J. Prosthet. Dent.* **2019**, *121*, 347–352. [CrossRef] [PubMed]

76. O'brien, W.J. *Dental Materials and Their Selection*; Quintessence: Carol Stream, IL, USA, 2002.
77. Yilmaz, B.; Karaagaclioglu, L. In vitro evaluation of color replication of metal ceramic specimens using visual and instrumental color determinations. *J. Prosthet. Dent.* **2011**, *105*, 21–27. [CrossRef] [PubMed]

inorganics

MDPI

Article

Portable Electrochemical Immunosensor Based on a Gold Microblobs-Optimized Screen-Printed Electrode for SARS-CoV-2 Diagnosis

Melissa M. Giacomet [1], Paulo H. M. Buzzetti [2], Oscar O. S. Junior [3], Alessandro F. Martins [4], Elton G. Bonafe [1] and Johny P. Monteiro [1,*]

1 Laboratório de Materiais, Macromoléculas e Compósitos—LAMMAC, Chemistry Department, Universidade Tecnológica Federal do Paraná (UTFPR), Apucarana 86812-460, PR, Brazil; melissagiacomet@gmail.com (M.M.G.); eltonbonafe@utfpr.edu.br (E.G.B.)

2 Département de Chimie Moleculaire (DCM), Université Grenoble-Alpes (UGA), 38000 Grenoble, France; pauloh.buzzetti@hotmail.com

3 Chemistry Department, Universidade Estadual de Maringá (UEM), Maringá 87020-900, PR, Brazil; oosjunior@uem.br

4 Department of Chemistry, Pittsburg State University, Pittsburg, KS 66762, USA; amartins@pittstate.edu

* Correspondence: johnymonteiro@utfpr.edu.br

Abstract: The development of biosensors for determining the most diverse biomolecules is a constant focus of many research groups. There is a latent need to propose sensors that combine portability, simple measurements, and good analytical performance. Here, we propose an electrochemical immunosensor that is fully portable and energy-independent for diagnosing antibodies against SARS-CoV-2 (the virus that causes COVID-19). Initially, disposable screen-printed carbon electrodes (SPEs) were covered by gold microblobs (AuMBs), which were synthesized amperometrically from Au^{3+} ions. Then, the SPE-AuMBs were coated with cysteamine, which allowed the N-hydroxysuccinimide-activated SARS-CoV-2 antigen (spike protein) to be immobilized. The antigen-activated electrode was used to detect COVID-19 antibodies from current measurements obtained by differential pulse voltammetry. The AuMBs synthesis time was optimized, and the presence of gold structures improved the electrochemical responses of the SPE. It was possible to quantitatively determine antibodies in the concentration range of 0.25 to 10 μg mL^{-1}. This range includes concentrations found in biological fluids from patients at any stage of the disease. An analysis took approximately the same time as traditional rapid nasal tests (20 min) and costed less, considering all the steps necessary to prepare a disposable antigen-functionalized SPE.

Keywords: biosensor; COVID-19; gold structure; electrochemistry; portable device; sensing

1. Introduction

Electrochemical sensors and biosensors have emerged as up-and-coming contenders for point-of-care detection owing to their advantageous traits, such as reliability and the potential for sensitive and selective detection [1]. These devices are frequently preferred due to their high portability, user-friendliness, short analysis time, and easy sample preparation. They offer a convenient and versatile platform for detecting viral antigens, antibodies, or genetic material in various biological specimens, emerging as a noteworthy specimen for infectious disease testing due to their non-invasive nature and practicality [2,3]. The development of electrodes for application in various electrochemical devices is constantly proposed by many research groups. The demand for electrodes modified in increasingly sophisticated ways has increased, seeking very high analytical performances [4–13]. Furthermore, advanced analytical techniques (such as electrochemical impedance spectroscopy (EIS)) have been used where the response analysis is more complex and requires deeper technical knowledge [14–16]. Combining these factors has allowed the development of electrochemical sensing devices with incredible performance, and the determination of analytes on the picomole scale has been possible [13]. Despite this, the complexity of the experimental arrangement has distanced the electrochemical devices from real, practical applicability.

Very low detection limits and very high sensitivity are not always essential requirements. Sometimes, simplicity, speed, low cost, and no technical analyst are the most important aspects of an analysis. In this sense, using screen-printed electrodes (SPEs) associated with simpler electrochemical techniques, such as voltammetric ones, can be advantageous. Electrochemical devices that use SPEs without or with little preparation have been reported by many groups, and they have been useful for proposing devices capable of determining molecules and biomolecules with important applications in the medical field [15–18]. Gold structures, such as gold nanoparticles, have been used to optimize the response of SPEs used in biosensors [18–20]. They are relatively simple elements that can be produced directly in the electrochemical analysis cell. However, the steps for optimizing gold structure synthesis on the SPE surface and their activation to receive biorecognition elements are not normally presented.

Since its initial manifestation in late 2019, the coronavirus disease 2019 (COVID-19), attributed to severe acute respiratory syndrome coronavirus 2 (SARS-CoV-2), has burgeoned into an international health emergency, inducing profound repercussions on public health and the global economy. An indispensable element in effectively managing an outbreak involves timely and expedient diagnostic modalities to curb infectious disease transmission [21]. To this end, SPE-based biosensors have been proposed and have become essential tools in general healthcare and analysis. For instance, Nascimento et al. [22] achieved a low detection limit in an electrochemical biosensor for the SARS-CoV-2 spike protein. The authors employed magnetic beads and gold nanoparticles conjugated to the human angiotensin-converting enzyme (ACE2) receptor on the surface of screen-printed carbon electrodes to recognize spike proteins in human samples using differential pulse voltammetry (DPV). In a sophisticated approach, Triastuti et al. [23] developed a SARS-CoV-2 aptasensor to detect the SARS-CoV-2 Spike Receptor Binding Domain (spike RBD). The authors functionalized a CeO_2-modified screen-printed carbon electrode with an aptamer for spike RBD protein and achieved detection by DPV in human saliva and oropharyngeal swab samples. Brazaca et al. developed gold-modified screen-printed carbon electrodes functionalized with anti-spike protein to determine the SARS-CoV-2 spike protein. A more complex electrochemical analysis based on impedance spectroscopy was used to assess the protein in biological fluids [24].

Although antigens (such as spike protein) have been used to diagnose infectious diseases in the early stages, human antibodies against SARS-CoV-2 spike protein have mainly been utilized to determine immune status or disease stage in COVID-19 diagnosis [25–28]. For example, the group of Drobysh et al. used gold nanostructure-activated SPEs to detect antibodies against recombinant SARS-CoV-2 spike protein (anti-rSpike). For this, the electrode received covalent immobilization of rSpike to recognize anti-rSpike in serum [29]. A study reported by Nature showed that the concentration of IgM antibody against SARS-CoV-2 spike (related to the disease's active phase) in the serum is 1.5 $\mu g\ mL^{-1}$ between 4 and 10 days after the disease and reaches a maximum value around 8.0 $\mu g\ mL^{-1}$ after 16–20 days of infection [30]. The relatively high biomarker levels in biological fluid make the use of SPE-based electrochemical biosensors interesting for COVID-19 diagnosis. The use of SPE-based electrochemical biosensors to determine antibodies against the SARS-CoV-2 virus is not a novelty, since numerous reports have been published over the last few years [22–24,31]. However, we wanted to combine the best elements to enable a novel electrode configuration to obtain a ready-for-use, simple, practical, and fully portable electrochemical device. This study introduces a gold microblob (AuMB)-modified SPE for electrochemical biosensing obtention using an antibody against SARS-CoV-2 as a detection model. An optimization study of electrochemical AuMB synthesis and SPE surface amination (surface activation by L-cysteine) is also presented. Using disposable SPE electrodes and a compact and energy-autonomous electrochemical apparatus enables obtaining a portable, straightforward analysis, quick response, and cost-effective platform. The biosensor evaluated the presence of antibodies against SARS-CoV-2-Spike both qualitatively and quantitatively using antigen-functionalized SPEs and simple DPV measurements.

2. Results and Discussion

2.1. Optimization of the AuMB Synthesis

The synthesis of AuMBs on carbon-based electrodes has already been proposed, and the electrode electrochemical signal has improved [31]. The synthesis of AuMBs on SPEs was optimized with respect to the electrochemical synthesis time. The potential was fixed at 1.2 V, which is the minimum to reduce gold to AuMBs in a controlled manner. Scanning electron microscopy (SEM) images of the AuMBs for each synthesis condition are shown in Figure 1. The AuMBs had a spherical shape and a rough surface. The average diameter of the particles was clearly larger, as the synthesis time increased up to 50 s. This is an expected result since applying a longer growth time results in larger particles. The average diameters were 5.5 ± 0.6 µm and 9.2 ± 0.9 µm (average of 50 particles) for times of 30 and 40 s, respectively. It was impossible to obtain the diameter accurately for 50 s as the particles are clearly agglomerated, without precisely defined contours and overlapping (formation of particle overlayers). A film of agglomerated particles without defined boundaries was observed in the delayed condition of 300 s.

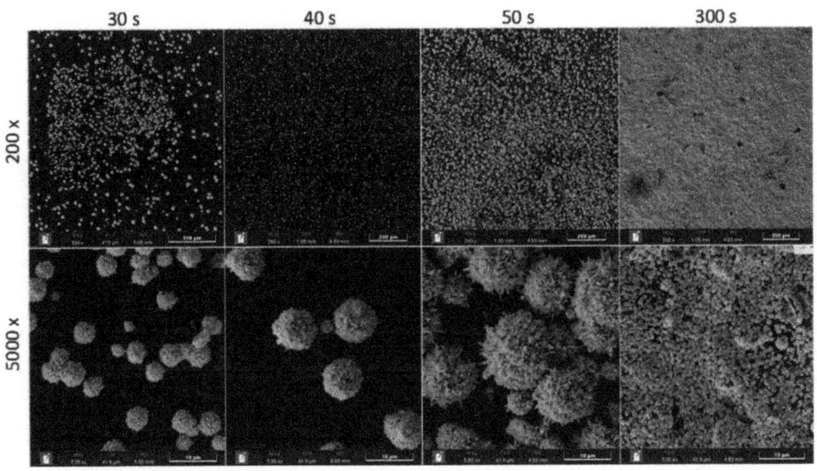

Figure 1. SEM images for SPE-AuMBs were obtained at different times for AuMB electrochemical synthesis.

The SPE-AuMBs produced were subjected to cyclic voltammetry (CV) measurements in the potassium ferrocyanide medium to evaluate the signal produced by the electrodes. The voltammograms are shown in Figure 2, and the SPE signals are compared before and after using different AuMBs synthesis times. The cyclic voltammograms show that the anodic peak moved to lower potential values for SPE-AuMBs produced between 30 and 50 s compared to the bare SPE. In this way, the voltammogram profile for potassium ferrocyanide better adjusted the reversible behavior of potassium ferrocyanide electrochemical reactions (mirrored anodic (I_{ap}) and cathodic (I_{cp}) current peaks). Furthermore, the I_{ap} increased in these electrodes compared to that obtained for the bare SPE. The I_{ap} increased by 10.2, 5.77, and 4.70 µA for 30, 40, and 50 s, respectively. For the SPE produced in an extrapolated time of 300 s, the voltammogram profile changed abruptly, indicating the occurrence of more complex reactions. The electrode produced at the 30 s condition presented an optimized electrochemical signal and was chosen to proceed with subsequent analyses. The presence of smaller AuMBs (which showed a greater surface area in relation to volume) played a fundamental role in the better electroactive performance.

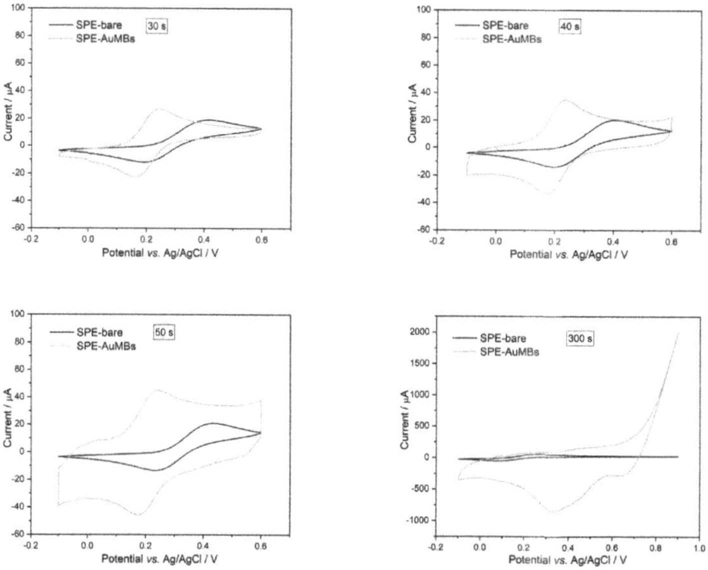

Figure 2. Cyclic voltammograms for bare SPE and SPE-AuMBs obtained at different electrochemical synthesis times. Measurements were performed for potassium ferrocyanide in PBS (pH = 7.43) using 25 mV s^{-1} and 10 mV step.

The electroactive area for SPE-AuMBs (with and without AuMBs) at the optimized synthesis time condition (30 s) was calculated by the Randles–Sevcik equation (Equation (1)) [32]. For this, cyclic voltammograms were obtained for the electrodes under different scanning speeds, as shown in Figure 3.

$$I_{ap} = 2.686 \times 10^5 A n^{3/2} C_0 D^{1/2} v^{1/2}, \quad (1)$$

where I_{ap} is the anodic peak current (A), A is the electroactive area (cm^2), C_0 is the concentration of the electroactive species (mol cm^{-3}), D is the diffusion coefficient of the electroactive species (6.39×10^{-06} cm^2 s^{-1} for potassium ferrocyanide in KCl 0.5 M), and v is the scan rate (V s^{-1}).

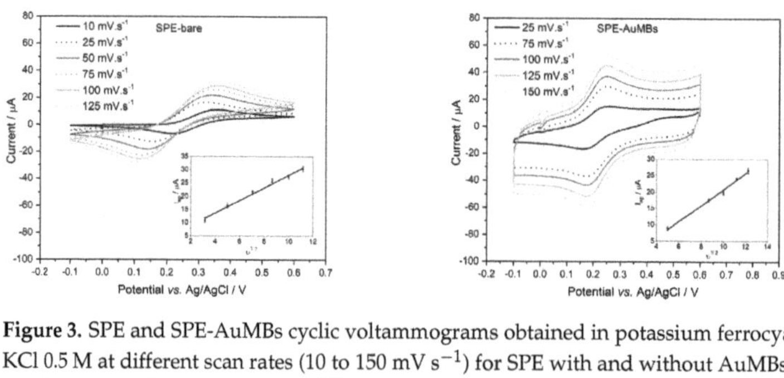

Figure 3. SPE and SPE-AuMBs cyclic voltammograms obtained in potassium ferrocyanide 5 mM in KCl 0.5 M at different scan rates (10 to 150 mV s^{-1}) for SPE with and without AuMBs. The insets in the figures are plots of I_{ap} versus $v^{1/2}$.

Using data from Figure 3, I_{ap} versus $v^{1/2}$ were plotted (inserts in Figure 3). The A for each electrode was determined from the slope of the straight line, and the values were 0.690

and 0.711 cm^2 for the SPE and SPE-AuMBs, respectively. This confirms that the presence of AuMBs on the electrode surface contributes to an increase in the electroactive area and justifies the improvement in the electroactive performance of the electrode.

2.2. Cysteamine-Activation of SPE-AuMBs

The immobilization of an antigen (Ant) on the electrode is necessary to obtain the immunosensor for the COVID-19 diagnosis proposed here. To achieve this, creating a good Ant immobilization strategy is important. Amination of the electrode surface is typically the strategy used to immobilize N-hydroxysuccinimide(NHS)-activated Ant. Typically, alkaneaminothiols (such as cysteamine) are used for the amination of gold-containing surfaces. Thiols show a good chemical affinity for gold and are useful for the amination of this surface type. Amination is crucial as it links the inorganic (electrode) and organic (Ant) parts and directly affects the quality of the immunosensor's performance. Sometimes, researchers do not pay due attention to this stage. Therefore, the immobilization of cysteamine on the electrode was evaluated, and two strategies were tested: drop coating and electrochemistry coating. Cyclic voltammograms (Figure 4) for potassium ferrocyanide were used to evaluate the aminated SPE-AuMBs.

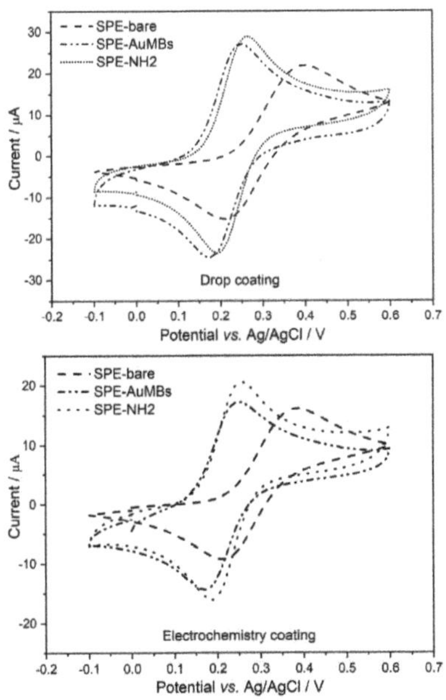

Figure 4. Cyclic voltammograms for NH_2-activated SPE-AuMBs that were aminated by drop coating and electrochemistry coating. Measurements were performed for potassium ferrocyanide in PBS (pH = 7.43) using 25 mV s^{-1} and 10 mV step.

The I_{ap} variation in SPE-AuMBs/-NH_2 with respect to SPE-AuMBs was 4.6 µA for drop coating and 4.9 µA for electrochemistry coating. A slightly greater advantage for electrochemical amination was observed. Therefore, this was the strategy chosen for the next steps.

The presence of cysteamine on the amino electrode through electrochemistry coating was evaluated by energy-dispersive X-ray spectroscopy (EDS). The elemental spectra obtained in the working electrode region in the SPE-AuMBs and SPE-AuMBs/-NH_2 are shown in Figure 5.

The nitrogen and sulfur signals that characterize cysteamine's presence occur at energies close to the expressive signals for carbon and gold, respectively. Furthermore, S and N should have a low occurrence because, theoretically, only a monolayer of cysteamine is adsorbed on the surface. This makes it difficult to identify signals for these elements. A small S signal was suggested for the EDS spectrum for SPE-AuMBs/-NH_2 and not for SPE-AuMBs. The presence of sulfur only in SPE-AuMBs/-NH_2 was corroborated through X-ray fluorescence (XRF) analysis, which is presented in Table 1.

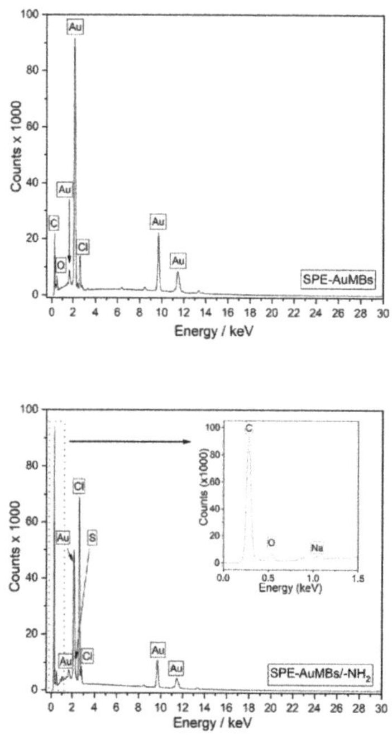

Figure 5. EDS spectrum for SPE-AuMBs (**top**) and SPE-AuMBs/-NH_2 (**bottom**).

Table 1. Elemental percentage occurrence obtained by XRF for SPE-AuMBs and SPE-AuMBs/-NH_2.

Elements	SPE-AuMBs	SPE-AuMBs/-NH_2
	Occurrence Percentage (%)	
Al	21.0	18.4
Si	47.9	43.3
S	--	0.2
Cl	11.9	19.8
Ti	16.4	17.3
Au	2.8	1.2

Some elements determined for the electrodes by XRF differ from those observed in the EDS spectra. EDS analyses were carried out, focusing only on the working electrode area. That was not possible using the XRF. Thus, regions of the reference electrode, counter-electrode, and SPE physical structure were also evaluated. Furthermore, XRF equipment is only able to detect elements heavier than Na. Thus, the C and O signals are not detected, but Al, Si, and Ti signals (present in the SPE physical structure) occurred in a large percentage.

Finally, EDS performed an elemental mapping for Au and S on the SPE-AuMBs/-NH_2 electrode (Figure 6). The orange and light green regions represent Au and S, respectively. It is evident that the clusters in the image have a chemical composition consisting of gold. Furthermore, it was observed that the small occurrence of S is concentrated in regions where AuMBs are present. This strongly suggests that cysteamine immobilization occurs through chemical adsorption onto the electrode by establishing Au—S-type bonds on the AuMBs.

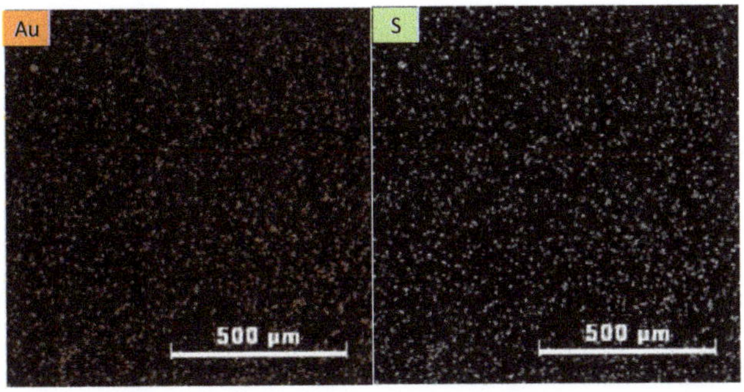

Figure 6. Elemental mapping of Au (orange region) and S (light green) by EDS in an image obtained by SEM for the SPE-AuMBs/-NH_2.

2.3. *AB Detection*

Ant was immobilized on NHS-activated SPE-AuMBs/-NH_2 to become able to receive AB. Voltammograms were obtained by CV and DPV after each modification step and are shown in Figure 7.

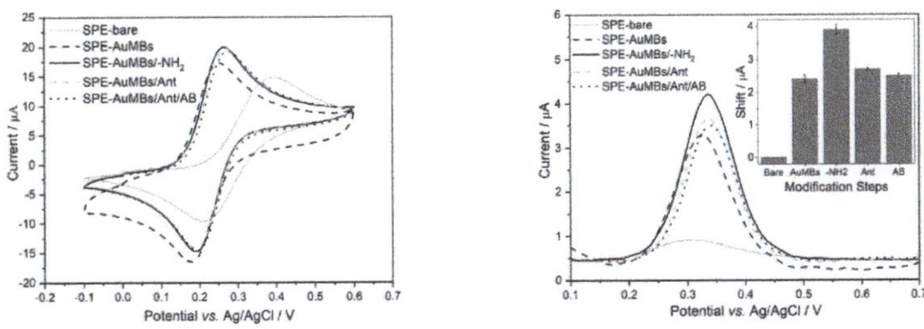

Figure 7. Voltammograms were obtained for potassium ferrocyanide 2 mM in PBS (pH = 7.43) by CV (**left**) and DPV (**right**) using unmodified SPE and SPE after each modification step. CV was performed with E_{step} = 2 mV and v of 25 mV s^{-1}. The parameters for DPV were E_{step} = 4 mV, E_{pulse} = 2 mV, t_{pulse} = 10 ms, and v = 10 mV s^{-1}. The AB concentration tested was 0.25 µg mL^{-1}.

As discussed previously, the cyclic voltammograms in Figure 7 (left) show that I_{ap} increases until cysteamine adsorption. Then, the I_{ap} of the electrode experiences two successive reductions due to the adsorptions of Ant and AB. Organic substances hinder electron transfer on the electrode surface. Thus, the presence of AB against the SARS-CoV-2 virus can be assessed by monitoring I_{ap} using an Ant-activated SPE.

Voltammetric measurements by DPV were performed to monitor the modification steps through the potassium ferrocyanide anode signal. DPV is a technique that minimizes background current and is more advantageous for quantitative analysis [32]. The same behavior was observed, that is, I_{ap} increases until cysteamine immobilization and shows successive reductions after Ant and AB immobilizations. The inset in Figure 7 shows a bar graph with the I_{ap} shift that occurred after each modification step. The shift observed after the immobilization of AB on Ant was 0.2 µA.

A calibration curve (Figure 8) was obtained for I_{ap} measurements from DPV voltammograms with respect to the application of different concentrations of AB.

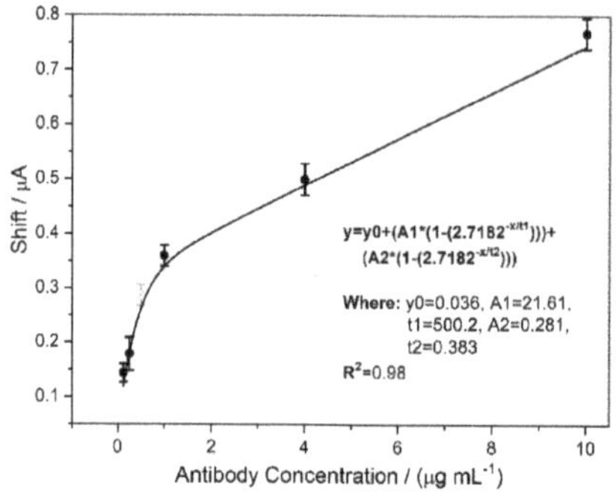

Figure 8. Calibration curve for I_{ap} variation as a function of AB concentration. The points were adjusted to exponential behavior.

The applied AB concentration range was from 0.250 to 10.0 µg mL^{-1}, and the device responses were adjusted to exponential behavior. The equation and its parameters are shown in Figure 8. The electrode responses increased rapidly up to a concentration of 1.00 µg mL^{-1} and clearly lost sensitivity for higher concentrations. This is a typical behavior of loss of analyte adsorption capacity (surface saturation). For concentrations greater than 10.0 µg mL^{-1}, the surface is completely saturated by AB molecules. This can be suggested because applying a concentration of 25.0 µg mL^{-1} caused a shift of 0.75 µA, which is approximately equal to the current variation observed for the concentration of 10.0 µg mL^{-1} of AB (0.77 ± 0.3 µA).

The analytical performance of the proposed device was compared with the recently reported SPE-based electrochemical biosensors. Table 2 presents the construction characteristics and the limit of detection (LOD) for the different sensor arrangements. The performance obtained here was comparable with another device with similar characteristics reported by Drobysch et al. [29] (first study in the list), considering a LOD obtained from the calibration curve's linear range. However, the LOD was higher than those of devices that used modified SPEs and/or more sophisticated electrochemical techniques (such as EIS). Despite this, the performance shown here is fully compatible with antibodies against SARS-CoV-2 virus for analysis. A study reported by Nature showed that the concentration of IgM antibody against SARS-CoV-2 Spike (related to the active phase of the disease) in the serum is 1.5 µg mL^{-1} between 4 and 10 days after the onset of the disease and reaches maximum values around 8.0 µg mL^{-1} after 16–20 days since infection [30]. Thus, the proposed device covers all concentration ranges achieved by AB in infected patients. Furthermore, the biosensor has some distinct characteristics as it is a portable and energy-autonomous device that could easily perform rapid qualitative testing of AB (positive/negative type) from biological fluids in point-of-care and rapid analysis. The required time is around 20 min using an electrode previously prepared with Ant.

Additionally, it was shown that the biosensor could be used to infer the infection evolution stage by quantitative AB measurements. The accuracy test of the proposed calibration curve of Figure 8 was carried out by applying an AB solution with an intermediate concentration close to the inflection point in the curve (gray point in Figure 8). The variation in I_{ap} after applying this solution was 0.287 µA. The fitting equation (insert in Figure 8) suggests that this response corresponds to a concentration of 0.574 µg mL^{-1} of AB. The actual AB concentration of the applied sample was 0.500 µg mL^{-1} (corresponds to half the AB concentration found in patients in the initial infection phase) and is close to that obtained by curve interpolation.

Table 2. Characteristics of electrochemical biosensors for COVID-19 diagnosis based on screen-printed electrodes.

Analyte	Bioreceptor	SPE Type	Electrochemical Technique	LOD	Reference
Anti-rSpike	rSpike	Carbon	DPV	0.14 nM	[29]
Anti-rSpike	rSpike	Carbon	CV	0.27 nM	[29]
RBD Spike	Monoclonal Anti-RBD Spike on cellulose nanocrystal	Graphene	DPV	2 fg/mL	[33]
RBD Spike	Anti-Horseradish peroxidase-conjugated SARS-CoV-2 RBD	poly(pyrrolepropionic acid)-modified graphene	Amperometry	2 pg/mL	[34]
N Protein	Anti-N protein	Carbon	EIS	1 ng/mL	[35]
RBD Spike	Anti-Spike	Graphene/carbon	EIS	0.25 fg/mL	[36]
Anti-Spike	Spike	Carbon	DPV	0.25 μg/mL (10 nM)	This Study

The estimated cost of electrode production for SARS-CoV-2 AB quantification is approximately 10.00 USD/unit. The disposable commercial SPE costs 6.00 USD/unit, and the reagents cost 4.00 USD/unit (including SARS-CoV-2 Ant, cysteamine, PBS, $HAuCl_4$, NHS, and 1-Ethyl-3-(3-dimethylaminopropyl)carbodiimide (EDC)). Costs could also be reduced with industrial-scale production. Comparatively, a serological or nasal test costs approximately USD 15.00. An additional advantage of the method proposed here is the possibility of acquiring a quantitative analysis of AB.

3. Materials and Methods

3.1. AuMBs Synthesis on SPE

Initially, water was dripped on a commercial carbon SPE (Metrohm, DPR-110, São Paulo, Brazil), and it was cleaned by applying a constant potential of 1.7 V vs. Ag/AgCl for 5 min. The AuMBs on the SPE were produced using the method reported in the literature [37]. For this, the SPE was rinsed with deionized water, and a 1.0 M $HAuCl_4$/0.5 M H_2SO_4 aqueous solution was dripped onto the cleaned SPE. Then, a constant potential of −0,4 V vs. Ag/AgCl was applied for 20, 30, 50, and 300 s. CV measurements were performed for the SPE and SPE-AuMBs. The parameters used for all CV tests were a potential range of −0.1 to +0.6 V (begin potential 0.0 V), a 25 mV s^{-1} scan rate, and a 10 mV step potential. A 2.0 mM potassium ferrocyanide solution in phosphate buffer (PBS, pH = 7.4) was consistently used to evaluate the CV measurements' electrochemical signal. The volume of liquids used to cover the SPEs fully was always 60 μL unless another volume was mentioned.

The electroactive area of the SPE and SPE-AuMBs was evaluated by acquiring cyclic voltammograms for potassium ferrocyanide 5.0 mM in KCl 0.5 M at different scan rates of 10, 25, 50, 75, 100, 125, and 150 mV s^{-1}. The electrodes were characterized by scanning SEM-EDS (TESCAN, Libušina tř., Brno, Czech Republic). All electrochemical measurements used a bipotentiostat/galvanostat (Metrohm, μStat 400, São Paulo, Brazil).

3.2. SPE-AuMBs Amination

A 20 mM cysteamine solution was dropped on rinsed SPE-AuMBs. Then, a constant potential of 1.2 V vs. Ag/AgCl was applied on the electrode under cysteamine solution for 20 min. The aminated electrode was rinsed with water. Alternatively, an amination obtained by drop coating was also tested [31]. An aliquot of 10 μL 60 mM cysteamine solution was dropped only onto the SPE's working electrode area. After incubation for 30 min, the aminated electrode was rinsed with water. The electrodes (SPE-AuMBs/-NH_2) were characterized by SEM-EDS and XRF (Malvern Panalytical, Epsilon 1, São Paulo, Brazil).

3.3. Immobilization of the Spike Protein (Antigen) on the Electrode

Solutions of EDC (Sigma-Aldrich, Barueri, Brazil) and NHS (Sigma-Aldrich) were prepared using ultrapure water. The antigen and antibody solutions were prepared using

PBS solution (pH = 7.4). Aliquots of 30 μL of 200 mM EDC and 20 μL of 200 mM NHS were mixed with 50 μL of 1 μg mL^{-1} SARS-CoV-2 spike antigen protein (Ant, Sigma-Aldrich) in an Eppendorf to activate the Ant carboxylic acid groups. After a 20 min reaction, the mixture was dropped (10 μL) onto the working electrode area of the SPE-AuMBs/-NH_2 and incubated for 15 min to obtain SPE-AuMBs/Ant. The reaction between NHS-activated carboxylic acid (Ant) and the amine (cysteamine) groups results in amide formation. A 10 μL aliquot of 1 g mL^{-1} bovine serum albumin (BSA, Sigma-Aldrich) PBS solution was also dropped on the SPE-AuMBs/Ant for 20 min to block the non-specific sites. All incubations were conducted at 21 ± 3 °C.

3.4. Antibody Detection

After the biosensor preparation, 20 μL of the 0.250 μg mL^{-1} SARS-CoV-2 spike antibody protein (AB, Sigma-Aldrich) for the receptor-binding domain region, prepared in PBS, was incubated on the SPE-AuMBs/Ant surface for 15 min for an initial test. After AB immobilization, CV and DPV measurements in the potassium ferrocyanide probe were obtained after rinsing the surface with water to remove non-immobilized AB. Sequentially, a calibration curve regarding AB detection was also obtained using the same methodology. AB solutions were prepared at concentrations 0.125, 0.250, 1.00, 4.00, 10.0, and 25.0 μg mL^{-1} in commercial human serum (Sigma-Aldrich, ref. SAE0012). Each AB solution was detected on an individual SPE-AuMBs/Ant, and the tests were carried out in duplicate. The SPE-AuMBs/Ant electrodes were also used to assess AB in commercial human serum from a 0.500 μg mL^{-1} AB-spiked solution. The assays were carried out using the portable experimental setup illustrated in Figure 9 by DPV measurements.

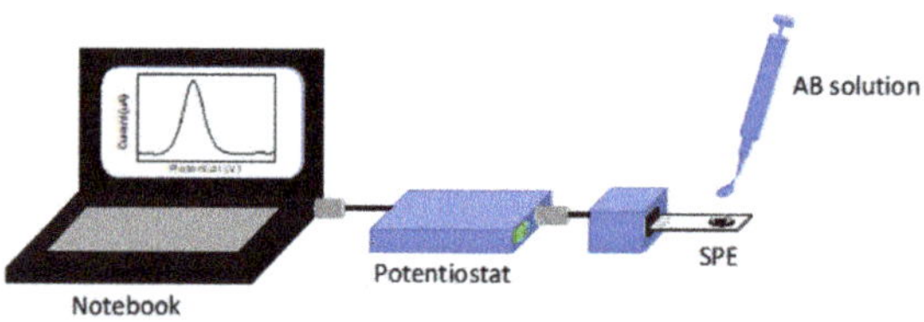

Figure 9. Fully portable experimental setup for SARS-CoV-2 AB detection assays by DPV measurements.

4. Conclusions

Here, we proposed a simple and cheap electrochemical method using disposable SPEs, which had optimized responses by applying AuMBs. AuMBs production on SPEs was optimized, and the shortest electrochemical synthesis time tested (30 s) was the condition that showed the best signal for the electrode. The AuMBs-improved electrodes were aminated using two strategies: electrochemical and drop coatings. The first proved to be more effective for immobilizing cysteamine on gold, exploiting the strong chemical affinity of the metal with -SH groups. Aminated AuMBs–electrodes were tested to detect COVID-19 antibodies. It was then demonstrated that an antigen-activated SPE could qualitatively and quantitatively evaluate the presence of SARS-CoV-2 antibodies. The antibody analysis took around 15 min and was cheaper than traditional rapid tests. Thus, the method proved to be a good alternative for both qualitative analyses (quick "yes" or "no" test) and proved to be capable of estimating the evolution of the infection through quantitative analyses.

Author Contributions: Conceptualization, J.P.M. and A.F.M.; methodology, M.M.G.; formal analysis, M.M.G.; investigation, O.O.S.J. and P.H.M.B.; resources, J.P.M., E.G.B. and A.F.M.; data curation, E.G.B.; writing—original draft preparation, J.P.M. and M.M.G.; writing—review and editing, J.P.M. and P.H.M.B.; visualization, O.O.S.J.; supervision, J.P.M.; project administration, J.P.M.; funding acquisition, J.P.M. All authors have read and agreed to the published version of the manuscript.

Funding: This research was funded by the National Council for Scientific and Technological Development—CNPq (Brazil), grant number 305290/2020-7.

Data Availability Statement: The original contributions presented in the study are included in the article, further inquiries can be directed to the corresponding author.

Acknowledgments: The authors thank the CNPq—Brazil and the Laboratório Multiusuário de Apoio à Pesquisa da UTFPR Campus Apucarana (LAMAP)—UTFPR, Apucarana, Brazil. J.P.M. thanks the CNPq's researcher grant.

Conflicts of Interest: The authors declare no conflicts of interest.

References

1. Yan, T.; Zhang, G.; Chai, H.; Qu, L.; Zhang, X. Flexible Biosensors Based on Colorimetry, Fluorescence, and Electrochemistry for Point-of-Care Testing. *Front. Bioeng. Biotechnol.* **2021**, *9*, 753692. [CrossRef] [PubMed]
2. de Eguilaz, M.R.; Cumba, L.R.; Forster, R.J. Electrochemical Detection of Viruses and Antibodies: A Mini Review. *Electrochem. Commun.* **2020**, *116*, 106762. [CrossRef] [PubMed]
3. Brazaca, L.C.; dos Santos, P.L.; de Oliveira, P.R.; Rocha, D.P.; Stefano, J.S.; Kalinke, C.; Abarza Muñoz, R.A.; Bonacin, J.A.; Janegitz, B.C.; Carrilho, E. Biosensing Strategies for the Electrochemical Detection of Viruses and Viral Diseases—A Review. *Anal. Chim. Acta* **2021**, *1159*, 338384. [CrossRef]
4. Wan, X.; Du, H.; Tuo, D.; Qi, X.; Wang, T.; Wu, J.; Li, G. UiO-66/Carboxylated Multiwalled Carbon Nanotube Composites for Highly Efficient and Stable Voltammetric Sensors for Gatifloxacin. *ACS Appl. Nano Mater.* **2023**, *6*, 19403–19413. [CrossRef]
5. Wang, T.; Xia, Y.; Wan, X.; Zhang, Y.; Chen, N.; Jin, Y.; Li, G. A Facile and Efficient Voltammetric Sensor for Marbofloxacin Detection Based on Zirconium-Based Metal-Organic Framework UiO-66/Nitrogen-Doped Graphene Nanocomposite. *Microchem. J.* **2024**, *201*, 110673. [CrossRef]
6. Safitri, E.; Heng, L.Y.; Ahmad, M.; Tan, L.L.; Nazaruddin, N.; Suhud, K.; Chiang, C.P.; Iqhrammullah, M. Electrochemical DNA Biosensor Based on Mercaptopropionic Acid-Capped ZnS Quantum Dots for Determination of the Gender of Arowana Fish. *Biosensors* **2022**, *12*, 650. [CrossRef]
7. Li, X.; Shi, F.; Wang, L.; Zhang, S.; Yan, L.; Zhang, X.; Sun, W. Electrochemical Biosensor Based on Horseradish Peroxidase and Black Phosphorene Quantum Dot Modified Electrode. *Molecules* **2023**, *28*, 6151. [CrossRef]
8. Shao, B.; Chen, W.; Yan, L.; Huang, Y.; Wang, B.; Zou, Q.; Xuan, Y.; Sun, W.; Niu, Y. Preparation and Electrocatalytic Study of Myoglobin Biosensor Based on Platinum-Gold-Three Dimensional Graphene Modified Electrode. *Int. J. Electrochem. Sci.* **2021**, *16*, 211039. [CrossRef]
9. Sheikholeslam, M.; Nanda, P.; Sanati, A.; Pritzker, M.; Chen, P. Direct Electrochemistry of Hemoglobin/Peptide-Carbon Nanotube Modified Electrode for Hydrogen Peroxide Biosensing. *Mater. Lett.* **2023**, *335*, 133799. [CrossRef]
10. Yan, L.; Shi, F.; Zhang, J.; Niu, Y.; Huang, L.; Huang, Y.; Sun, W. Electrochemical DNA Biosensor Based on Platinum-Gold Bimetal Decorated Graphene Modified Electrode for the Detection of Vibrio Parahaemolyticus Specific Tlh Gene Sequence. *Curr. Anal. Chem.* **2022**, *18*, 781–789. [CrossRef]
11. Sun, Y.; Wang, B.; Deng, Y.; Cheng, H.; Li, X.; Yan, L.; Li, G.; Sun, W. Reduced Graphene Oxide/Titanium Carbide MXene Nanocomposite-Modified Electrode for Electrochemical Hemoglobin Biosensor. *J. Chin. Chem. Soc.* **2021**, *68*, 2326–2336. [CrossRef]
12. Wen, Z.; Niu, X.; Li, X.; Zhao, W.; Li, X.; Ma, D.; Deng, Y.; Sun, X.; Sun, W. Application of Nanosized $LiFePO_4$ Modified Electrode to Electrochemical Sensor and Biosensor. *Curr. Anal. Chem.* **2018**, *14*, 452–457. [CrossRef]
13. Zheng, D.; Zhang, R.; Chen, J.; Feng, Z.; Lin, S. A Highly Sensitivity Electrochemistry Biosensor for *E. coli* DNA Detection Based on Hollow and Porous DCuO@rGO Composite. *Inorg. Chem. Commun.* **2024**, *166*, 112621. [CrossRef]
14. Pusomjit, P.; Teengam, P.; Thepsuparungsikul, N.; Sanongkiet, S.; Chailapakul, O. Impedimetric Determination of Cortisol Using Screen-Printed Electrode with Aptamer-Modified Magnetic Beads. *Microchim. Acta* **2021**, *188*, 41. [CrossRef] [PubMed]
15. Moon, J.; Jiang, H.; Lee, E.C. Physical Surface Modification of Carbon-Nanotube/ Polydimethylsiloxane Composite Electrodes for High-Sensitivity Dna Detection. *Nanomaterials* **2021**, *11*, 2661. [CrossRef] [PubMed]
16. Ye, X.; Li, J.; Jiang, T.; Li, B.; Chen, J. A Portable Electrochemical Impedance Spectroscopy Lab-on-Chip System for Biosensing Applications. In Proceedings of the IEEE International Symposium on Circuits and Systems, Austin, TX, USA, 27 May–1 June 2022; pp. 1347–1351. [CrossRef]
17. Roberto de Oliveira, P.; Crapnell, R.D.; Garcia-Miranda Ferrari, A.; Wuamprakhon, P.; Hurst, N.J.; Dempsey-Hibbert, N.C.; Sawangphruk, M.; Janegitz, B.C.; Banks, C.E. Low-Cost, Facile Droplet Modification of Screen-Printed Arrays for Internally Validated Electrochemical Detection of Serum Procalcitonin. *Biosens. Bioelectron.* **2023**, *228*, 115220. [CrossRef]
18. Elewi, A.S.; Al-Shammaree, S.A.W.; AL Sammarraie, A.K.M.A. Hydrogen Peroxide Biosensor Based on Hemoglobin-Modified Gold Nanoparticles–Screen Printed Carbon Electrode. *Sens. Biosensing Res.* **2020**, *28*, 100340. [CrossRef]
19. Koudelková, Z.; Sedláčková, E.; Richtera, L.; Švec, P.; Adam, V. Gold Nanoparticles Modified Screen Printed Carbon Electrode as a Tool for Detection of TP53. In Proceedings of the NANOCON 11th International Conference on Nanomaterials—Research & Application, Brno, Czech Republic, 16–18 October 2020; pp. 402–405. [CrossRef]
20. Mulyani, R.; Yumna, N.; Maksum, I.P.; Subroto, T.; Hartati, Y.W. Optimization of Aptamer-Based Electrochemical Biosensor for ATP Detection Using Screen-Printed Carbon Electrode/Gold Nanoparticles (SPCE/AuNP). *Indones. J. Chem.* **2022**, *22*, 1256–1268. [CrossRef]
21. Hodson, R. Preparing the World for the next Pandemic. *Nature* **2022**, *610*, S33. [CrossRef]

22. Nascimento, E.D.; Fonseca, W.T.; de Oliveira, T.R.; de Correia, C.R.S.T.B.; Faça, V.M.; de Morais, B.P.; Silvestrini, V.C.; Pott-Junior, H.; Teixeira, F.R.; Faria, R.C. COVID-19 Diagnosis by SARS-CoV-2 Spike Protein Detection in Saliva Using an Ultrasensitive Magneto-Assay Based on Disposable Electrochemical Sensor. *Sens. Actuators B Chem.* **2022**, *353*, 131128. [CrossRef]
23. Triastuti, A.; Zakiyyah, S.N.; Gaffar, S.; Anshori, I.; Surawijaya, A.; Hidayat, D.; Wiraswati, H.L.; Yusuf, M.; Hartati, Y.W. CeO_2@NH_2 Functionalized Electrodes for the Rapid Detection of SARS-CoV-2 Spike Receptor Binding Domain. *RSC Adv.* **2023**, *13*, 5874–5884. [CrossRef] [PubMed]
24. Brazaca, L.C.; Imamura, A.H.; Gomes, N.O.; Almeida, M.B.; Scheidt, D.T.; Raymundo-Pereira, P.A.; Oliveira, O.N.; Janegitz, B.C.; Machado, S.A.S.; Carrilho, E. Electrochemical Immunosensors Using Electrodeposited Gold Nanostructures for Detecting the S Proteins from SARS-CoV and SARS-CoV-2. *Anal. Bioanal. Chem.* **2022**, *414*, 5507–5517. [CrossRef] [PubMed]
25. Karakuş, E.; Erdemir, E.; Demirbilek, N.; Liv, L. Colorimetric and Electrochemical Detection of SARS-CoV-2 Spike Antigen with a Gold Nanoparticle-Based Biosensor. *Anal. Chim. Acta* **2021**, *1182*, 338939. [CrossRef] [PubMed]
26. Zamzami, M.A.; Rabbani, G.; Ahmad, A.; Basalah, A.A.; Al-Sabban, W.H.; Nate Ahn, S.; Choudhry, H. Carbon Nanotube Field-Effect Transistor (CNT-FET)-Based Biosensor for Rapid Detection of SARS-CoV-2 (COVID-19) Surface Spike Protein S1. *Bioelectrochemistry* **2022**, *143*, 107982. [CrossRef]
27. Yang, X.; Yin, Z.Z.; Zheng, G.; Zhou, M.; Zhang, H.; Li, J.; Cai, W.; Kong, Y. Molecularly Imprinted Miniature Electrochemical Biosensor for SARS-CoV-2 Spike Protein Based on Au Nanoparticles and Reduced Graphene Oxide Modified Acupuncture Needle. *Bioelectrochemistry* **2023**, *151*, 108375. [CrossRef]
28. Ayankojo, A.G.; Boroznjak, R.; Reut, J.; Öpik, A.; Syritski, V. Molecularly Imprinted Polymer Based Electrochemical Sensor for Quantitative Detection of SARS-CoV-2 Spike Protein. *Sens. Actuators B Chem.* **2022**, *353*, 131160. [CrossRef]
29. Drobysh, M.; Liustrovaite, V.; Baradoke, A.; Viter, R.; Chen, C.F.; Ramanavicius, A.; Ramanaviciene, A. Determination of RSpike Protein by Specific Antibodies with Screen-Printed Carbon Electrode Modified by Electrodeposited Gold Nanostructures. *Biosensors* **2022**, *12*, 593. [CrossRef]
30. Zeng, W.; Ma, H.; Ding, C.; Yang, Y.; Sun, Y.; Huang, X.; He, W.; Xiang, Y.; Gao, Y.; Jin, T. Characterization of SARS-CoV-2-Specific Antibodies in COVID-19 Patients Reveals Highly Potent Neutralizing IgA. *Signal Transduct. Target. Ther.* **2021**, *6*, 35. [CrossRef]
31. Liv, L.; Kayabay, H. An Electrochemical Biosensing Platform for the SARS-CoV-2 Spike Antibody Detection Based on the Functionalised SARS-CoV-2 Spike Antigen Modified Electrode. *ChemistrySelect* **2022**, *7*, e202200256. [CrossRef]
32. Klein, R.S.; Taniguchi, M.M.; dos Santos, P.D.; Bonafe, E.G.; Martins, A.F.; Monteiro, J.P. Trans-Resveratrol Electrochemical Detection Using Portable Device Based on Unmodified Screen-Printed Electrode. *J. Pharm. Biomed. Anal.* **2022**, *207*, 114399. [CrossRef]
33. Jaewjaroenwattana, J.; Phoolcharoen, W.; Pasomsub, E.; Teengam, P.; Chailapakul, O. Electrochemical Paper-Based Antigen Sensing Platform Using Plant-Derived Monoclonal Antibody for Detecting SARS-CoV-2. *Talanta* **2023**, *251*, 123783. [CrossRef] [PubMed]
34. Primpray, V.; Kamsong, W.; Pakapongpan, S.; Phochakum, K.; Kaewchaem, A.; Sappat, A.; Wisitsoraat, A.; Lomas, T.; Tuantranont, A.; Karuwan, C. An Alternative Ready-to-Use Electrochemical Immunosensor for Point-of-Care COVID-19 Diagnosis Using Graphene Screen-Printed Electrodes Coupled with a 3D-Printed Portable Potentiostat. *Talanta Open* **2022**, *6*, 100155. [CrossRef] [PubMed]
35. Gevaerd, A.; Carneiro, E.A.; Gogola, J.L.; Nicollete, D.R.P.; Santiago, E.B.; Riedi, H.P.; Timm, A.; Predebon, J.V.; Hartmann, L.F.; Ribeiro, V.H.A.; et al. Utilizing COVID-19 as a Model for Diagnostics Using an Electrochemical Sensor. *Sensors* **2024**, *24*, 3772. [CrossRef] [PubMed]
36. Ehsan, M.A.; Khan, S.A.; Rehman, A. Screen-Printed Graphene/Carbon Electrodes on Paper Substrates as Impedance Sensors for Detection of Coronavirus in Nasopharyngeal Fluid Samples. *Diagnostics* **2021**, *11*, 1030. [CrossRef]
37. Malvano, F.; Albanese, D.; Crescitelli, A.; Pilloton, R.; Esposito, E. Impedimetric Label-Free Immunosensor on Disposable Modified Screen-Printed Electrodes for Ochratoxin A. *Biosensors* **2016**, *6*, 33. [CrossRef]

inorganics

MDPI

Article

Preparation of Ca-Mg Double-Doped Mesoporous Silica Nanoparticles and Their Drug-Loading and Drug-Releasing Properties

Qian Zhang [1], Jiamin Huang [1], Chao Liu [2], Ruihua Chen [3], Tao Jiang [3], Yusufu Hailili [1], Telieke Bahetibieke [1], Xiaohui Tang [4,*] and Mei Wang [1,5,6,*]

1 School of Pharmacy, Xinjiang Medical University, Urumqi 830011, China; zhangqianxjmu@163.com (Q.Z.); woahjmjy@163.com (J.H.); 2790610460@163.com (Y.H.); 3549415558@163.com (T.B.)
2 School of Medical Engineering and Technology, Xinjiang Medical University, Urumqi 830011, China; cliu20@sohu.con
3 Animal Centre, Xinjiang Medical University, Urumqi 830011, China; xjchenrh@xjmu.edu.cn (R.C.); 13579977272@163.com (T.J.)
4 Central Laboratory, Xinjiang Medical University, Urumqi 830011, China
5 Xinjiang Key Laboratory of Natural Medicines Active Components and Drug Release Technology, Xinjiang Medical University, Urumqi 830011, China
6 Engineering Research Center of Xinjiang and Central Asian Medicine Resources, Ministry of Education, Urumqi 830011, China
* Correspondence: tangxh0429@163.com (X.T.); meiwang@xjmu.edu.cn (M.W.)

Abstract: Breast cancer is a common clinical malignant tumor that seriously threatens women's physical and mental health. Chemotherapy, as the first choice of breast cancer treatment, has limited its application in the clinic due to problems of poor stability, short half-life, and serious toxic side effects. With the emergence of nanotechnology, inorganic materials to prepare mesoporous silica nanoparticles (MSNs) have been widely used in anti-tumor drug carriers. However, their slow degradation rate limits their application in the biomedical field. Therefore, developing low-toxicity MSNs with good biocompatibility, biodegradability, and rapid release at the tumor site is a key scientific issue to be addressed. Here, we prepared DOX-loaded Ca-Mg-doped MSNs by electrostatic adsorption to obtain Ca-Mg@DOX@MSNs with suitable particle sizes and zeta potential, and the incorporation of calcium and magnesium also led to an increase in the degradation rate under acidic conditions and an accelerated release, which reduced the toxicity of DOX and promoted cellular uptake with good anti-tumor effects. This study provides a new idea for the clinical treatment of breast cancer.

Keywords: Ca-Mg double-doped mesoporous silica; Doxorubicin; nanoparticles; breast cancer

1. Introduction

As a significant global public health problem, cancer is currently one of the diseases with the highest morbidity and mortality rates in the world; it will be the most significant obstacle to the growth of life expectancy of the world population in the 21st century [1–3]. According to the World Cancer Report, the annual number of new cancer cases worldwide is expected to increase from 18 million in 2018 to 27 million in 2040, which is a 50% rise [4]. Among them, breast cancer is the most common clinical malignancy, which seriously threatens women's physical and mental health. According to statistics, the incidence rate of female breast cancer has now exceeded that of lung cancer, becoming the leading cause of global cancer incidence in recent years, accounting for 11.7% of all global cancer cases and 6.9% of total cancer deaths [1]. Current clinical treatments for breast cancer mainly include surgery, radiotherapy, chemotherapy, endocrine therapy, molecular targeted therapy, and immunotherapy [5–9]. Chemotherapy, as the first choice of treatment for breast cancer, has been limited in clinical applications due to the disadvantages of poor stability, short half-life, and the lack of tissue specificity of chemotherapeutic drugs [10,11]. In addition to this, it has also been shown that chemotherapy can lead to complications, such as anthracycline-induced cardiotoxicity [12]. The emergence of nanotechnology provides a powerful tool for effective delivery, responsive release, and reductions in the toxicity of chemotherapeutic drugs [13,14]. Among them, inorganic nanoparticles can be applied to the fields of catalysis and environmental protection [15,16], in addition to their use as drug carriers to deliver anti-tumor medications to the target site and achieve the controlled release of drugs in response to specific internal stimuli in the tumor microenvironment [17–20].

Silica nanoparticles (SiO_2 NPs), the most prevalent inorganic nanoparticles, are widely used in biomedical applications [21,22]. There are two main types of SiO_2 NPs: non-porous and mesoporous silica nanoparticles (MSNs), in which the enormous specific surface area, the ordered and tunable pore size, and the easily modifiable surface of MSNs [23] have led to better applications in the field of anti-tumor drug carriers [24–28]. However, pure silica degrades very slowly due to its stable network structure, which limits its biomedical applications [29,30]. Therefore, the preparation of degradable mesoporous silica drug carriers is significant. In addition, the pores and surface of SiO_2 NPs contain silica hydroxyl (Si-OH) groups that make their zeta potential very damaging, and positively charged drugs can be readily adsorbed to the pores and surface of SiO_2 NPs by electrostatic solid interaction [31], which also lays the foundation for the successful loading of Doxorubicin (DOX).

Recently, in order to overcome the incomplete drug release and toxic bioaccumulation phenomenon of MSNs, researchers have proposed methodological strategies, such as reconfiguring silica nanoparticle structures by organic–inorganic hybridization and metal ion doping [32], to improve the biodegradation properties of MSNs. Academician Shi Jianlin's group at the Shanghai Institute of Silicate Research prepared MSNs doped with manganese ions and organic fragments by sol–gel and hydrothermal methods [33], respectively, successfully achieved the rapid degradation of MSNs in the tumor microenvironment (TME), and also significantly improved the therapeutic effect and the resolution of MRI-weighted images. Aizheng Chen's group at Huaqiao University prepared copper and iron double-doped MSNs by the solvothermal method [34], which achieved the specific biodegradation of MSNs under microacidic conditions. The therapeutic effect of Cu and Fe double-doped drug-carrying MSNs was also superior to that of other types of drug-carrying MSNs. In addition, the intracellular reactive oxygen species level was increased by the Fenton-like reaction catalyzed by Fe and Cu ions. The killing power of the tumor cells was significantly improved with the help of photodynamic therapy, which further enhanced the tumor's therapeutic effect. However, this kind of metal-doped MSN will release many metal ions in the body or the cells during descending explanatory drugs, and the released Mn, Cu, and Fe will cause potential hazards to the human body at higher concentrations. Therefore, it is essential to dope non-toxic and crucial elements in MSN carriers.

In addition, the insufficient intracellular drug release from nanoparticles reduces the amount of anticancer drug that actually reaches the cancer cells, which not only hinders the efficacy of cancer chemotherapy but also induces toxic side effects in vivo. Therefore, the development of low-toxicity MSNs with good biocompatibility, biodegradability, and rapid release at the tumor site is a critical scientific issue that needs to be solved [35]. Ca and Mg are essential minerals in the human body, playing crucial roles in various physiological processes [33]. Ca is primarily involved in the formation and maintenance of bones, nerve transmission, and muscle contraction. In contrast, Mg, as an important mineral, is widely involved in enzymatic reactions within the body, particularly in energy metabolism, DNA synthesis, protein synthesis, and cellular function regulation. Mg acts as a cofactor for numerous enzymes, regulating intracellular Ca concentrations and supporting normal cellular metabolism and function. Compared with elemental Si, Ca^{2+} reacts more easily with Si-OH to form Si-O-Ca, obtaining the O-Ca bond, which is less stable than the Si-O bond [36], while Mg^{2+} replaces part of the Si in Si-O-Si to form Mg-O-Si. The acid-responsive Mg-O bond obtained combines the advantages of both substances and synergistically exerts an accelerated biodegradation effect on MSNs. The synergistic effects of Ca and Mg not only optimize the performance of MSNs but also improve their prospects in biomedical applications. By adjusting the ratio of these two elements, precise material design can be achieved, facilitating accelerated biodegradation and enhancing their application in the biomedical field.

Doxorubicin (DOX) belongs to the anthracycline glycoside class of drugs and is a commonly used clinical cycle non-specific anti-tumor drug [37]. It kills tumor cells of various growth cycles, but the obvious toxic side effects make its clinical application severely limited. Therefore, in order to promote the further application of DOX in the clinic, exploring a novel therapeutic strategy that can reduce the toxic side effects of DOX [38] while preserving its tumor tissue-killing effect has become a hot issue in breast cancer treatment.

Given this, in this study, DOX was used as a model drug. It was proposed that the Ca-O and Mg-O bonds, which are unstable under acidic conditions, were formed in MSNs by doping the essential Ca-Mg elements in MSNs (Scheme 1). It was proposed to increase the biodegradation rate and drug release rate of MSNs under acidic conditions to make the Ca-Mg-doped MSNs have a higher drug release and drug loading rate in the tumor microenvironment and to promote the killing effect of DOX on tumor cells and exert the excellent performance of reducing the toxicity of DOX at the same time.

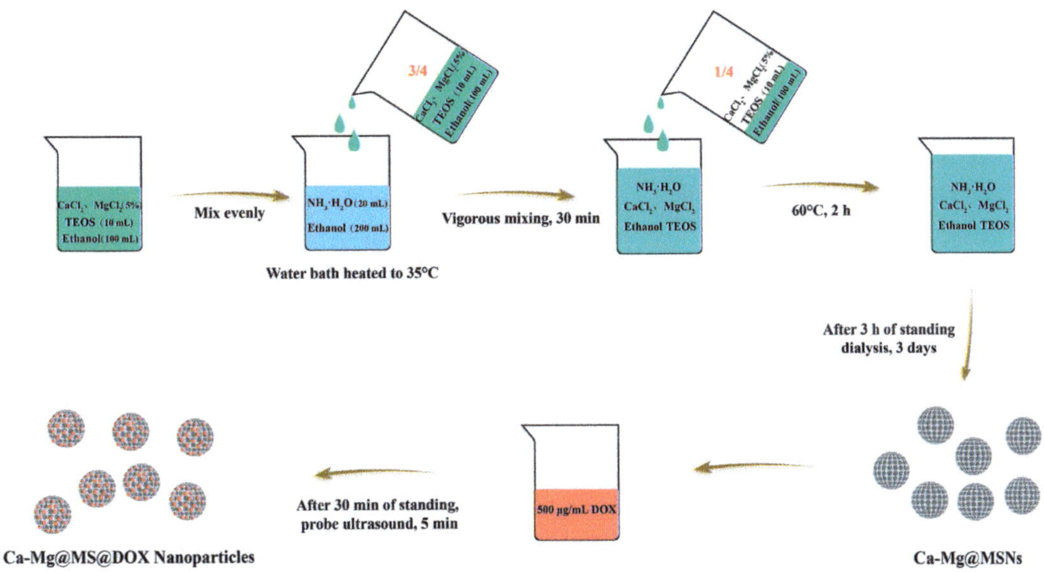

Scheme 1. Schematic illustration of Ca-Mg@DOX@MSNs preparation process.

2. Results

2.1. Characterization of Ca-Mg@MSNs

The MSN subs were dispersed in different concentrations of $CaCl_2$ and $MgCl_2$ solutions, keeping Ca^{2+}:Mg^{2+} = 1:1, and determined the particle size, the PDI, and potential. The results are shown in Figure 1a. The particle sizes of Ca-Mg@MSNs in different concentrations of $CaCl_2$ and $MgCl_2$ solutions were around 200 nm and negatively charged, and the PDI was more significant than 3%. The dispersion was reduced when the concentration of $CaCl_2$ and $MgCl_2$ solutions was more significant than 5%. Therefore, the concentration of $CaCl_2$ and $MgCl_2$ was determined to be 5%.

Ca-Mg@MSNs were prepared with a 5% $CaCl_2$ and $MgCl_2$ concentration. The morphology of MSNs and Ca-Mg@MSNs was observed by scanning electron microscopy; from the results, we can see that the MSNs are regularly spherical-shaped, while Ca-Mg@MSNs are irregularly spherical, which is due to the fact that Mg^{2+} and Ca^{2+} can replace some of the Si in Si-O-Si to form Si-O-M (M = calcium, magnesium), which leads to the morphology change, while, at the same time, it also imparts Ca-Mg@MSN new properties [39,40].

The elemental EDX mapping of scanning electron microscopy was used to observe the content and distribution of $CaCl_2$ and $MgCl_2$. It can be seen from Figure 1d that Ca and Mg elements are indeed present in the Ca-Mg@MSNs, which were successfully synthesized with a uniform distribution of elements, with the content of Ca being 2.92% and that of Mg being 1.19%.

a

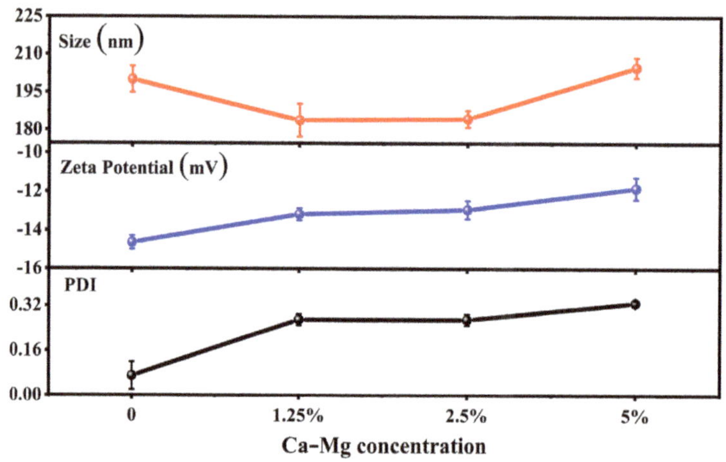

b

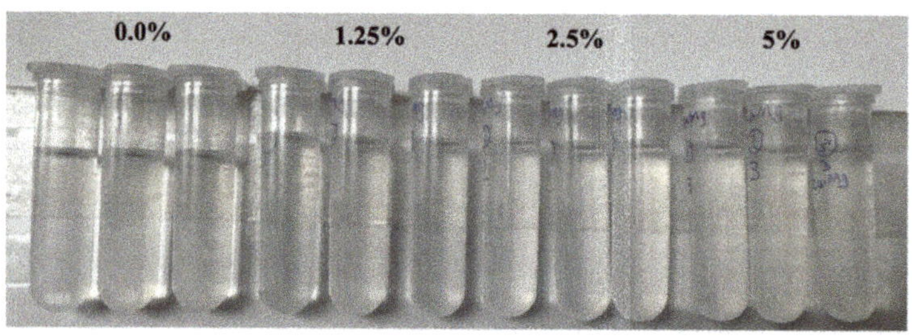

Figure 1. *Cont.*

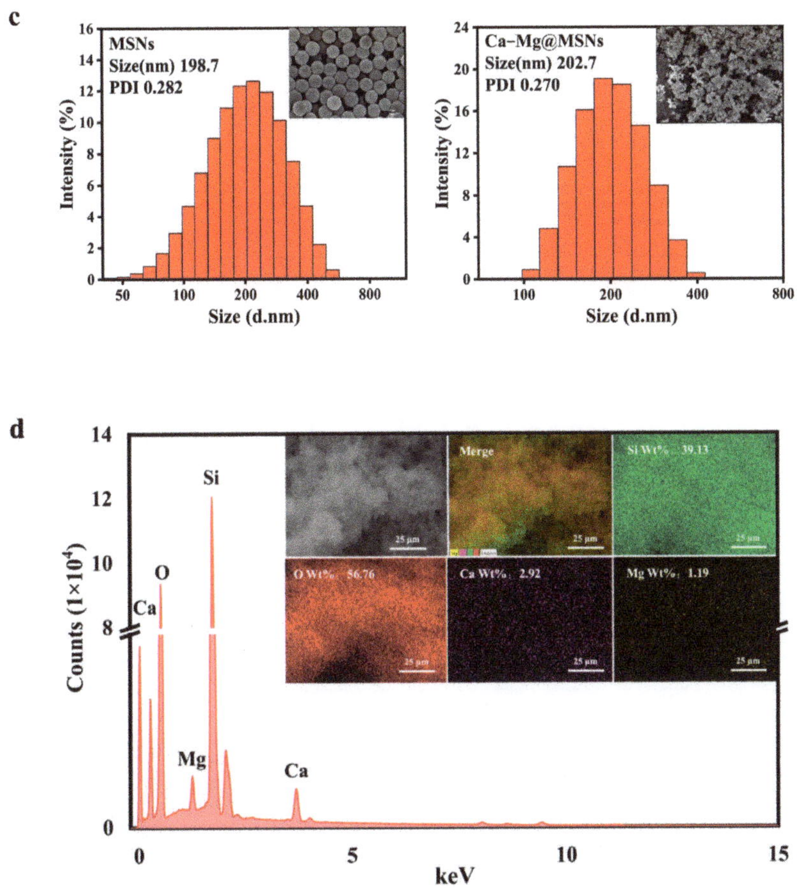

Figure 1. Preparation and characterization of Ca-Mg@MSNs: (**a**) Particle size, PDI, and potential of MSNs with different $CaCl_2$ and $MgCl_2$ concentrations. (**b**) Appearance of MSNs with different $CaCl_2$ and $MgCl_2$ concentrations. (**c**) Particle size, PDI, and morphology of MSNs and Ca-Mg@MSNs. (**d**) Elemental distribution of Ca-Mg@MSNs.

2.2. *Characterization of DOX@MSNs and Ca-Mg@DOX@MSNs*

2.2.1. Specific Surface Area and Pore Size Analysis (BET)

The specific surface area and the pore size distribution are determined by the BET method and BJH model, respectively. The results are shown in Figure 2a. Both MSNs and Ca-Mg@MSNs had typical type IV isotherms and H3 hysteresis lines, which prove that both materials have the structural characteristics of mesoporous materials, the specific surface areas of 159.07 m^2/g and 32.09 m^2/g, respectively, the mesoporous pore volumes of 0.85 cm^3/g and 0.14 cm^3/g, respectively, and average pore sizes of 23.85 nm and 29.73 nm. In comparison, it can be seen that the BET surface area of Ca-Mg@MSNs decreased, and the pore size increased, indicating that the Ca-Mg@MSNs samples have a complex porous structure. The large size of the pores makes it suitable for high drug-loading efficiency, which also lays a solid foundation for the further loading of DOX.

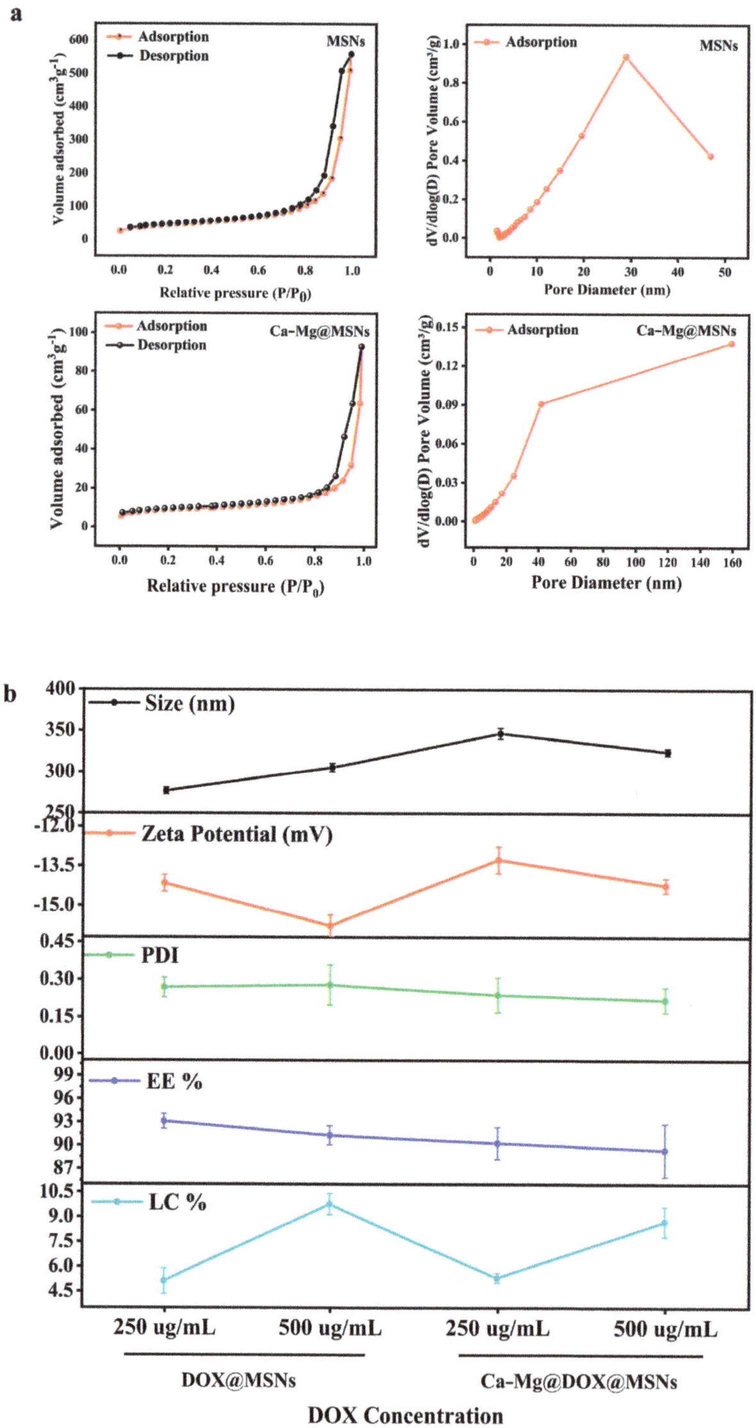

Figure 2. *Cont.*

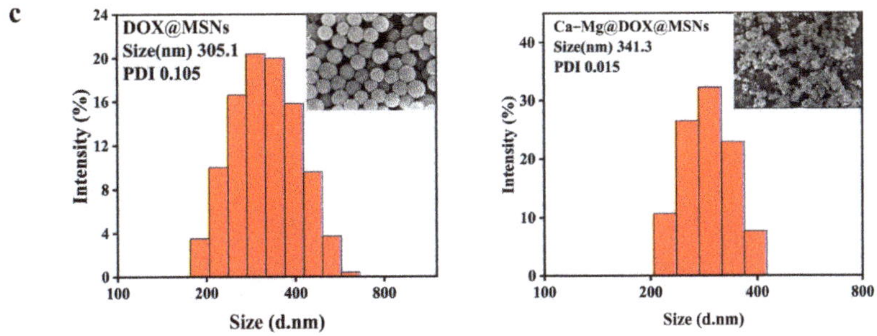

Figure 2. Preparation and characterization of Ca-Mg@DOX@MSNs: (**a**) Results of the BET experiment. (**b**) The particle size, PDI, potential, encapsulation efficiency (%), and loading capacity (%) of DOX@MSNs and Ca-MgDOX@MSNs with different DOX concentrations. (**c**) The particle size, PDI, and morphology of DOX@MSNs and Ca-MgDOX@MSNs.

2.2.2. Determination of Particle Size, Potential, PDI, Drug Loading and Encapsulation Efficiency

MSNs and Ca-Mg@MSNs with different concentrations of DOX were prepared. DOX@MSNs and Ca-Mg@DOX@MSNs had particle sizes of about 200–300 nm, which were negatively charged, and the presence of the electrostatic repulsive force made the prepared nanocarriers more stable and able to be stored for a longer time. It provides the experimental basis for the subsequent in vivo and ex vivo experiments and the regular use of each group of nanoparticles. The encapsulation efficiency was greater than 90%, and the drug loading capacity was more significant than 5%. It lays the foundation for the subsequent exertion of excellent anti-tumor effects (Figure 2b).

Observing the morphology of DOX@MSNs and Ca-Mg@DOX@MSNs by scanning electron microscopy is one of the essential ways of characterization, which can present the results intuitively, objectively, and realistically. From the results, we can see that the distribution of the prepared groups of nanocarriers is uniform, the MSN sub is a regular spherical shape, and the surface of the MSN sphere becomes non-smooth after adding DOX, which proves that DOX is successfully loaded on the surface of MSNs (Figure 2c).

2.3. *Release of Ca-Mg@DOX@MSNs*

2.3.1. Calculated Results of Molecular Dynamic Simulations of Ca-Mg@DOX@MSNs

Molecular dynamic simulation can be used to study intermolecular interactions and physicochemical properties on the atomic scale, which is of high significance as a guide in the study of the structure and properties of novel mesoporous silica. In molecular dynamic simulation, the magnitude of free energy can reflect the stability of molecules and the strength of interactions under specific conditions. Specifically, the smaller the free energy, the more stable the system is and the stronger the interactions; conversely, the larger the free energy, the more unstable the system is and the weaker the interactions. From Figure 3a, it can be seen that the free energy of the system becomes larger after doping calcium and magnesium, and the free energy of the Ca_2Mg_2@MSNs group is the largest, which indicates that the stability of the MSN system decreases after doping calcium and magnesium and shows a trend of gradual decrease with the increase in theincrease of the amount of calcium and magnesium doped, which also predicts that the doping of calcium and magnesium can improve the degradation rate of MSNs to increase the release of the loaded drug. The cumulative release rate will be examined subsequently to verify the simulation results.

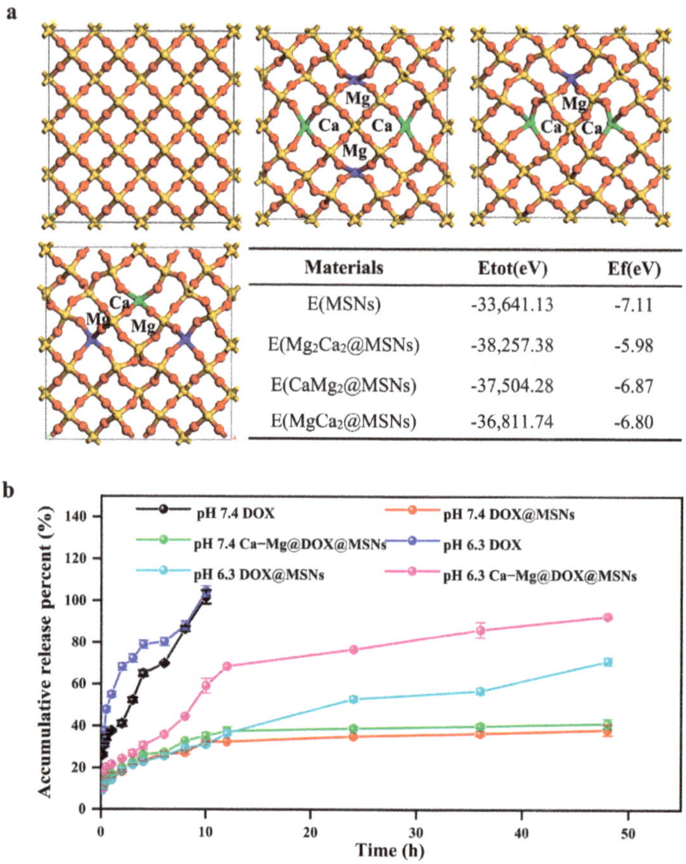

Materials	Etot(eV)	Ef(eV)
E(MSNs)	-33,641.13	-7.11
E(Mg_2Ca_2@MSNs)	-38,257.38	-5.98
E($CaMg_2$@MSNs)	-37,504.28	-6.87
E($MgCa_2$@MSNs)	-36,811.74	-6.80

Figure 3. Release of Ca-Mg@DOX@MSNs: (**a**) Calculated results of molecular dynamic simulations of Ca-Mg@DOX@MSNs. (**b**) Cumulative release of Ca-Mg@DOX@MSNs.

2.3.2. Cumulative Release of Ca-Mg@DOX@MSNs

The drug's ability in the nanoparticles to be released sustainably is a critical factor in achieving therapeutic effects. In this experiment, the in vitro release was investigated using the positive dynamic dialysis method. The results are shown in Figure 3b; free DOX could be rapidly released in 10 h under different pH conditions, while the release rate of the Ca-Mg@DOX@MSNs is higher than that of DOX@MSN under both pH 7.4 and pH 6.3. This result is consistent with molecular dynamic simulations, as calcium and magnesium doping decreases the stability of the MSN system, which in turn accelerates the rate of drug release. We also found that Ca-Mg@DOX@MSNs were relatively stable in the PBS solution at pH 7.4, with less than 50% release in 48 h, while the release rate was significantly increased to 92.73 $\pm$ 0.76% under acidic conditions at pH 6.3. This result also indicates that the prepared nanoparticles release better in the tumor environment than in normal tissues, which can exert a better anti-tumor effect. The results of fitting the cumulative release and time (Tables 1 and 2) showed that MSNs, DOX@MSNs, and Ca-Mg@DOX@MSNs conformed to the first-order kinetic model, Weibull's distribution function, and Higuchi's model, respectively, at pH 7.4, and MSNs conformed to the Higuchi's model at pH 6.3. In contrast, DOX@MSNs conformed to Higuchi's model, and DOX@MSNs were more consistent with the mono-exponential and first-order kinetic models. In comparison, Ca-Mg@DOX@MSNs were more consistent with the first-order kinetic model.

Table 1. Dissolution curve fitting of DOX@MSNs and Ca-Mg@DOX@MSNs in pH 7.4.

Model	DOX	r	DOX@MSNs	r	Ca-Mg @DOX@MSNs	r
Single index model	lg(y∞ − y) = lgy∞ − 0.195t/2.303	0.976	lg(y∞ − y) = lgy∞ − 0.007t/2.303	0.848	lg(y∞ − y) = lgy∞ + 0.007t/2.303	0.683
Zero-order kinetic	Q = 7.3056t + 28.983	0.993	Q = 0.523t + 18.896	0.825	Q = 0.5902t + 20.293	0.795
First-order kinetics	ln(1 − Q) = −0.195t − 0.2463	0.976	ln(1 − Q) = −0.0072t − 0.2112	0.848	ln(1 − Q) = −0.0083t − 0.2305	0.819
Ritger–Peppas	$M\infty/Mt = 0.415t^{0.312}$	0.965	$M\infty/Mt = 0.157t^{0.2392}$	0.844	$M\infty/Mt = 0.171t^{0.2642}$	0.952
Weibull distribution function	ln(1/(1 − Q)) = 0.3361lnt + 0.6602	0.831	ln(1/(1 − Q)) = 0.0757lnt + 0.1731	0.913	ln(1/(1 − Q)) = 0.0834lnt + 0.2106	0.976
Higuchi	$Q = 1.366t^{1/2}$	0.966	$Q = 1.270t^{1/2}$	0.844	$Q = 1.302t^{1/2}$	0.985

Table 2. Dissolution curve fitting of DOX@MSNs and Ca-Mg@DOX@MSNs in pH 6.3.

Model	DOX	r	DOX@MSNs	r	Ca-Mg @DOX@MSNs	r
Single index model	lg(y∞ − y) = lgy∞ − 0.206t/2.303	0.968	lg(y∞ − y) = lgy∞ − 0.022t/2.303	0.991	lg(y∞ − y) = lgy∞ + 0.010t/2.303	0.768
Zero-order kinetic	Q = 6.2226t + 44.642	0.932	Q = 1.2312t + 16.032	0.973	Q = 1.715t + 24.906	0.918
First-order kinetics	ln(1 − Q) = −0.2066t − 0.5414	0.968	ln(1 − Q) = −0.0222t − 0.1499	0.991	ln(11 − Q) = −0.0512t − 0.2081	0.989
Ritger–Peppas	$M\infty/Mt = 0.535t^{0.2684}$	0.99	$M\infty/Mt = 0.140t^{0.3691}$	0.906	$M\infty/Mt = 0.217t^{0.3745}$	0.974
Weibull distribution function	ln(1/(1 − Q)) = 0.4169lnt + 0.957	0.966	ln(1/(1 − Q)) = 0.185lnt + 0.1078	0.845	ln(1/(1 − Q)) = 0.364lnt + 0.2856	0.837
Higuchi	$Q = 1.308t^{1/2}$	0.991	$Q = 1.411t^{1/2}$	0.985	$Q = 1.454t^{1/2}$	0.975

2.4. Safety Evaluation of Ca-Mg@DOX@MSNs

Nanoparticles have the danger of hemolysis, platelet aggregation, or blood coagulation; hemolysis experiments of medical materials, cytotoxicity of the antecedent experiments, and the necessary characterization of blood-contacting materials in the biocompatibility characterization of this piece have a unique and vital position. The results of hemolysis experiments showed that free DOX (30 μg/mL) had a hemolytic effect, and the hemolysis of MSNs, Ca-Mg@MSNs, DOX@MSNs, and Ca-Mg@DOX@MSNs was less than 5%, which proved that MSNs could be used as a carrier for medical purposes. It could also reduce the hemolysis of DOX. It provides an experimental basis for its further application (Figure 4).

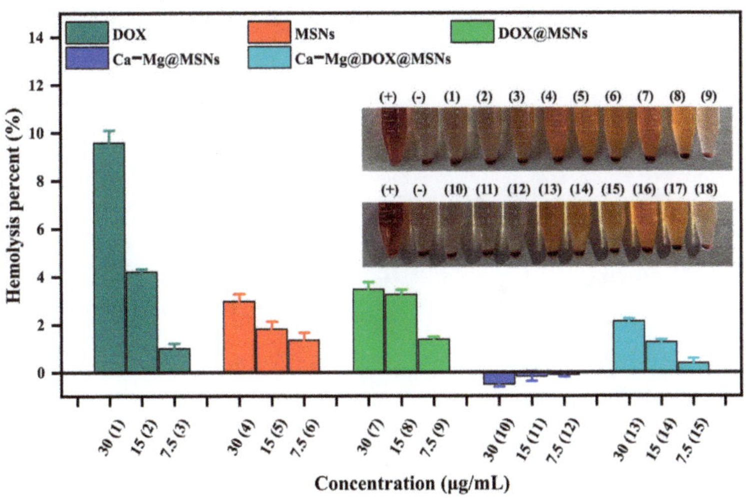

Figure 4. Safety evaluation of Ca-Mg@DOX@MSNs.

2.5. Uptake and In Vitro Anti-Tumor Activity of Ca-Mg@DOX@MSNs

2.5.1. Cellular Uptake of Ca-Mg@DOX@MSNs

FCM measured the uptake of DOX@MSNs and Ca-Mg@DOX@MSNs by 4T1 cells. The uptake of DOX@MSNs and Ca-Mg@DOX@MSNs by 4T1 cells was significantly increased when compared with free DOX, indicating that the prepared nanoparticles can be taken up well by the tumor cells, thus exerting an excellent anti-tumor effect. The prolonged incubation time did not significantly enhance the uptake effect, indicating that the 4T1 cells were able to uptake the DOX@MSNs and Ca-Mg@DOX@MSNs well within a short period (Figure 5a). This result further demonstrates that the prepared DOX@MSNs and Ca-Mg@DOX@MSNs can accumulate in the tumor cells and thus kill the tumor cells to exert excellent anti-tumor effects.

2.5.2. Cellular Toxicity of Ca-Mg@DOX@MSNs

DOX is currently a commonly used drug in breast cancer treatment in clinical practice. As DOX can quickly enter the tumor cells through the cell membranes and act on DNA, it has anti-tumor solid pharmacological activity. As a broad-spectrum anti-tumor drug, it has an inhibitory effect on various tumor cells. However, while killing tumor cells, DOX also produces toxic side effects such as myelosuppression, cardiotoxicity, and liver and kidney functions, which severely limits the clinical application of the drug [41]. Loading DOX with nanocarriers can avoid these drawbacks; to explore this role of the prepared nanocarriers, the toxicity of DOX@MSNs and Ca-Mg@DOX@MSNs on 4T1 cells was determined by the CCK-8 method in this paper. The results at both 24 h and 48 h showed that, with the increasing concentration of the administered drug, the toxicity of DOX, DOX@MSNs, and Ca-Mg @DOX@MSNs significantly enhanced the killing ability of 4T1 cells. Still, the toxicity of DOX@MSNs and Ca-Mg@DOX@MSNs was slightly weaker than that of DOX, which may be attributed to the fact that calcium and magnesium, as trace elements, have the effect of reducing the toxicity of DOX and the property of slightly mitigating the toxicity of DOX, which makes the prepared DOX@MSNs and Ca-Mg@DOX@MSNs reduce the chemotherapeutic toxicity of DOX while exerting anti-tumor effects (Figure 5b).

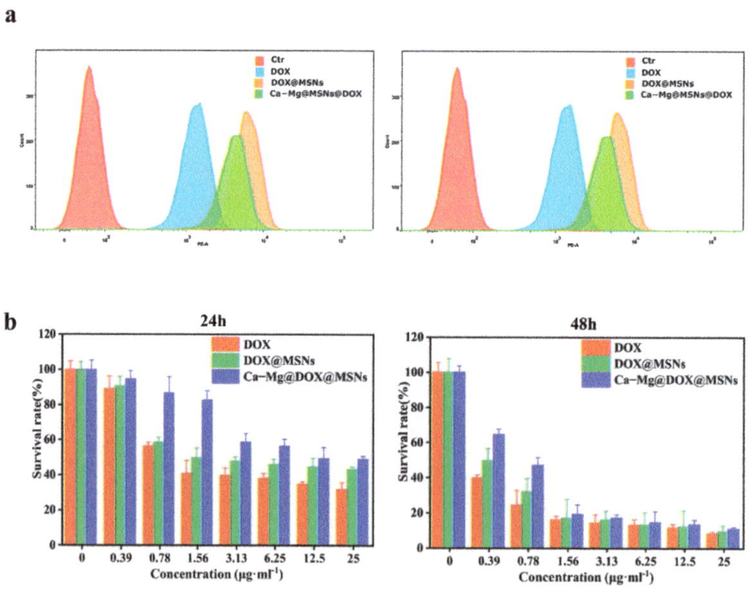

Figure 5. *Cont.*

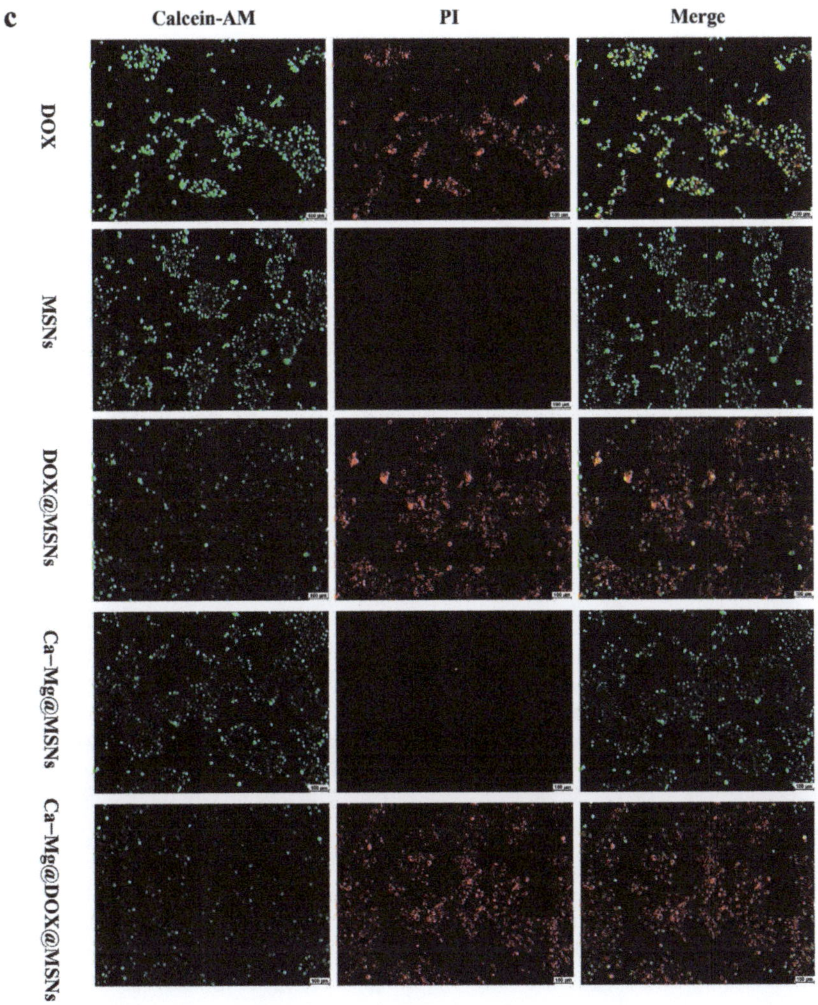

Figure 5. Uptake and in vitro anti-tumor activity of Ca-Mg@DOX@MSNs: (**a**) Results of uptake experiments. (**b**) CCK-8 assay to determine the effects on 4T1 cells. (**c**) Results of live and dead cell staining.

2.5.3. Living/Dead Cell Staining of Ca-Mg@DOX@MSNs

The effects of DOX@MSNs and Ca-Mg@DOX@MSNs on 4T1 cells were observed by the laser confocal staining of live and dead cells. Calcein-AM can pass through the intact cell membrane and embed in the nuclei of all cells (live and dead), showing green fluorescence. PI can only pass through the incomplete cell membrane of the dead cells and be embedded in the nuclei of all dead cells. The nuclei of all dead cells show red fluorescence. From the experimental results, we can see that the red fluorescence in the nuclei of 4T1 cells after the action of DOX@MSNs and Ca-Mg@DOX@MSNs was significantly higher than that of MSNs and Ca-Mg@MSN groups, which indicated that the dead cells were increased by loading DOX. The results were in agreement with the results of CCK-8 (Figure 5c).

3. Discussion

As an emerging nanocarrier, the MSN has potential applications in tumor-targeted drug delivery systems [42–47]. However, the poor biodegradability of MSNs severely hinders its practical biological applications [48,49]. The biodegradation process of conventional MSNs takes days [50] and weeks for excretion from the body [51]. Many studies have shown that poor biodegradability can lead to severe particle bioaccumulation, which may cause [52,53] a range of biosafety issues. Therefore, improving the biodegradability of MSNs is crucial to facilitate their in vivo studies and clinical applications.

Metal ion doping has been reported as an effective strategy to promote MSN biodegradation in the tumor microenvironment. Doping metal ions, such as Mn and Fe, into the silica skeleton to form Si-O-M (M = Mn, Fe, etc.) significantly improved the biodegradability because the M-O bond is more fragile than the Si-O bond in an acidic environment [54]. However, large amounts of Mn, Fe, etc., can harm the human body. Calcium, an essential nutrient element, is commonly used to improve the biodegradability of MSNs compared to other metallic elements by doping into a silica matrix without toxicity. Zhang et al. [55] synthesized MSN/HAP hybrid nanoparticles with good acid-responsive properties by adding calcium salts ($CaCl_2$) and $Na_2HPO_4 \cdot 12\ H_2O$ to the reaction system to make them biodegradable. Similarly, Hu et al. [56] successfully prepared mesoporous silica-calcium phosphate (MS-CAP) hybrid nanoparticles by employing $CaCl_2$ as a Ca^{2+} source, which reacts more readily with Si-OH to form Si-O-Ca than elemental Si. In addition, by TEM observation and ICP-OES analysis, we found that the MS-CAP particles have good stability in neutral and weakly acidic environments (pH 7.4 and pH 6.5). However, in a strongly acidic environment (pH 4.5), the MS-CAP particles were almost completely degraded within 24 h. The MS-CAP particles were also stable in the neutral and weakly acidic environment (pH 7.4 and pH 6.5). In addition, as an essential element, Mg has been used to enhance the biodegradability of MSNs. Yu et al. [40] prepared a pH-responsive biodegradable Mg-doped MSN by the dissolution–re-growth method, in which Mg^{2+}could replace some of the Si in the Si-O-Si to form Mg-O-Si. The Mg-O bond obtained was acid-responsive to fracture, which helped to accelerate the silica particles for biodegradation.

However, most MSNs are doped with only one substance and usually need to exhibit satisfactory biodegradation efficiency. Therefore, multi-substance-doped mesoporous silica with better biodegradability is essential. In this paper, we simultaneously doped two substances, Ca and Mg, into the MSN. The cumulative release of Ca-Mg@DOX@MSNs for 48 h was as high as $92.73 \pm 0.76\%$, which demonstrated that the simultaneous doping of Ca and Mg could significantly enhance the biodegradability of MSNs and thus could exert an excellent anti-tumor effect. The mechanism may be that pure silica materials are composed of tetrahedra with bridging oxygens with a stable Si-O-Si network structure. For Ca- and Mg-doped mesoporous silica, the presence of Ca and Mg in the silica network structure introduces non-bridging oxygen, resulting in relatively weak Si-O-M (M = Ca, Mg) connections. Therefore, the presence of non-bridging oxygen MS-Ca and MS-Mg in the silica network accelerates the hydrolysis of silica. In addition, the amorphous structure formed by Ca and Mg doping is also believed to favor its degradation [55,57].

Meanwhile, from the cytotoxicity experiments, it can be seen that the doping of Ca and Mg as trace elements into MSNs makes the nanoparticles show the utility of reducing DOX toxicity while killing the tumor cells. This result indicates that the prepared nanoparticles are expected to reduce the toxic side effects of DOX and provide a new idea for the further application of DOX, a broad-spectrum anti-tumor drug, in the clinic.

In summary, the prepared Ca-Mg double-doped MSNs not only significantly enhanced the biodegradation rate of MSNs under acidic conditions, providing a feasible solution to the drawbacks of MSNs in clinical applications at this stage but also showed, due to

Ca-Mg doping of the essential human element, a reduction in DOX toxicity. This surprising discovery provides a basis for the in-depth study of DOX in the clinic, which is expected to be a new strategy for tumor therapy.

4. Materials and Methods

4.1. Materials

The following materials were used: adriamycin hydrochloride (Shanghai McLean Biochemical Technology Co., Ltd., Shanghai, China); tetraethyl silicate-TEOS (Shanghai McLean Biochemical Technology Co., Ltd.); anhydrous ethanol (Tianjin Zhiyuan Chemical Reagent Co., Ltd., Tianjin, China); calcium chloride (Tianjin Xinbute Chemical Co., Ltd., Tianjin, China); magnesium chloride (Tianjin Zhiyuan Chemical Reagent Co., Ltd.); CCK8 (Omer Biotechnology (Shanghai) Co., Ltd., Shanghai, China); live dead cell staining kit (Beijing Solebo Technology Co., Ltd., Beijing, China); and 4T1 cells (Procell Life Science & Technology Co., Ltd, Wuhan, China).

4.2. Preparation of MSNs and Ca-Mg@MSNs

4.2.1. MSNs

After mixing 20 mL of concentrated ammonia and 200 mL of ethanol and heating the water bath to a temperature of 35 °C, 10 mL of TEOS and 100 mL of ethanol were mixed well, and 3/4 of the TEOS ethanol solution was added dropwise to the ammonia–ethanol mixture. The mixture was stirred vigorously in a magnetic stirrer (85-2 temperature-controlled magnetic stirrer, Jintan Medical Instrument Factory) for 30 min; the remaining 1/4 of the TEOS solution was added dropwise, and the reaction was carried out at a temperature of 60 °C for 2 h. After cooling the system to room temperature, it was left to age for 3 h. The mixture was poured into a treated dialysis bag with a cut-off relative molecular mass of 8000~14,000 kDa, and the bag was submerged. The dialysis bag was immersed in deionized water for 3 d, changing the water every 6 h. After dialysis, the bag was freeze-dried and stored at −20 °C [58,59].

4.2.2. Ca-Mg@MSNs

After mixing 20 mL of concentrated ammonia and 200 mL of ethanol and heating the water bath to a temperature of 35 °C, a certain amount of $CaCl_2$ and $MgCl_2$ (specifically, the dosage forms were prepared under Ca^{2+}:Mg^{2+} = 1:1 conditions with 1.25%, 2.5%, and 5% $CaCl_2$ and $MgCl_2$ concentration gradients, respectively), 10 mL of TEOS, and 100 mL of ethanol were mixed homogeneously, and 3/4 of the TEOS ethanol solution was taken and added dropwise to the ammonia–ethanol mixed solution. Stir vigorously for 30 min; add the remaining 1/4 of TEOS solution dropwise; and react for 2 h at a temperature of 60 °C. After the system was cooled down to room temperature and left to age for 3 h, the mixture was poured into a treated dialysis bag with a cut-off relative molecular mass of 8000–14,000 kDa. The bag was submerged in deionized water for dialysis for 3 d. During this period, the water was changed every 6 h, and, after the end of dialysis, the resulting samples were freeze-dried and stored at −20 °C [55].

4.3. Characterization of MSNs and Ca-Mg@MSNs

4.3.1. Particle Size, Potential, and Morphology

A certain amount of MSNs and Ca-Mg@MSNs was taken in a specific volume of deionized water. Ultrasonic oscillation (KQ5200DE, CNC ultrasonic cleaner, Jiangsu, China) was used to disperse the nanoparticles nicely. The particle size was determined by using a laser particle size analyzer (Nano-2S90, Malvern Laser Particle Sizing, Malvern, UK), and the potential and PDI were determined using a laser particle size analyzer (Nano-2S90,

Malvern Laser Particle Sizing, Malvern, UK) [60]. After coating the MSNs at an accelerating voltage of 5.0 kV, their morphology and dimensions were analyzed by scanning electron microscopy [57].

4.3.2. EDX Mapping Analysis

A certain amount of Ca-Mg@MSNs was placed on the scanning electron microscope (JSM-7610FPlus, thermal field emission scanning electron microscope, Tokyo, Japan), and nanoparticles were irradiated by X-rays. The element's intra-atomic shell electrons were excited, produced shell electron leaps, and emitted the element's characteristics of X-rays. Through the detector to measure the wavelength of the characteristic X-rays of the element (energy) of the intensity of the concentration of the ratio of the relationship between the wavelength of the X-rays, it will be able to carry out a quantitative analysis of the wavelength dispersive X-ray fluorescence spectrometry.

4.4. Preparation of DOX@MSNs and Ca-Mg@DOX@MSNs

The DOX was adsorbed on the surfaces of MSNs and Ca-Mg@MSNs by non-covalent bonding. MSNs and Ca-Mg@MSNs were dispersed in the DOX aqueous solution with concentrations of 250, 500, and 750 ug/mL, respectively, and, after standing for 30 min, the probes were ultrasonicated for 5 min, which led to the preparation of DOX@MSNs and Ca-Mg@DOX@MSNs [61].

4.5. Characterization of DOX@MSNs and Ca-Mg@DOX@MSNs

4.5.1. Specific Surface Area and Pore Size Analysis (BET)

N_2 adsorption–desorption isotherms were carried out on a Micromeritics ASAP2460 analyzer (ASAP 2460, Micromeritics Instrument Corporation, Norfolk, USA) at 77.3 K under continuous adsorption conditions. The samples were processed under vacuum at 200 °C for 4 h prior to nitrogen adsorption. The specific surface area was calculated from the adsorption and desorption isotherms of the samples according to the BET formula, and the pore size distribution was calculated using the BJH equivalent cylindrical model.

4.5.2. Particle Size, Potential, Morphology, Loading, and Encapsulation Efficiency of DOX@MSNs and Ca-Mg@DOX@MSNs

A certain amount of DOX@MSNs and Ca-Mg@DOX@MSNs were taken in a specific volume of deionized water. Ultrasonic oscillation was used to make the nanoparticles fully dispersed, and a laser particle size analyzer was used to determine the particle size, potential, and PDI [62]. After coating the MSNs at an accelerating voltage of 5.0 kV, their morphology and dimensions were analyzed by scanning electron microscopy.

Take a certain amount of the ultrafiltered off DOX small molecule solution, measure the absorbance with the UV2550 UV–visible spectrophotometer (Shanghai Optical Instrument Factory Co., Ltd., Shanghai, China), calculate the drug concentration by substituting into the standard curve equation, and calculate the encapsulation efficiency according to Equation (1). Weigh the same volume of DOX@MSNs and Ca-Mg@DOX@MSNs, freeze dry, and then calculate the drug loading efficiency according to Equation (2):

$$\text{Encapsulationefficiency}(\%) = \frac{\begin{array}{c}\text{Amountofdrugina definite}\\ \text{massofthepreparedparticle (mg)}\end{array}}{\begin{array}{c}\text{Theoreticalamountof}\\ \text{druginthesamemass (mg)}\end{array}} \times 100\% \quad (1)$$

$$\text{Loadingefficiency}(\%) = \frac{\begin{array}{c}\text{Amountofdrugina definite}\\ \text{massofthepreparedparticle (mg)}\end{array}}{\text{Totalmassoftheparticles (mg)}} \times 100\% \quad (2)$$

4.6. *Release of Ca-Mg@DOX@MSNs*

4.6.1. Calculated Results of Molecular Dynamic Simulations of Ca-Mg@DOX@MSNs

We have employed molecular dynamic computer simulations to investigate the structure and properties of Ca-Mg-doped MSNs. All calculations were carried out using the pseudopotential plane wave method based on first principles, performed by the CASTEP software package (V23.1.1) [63], and their surface charges were calculated using density-functional theory (DFT) [64].

4.6.2. Cumulative Release of Ca-Mg@DOX@MSNs

A total of 2 mL of DOX@MSNs, Ca-Mg@DOX@MSNs, and the DOX solution (the amount of DOX was 250 ug/mL) were taken in a dialysis bag (molecular weight cut-off 3500), clamped at both ends, and then placed in the dissolution medium containing 200 mL of isotonic PBS. The temperature was kept at (37 ± 0.5) °C, and the rate was 500 r/min (ZWYR-D2402 Zhicheng constant temperature culture oscillator, Shanghai Zhicheng Analytical Instrument Manufacturing Co., Shanghai, China). Samples of 2 mL were taken at 10 min, 20 min, 30 min, 1.0 h, 2.0 h, 3.0 h, 4.0 h, 6.0 h, 8.0 h, 24 h, and 48 h (immediately supplemented with isotonic PBS at the same temperature and same volume). The drug concentration was determined using a UV–visible spectrophotometer at 488 nm, and the cumulative release rate was calculated and fitted to cumulative releases and time. The test was repeated three times to obtain the mean value [65,66].

4.7. *Safety Evaluation of Ca-Mg@DOX@MSNs*

Hemolysis Experiment of Ca-Mg@DOX@MSNs

Using saline as the negative control and distilled water as the positive control, dilute 10 mL each of MSNs, DOX@MSNs, Ca-Mg@MSNs, Ca-Mg@DOX@MSNs, and DOX (DOX concentrations of 30 μg/mL, 15 μg/mL, and 7.5 μg/mL) with saline and incubate for 30 min in a 37 °C water bath, adding freshly diluted anticoagulated blood of 0.2 mL, mixing them evenly. After homogenization, incubate at 37 °C for 60 min, centrifuge at 3000 r/min for 10 min, take the supernatant, and measure the OD value at 545 nm, three times in parallel, for the evaluation of hemolysis efficiency according to Equation (3).

$$\text{Hemolysisefficiency}(\%) = \frac{OD_{\text{Sample}} - OD_{\text{Negative}}}{OD_{\text{Positive}} - OD_{\text{Negative}}} \times 100\% \quad (3)$$

The OD value of the negative control group is less than 0.03, and that of the positive control group is 0.8 ± 0.3. If the hemolysis efficiency of the material is less than 5%, it conforms to the requirements of the hemolysis experiment of medical materials; if the hemolysis efficiency is more than 5%, it fails to conform to the requirements.

4.8. *Uptake and In Vitro Anti-Tumor Activity of Ca-Mg@DOX@MSNs*

4.8.1. Cellular Uptake of Ca-Mg@DOX@MSNs

4T1 cells were inoculated into six-well plates (5×10^5/well) and incubated overnight at 37 °C, 5% CO_2, and suitable humidity. DOX, DOX@MSNs, and Ca-Mg@DOX@MSNs (DOX content 1 ug/mL) were added, respectively, along with a blank control, and, after 2 h and 24 h of action, the cells were washed three times with PBS, digested, centrifuged at

1000 r/min, and resuspended with PBS. Sieving was performed, flow cytometry (LSR-II, BD, USA) was performed, and FlowJo analyzed the data [67].

4.8.2. Cellular Toxicity of Ca-Mg@DOX@MSNs

4T1 cells were cultured in DMEM medium with 10% FBS and 1% antibiotics (100 U/mL of penicillin and 100 μg/mL of streptomycin). After the cells grow to the logarithmic growth phase, inoculate into 96-well culture plates (5000/well) and incubate overnight at 37 °C, 5% CO_2, and suitable humidity. After adding DOX@MSNs and Ca-Mg@DOX@MSNs (DOX concentrations of 0, 3.12, 5.25, 12.50, 25.00, and 50.00 μg/mL), respectively, along with a blank control and negative control, with five replicate wells in each group, and after 24 h and 48 h of action, add 20 μL of the CCK-8 solution (5 mg/mL) and incubate for 2 h. Absorbance was measured at 450 nm by an enzyme labeling instrument (Multiskan FC, Enzyme-Linked Immunosorbent Assay Analyzer, Waltham, US.), and the cell survival rate was calculated according to Equation (4) [68]:

$$\text{Cellsurvivalrate}(\%) = \frac{\text{Experimentalgroup} - \text{Blankcontrol}}{\text{Negativecontrol} - \text{Blankcontrol}} \times 100\% \quad (4)$$

4.8.3. Living/Dead Cell Staining of Ca-Mg@DOX@MSNs

The 4T1 cells were inoculated onto confocal culture dishes (5000/well) and incubated overnight at 37 °C, 5% CO_2, and suitable humidity. DOX@MSNs and Ca-Mg@DOX@MSNs (DOX concentration of 1 μg/mL) were added. At the same time, a blank control was set up, and, after 24 h of action, the cells were washed three times with PBS and added 100 μL of a 2 μM Calcein-AM working solution and a 4.5 μM PI working solution to each well. The cells were incubated for 30 min, and the cells were washed three times with PBS. The 100 μL PBS in LSCM (Nanjing Henqiao Instrument Co., Ltd., Nanjing, China) observed cell status [69].

5. Conclusions

In this study, Ca-Mg-doped MSNs were prepared, significantly increasing their biodegradation rate because Ca-O and Mg-O bonds are more fragile than Si-O bonds in an acidic environment. The MSNs and DOX@MSNs were successfully synthesized with a particle size of approximately 200–300 nm and exhibited a negative charge. Both MSNs and DOX@MSNs exhibited a regular spherical morphology, whereas the Ca and Mg-doped nanoparticles displayed an irregular spherical shape. Elemental mapping revealed that the Ca and Mg contents were 2.92% and 1.19%, respectively. The DOX@MSNs and Ca-Mg@DOX@MSNs were also successfully prepared, with a particle size ranging from 300 to 400 nm and a negative charge. The encapsulation efficiency was approximately 90%, and the drug loading efficiency was around 10%. Notably, these nanoparticles demonstrated a slower drug release compared to free DOX in solution. The release of Ca-Mg@DOX@MSNs was notably accelerated under acidic conditions. All samples exhibited a hemolysis rate of less than 5%. Cellular toxicity assays revealed that DOX@MSNs and Ca-Mg@DOX@MSNs exhibited lower toxicity compared to free DOX. Similar results were observed in live/dead cell staining experiments. Furthermore, cellular uptake studies indicated that DOX@MSNs were taken up more efficiently than Ca-Mg@DOX@MSNs and free DOX. In conclusion, this study provides new insights into the treatment of breast cancer by selecting essential and non-toxic elements to optimize the properties of MSN materials, which significantly increased drug release and reduced chemotherapeutic drug toxicity.

It is important to acknowledge that, while this study offers promising insights into breast cancer treatment, there are several limitations that require further attention in future research. First, although this study highlights that DOX@MSNs and Ca-Mg@DOX@MSNs can effectively reduce the toxicity of DOX and exhibit anti-tumor effects, the experiments were primarily conducted in vitro using 4T1 cell lines. This study did not include other tumor cell lines or clinical samples, which limits the broader applicability of these findings. Tumor types and individual patient variability may result in differences in the performance of these materials. Therefore, expanding research to include various tumor models and preclinical animal experiments is crucial to better assess the clinical translational potential of these materials. Moreover, the preparation method for calcium and magnesium-doped MSN materials used in this study has not been validated in vivo. While the method has shown some success in vitro, its reproducibility and scalability have not been fully confirmed. Although relevant preparation techniques are available in the literature, industrial-scale applications may face challenges such as high production costs and potential material instability. Hence, future research should focus on developing methods for controllable synthesis, large-scale production, cost reduction, and ensuring the long-term stability of these materials.

Author Contributions: Conceptualization, Q.Z., X.T. and M.W.; methodology, Q.Z., X.T. and M.W.; investigation, Q.Z., J.H., R.C., T.J., Y.H., T.B. and X.T.; resources, M.W.; data curation, Q.Z., X.T. and M.W.; writing—original draft preparation, Q.Z., C.L. and X.T.; writing—review and editing, Q.Z., C.L., X.T. and M.W.; formal analysis, J.H., R.C., Y.H., T.B. and T.J.; supervision, C.L. and M.W.; project administration, C.L. and M.W.; funding acquisition, C.L. and M.W. All authors have read and agreed to the published version of the manuscript.

Funding: This study was funded by the National Science Foundation of China (No. 82460366), the Particular Project of Innovation Environment Construction in Autonomous Region of Xinjiang-Construction of Science and Technology Innovation Base (PT2304), the Special Funds for Xinjiang Key Laboratory of Natural Medicines Active Components and Drug Release Technology, Engineering Research Center of Xinjiang and Central Asian Medicine Resources, Ministry of Education.

Data Availability Statement: Data is contained within the article.

Conflicts of Interest: The authors have no financial or proprietary interests in any material discussed in this article.

References

1. Sung, H.; Ferlay, J.; Siegel, R.L.; Laversanne, M.; Soerjomataram, I.; Jemal, A.; Bray, F. Global Cancer Statistics 2020: GLOBOCAN Estimates of Incidence and Mortality Worldwide for 36 Cancers in 185 Countries. *CA Cancer J. Clin.* **2021**, *71*, 209–249. [CrossRef] [PubMed]
2. Cainap, C.; Crisan, N. Advances in Cancer Therapy from Research to Clinical Practice—Surgical, Molecular or Systemic Management of Cancer. *Medicina* **2023**, *59*, 1309. [CrossRef] [PubMed]
3. Qi, J.; Li, M.; Wang, L.; Hu, Y.; Liu, W.; Long, Z.; Zhou, Z.; Yin, P.; Zhou, M. National and Subnational Trends in Cancer Burden in China, 2005-20: An Analysis of National Mortality Surveillance Data. *Lancet Public Health* **2023**, *8*, e943–e955. [CrossRef] [PubMed]
4. Bray, F.; Ferlay, J.; Soerjomataram, I.; Siegel, R.L.; Torre, L.A.; Jemal, A. Global Cancer Statistics 2018: GLOBOCAN Estimates of Incidence and Mortality Worldwide for 36 Cancers in 185 Countries. *CA Cancer J. Clin.* **2018**, *68*, 394–424. [CrossRef] [PubMed]
5. Wang, J.; Wu, S.-G. Breast Cancer: An Overview of Current Therapeutic Strategies, Challenge, and Perspectives. *Breast Cancer Targets Ther.* **2023**, *15*, 721–730. [CrossRef]
6. Debien, V.; De Caluwé, A.; Wang, X.; Piccart-Gebhart, M.; Tuohy, V.K.; Romano, E.; Buisseret, L. Immunotherapy in Breast Cancer: An Overview of Current Strategies and Perspectives. *NPJ Breast Cancer* **2023**, *9*, 7. [CrossRef] [PubMed]
7. Zarenezhad, E.; Kanaan, M.H.G.; Abdollah, S.S.; Vakil, M.K.; Marzi, M.; Mazarzaei, A.; Ghasemian, A. Metallic Nanoparticles: Their Potential Role in Breast Cancer Immunotherapy via Trained Immunity Provocation. *Biomedicines* **2023**, *11*, 1245. [CrossRef] [PubMed]
8. Yu, L.-Y.; Tang, J.; Zhang, C.-M.; Zeng, W.-J.; Yan, H.; Li, M.-P.; Chen, X.-P. New Immunotherapy Strategies in Breast Cancer. *Int. J. Environ. Res. Public Health* **2017**, *14*, 68. [CrossRef]
9. Zhang, R.; Clark, S.D.; Guo, B.; Zhang, T.; Jeansonne, D.; Jeyaseelan, S.J.; Francis, J.; Huang, W. Challenges in the Combination of Radiotherapy and Immunotherapy for Breast Cancer. *Expert Rev. Anticancer Ther.* **2023**, *23*, 375–383. [CrossRef]

10. Du, Y. Research Progress of Tumor Microenvironment-Responsive Anti-Tumor Nano-Drug Carriers. *BIO Web Conf.* **2023**, *61*, 01031. [CrossRef]
11. Pal, S.; Mehta, D.; Dasgupta, U.; Bajaj, A. Advances in Engineering of Low Molecular Weight Hydrogels for Chemotherapeutic Applications. *Biomed. Mater. Bristol Engl.* **2021**, *16*, 024102. [CrossRef] [PubMed]
12. Maksymowicz, M.; Machowiec, P.; Korzeniowska, A.; Baran, N.; Bielak, A.; Nowak, A.; Cywka, Ł.; Szwed, W.; Nowak, A.; Bogusz, K. Adverse Effects in the Management of Breast Cancer—Recent Studies. *J. Educ. Health Sport* **2023**, *37*, 11–24. [CrossRef]
13. Li, L.; Yang, W.-W.; Xu, D.-G. Stimuli-Responsive Nanoscale Drug Delivery Systems for Cancer Therapy. *J. Drug Target.* **2019**, *27*, 423–433. [CrossRef] [PubMed]
14. Xiang, J.; Zhao, R.; Wang, B.; Sun, X.; Guo, X.; Tan, S.; Liu, W. Advanced Nano-Carriers for Anti-Tumor Drug Loading. *Front. Oncol.* **2021**, *11*, 758143. [CrossRef] [PubMed]
15. Ta, Q.T.H. Highly Surface-Active Si-Doped $TiO_2/Ti_3C_2T_x$ Heterostructure for Gas Sensing and Photodegradation of Toxic Matters. *Chem. Eng. J.* **2021**, *425*, 131437. [CrossRef]
16. Tran, M.N.; Moreau, M.; Addad, A.; Teurtrie, A.; Roland, T.; de Waele, V.; Dewitte, M.; Thomas, L.; Levêque, G.; Dong, C.; et al. Boosting Gas-Phase TiO_2 Photocatalysis with Weak Electric Field Strengths of Volt/Centimeter. *ACS Appl. Mater. Interfaces* **2024**, *16*, 14852–14863. [CrossRef]
17. Deng, S.; Cui, C.-X.; Duan, L.; Hu, L.; Yang, X.; Wang, J.-C.; Qu, L.-B.; Zhang, Y. Anticancer Drug Release System Based on Hollow Silica Nanocarriers Triggered by Tumor Cellular Microenvironments. *ACS Omega* **2021**, *6*, 553–558. [CrossRef] [PubMed]
18. Yang, W.; Yang, S.; Jiang, L.; Zhou, Y.; Yang, C.; Deng, C. Tumor Microenvironment Triggered Biodegradation of Inorganic Nanoparticles for Enhanced Tumor Theranostics. *RSC Adv.* **2020**, *10*, 26742–26751. [CrossRef] [PubMed]
19. Chiou, A.E.; Hinckley, J.A.; Khaitan, R.; Varsano, N.; Wang, J.; Malarkey, H.F.; Hernandez, C.J.; Williams, R.M.; Estroff, L.A.; Weiner, S.; et al. Fluorescent Silica Nanoparticles to Label Metastatic Tumor Cells in Mineralized Bone Microenvironments. *Small* **2021**, *17*, e2001432. [CrossRef]
20. Zhao, H.; Li, Y.; Chen, J.; Zhang, J.; Yang, Q.; Cui, J.; Shi, A.; Wu, J. Environmental Stimulus-Responsive Mesoporous Silica Nanoparticles as Anticancer Drug Delivery Platforms. *Colloids Surf. B Biointerfaces* **2024**, *234*, 113758. [CrossRef]
21. Hosseinpour, S.; Walsh, L.J.; Xu, C. Biomedical Application of Mesoporous Silica Nanoparticles as Delivery Systems: A Biological Safety Perspective. *J. Mater. Chem. B* **2020**, *8*, 9863–9876. [CrossRef]
22. Croissant, J.G.; Fatieiev, Y.; Khashab, N.M. Degradability and Clearance of Silicon, Organosilica, Silsesquioxane, Silica Mixed Oxide, and Mesoporous Silica Nanoparticles. *Adv. Mater.* **2017**, *29*, 1604634. [CrossRef] [PubMed]
23. Niculescu, V.-C. Mesoporous Silica Nanoparticles for Bio-Applications. *Front. Mater.* **2020**, *7*, 36. [CrossRef]
24. Manzano, M.; Vallet-Regí, M. Mesoporous Silica Nanoparticles in Nanomedicine Applications. *J. Mater. Sci. Mater. Med.* **2018**, *29*, 65. [CrossRef] [PubMed]
25. Yue, J.; Luo, S.-Z.; Lu, M.-M.; Shao, D.; Wang, Z.; Dong, W.-F. A Comparison of Mesoporous Silica Nanoparticles and Mesoporous Organosilica Nanoparticles as Drug Vehicles for Cancer Therapy. *Chem. Biol. Drug Des.* **2018**, *92*, 1435–1444. [CrossRef] [PubMed]
26. Zhang, W.; Liu, H.; Qiu, X.; Zuo, F.; Wang, B. Mesoporous Silica Nanoparticles as a Drug Delivery Mechanism. *Open Life Sci.* **2024**, *19*, 20220867. [CrossRef] [PubMed]
27. Zhou, J.; Liu, C.; Zhong, Y.; Luo, Z.; Wu, L. A Review of Current Developments in Functionalized Mesoporous Silica Nanoparticles: From Synthesis to Biosensing Applications. *Biosensors* **2024**, *14*, 575. [CrossRef]
28. Kovtareva, S.; Kusepova, L.; Tazhkenova, G.; Mashan, T.; Bazarbaeva, K.; Kopishev, E. Surface Modification of Mesoporous Silica Nanoparticles for Application in Targeted Delivery Systems of Antitumour Drugs. *Polymers* **2024**, *16*, 1105. [CrossRef]
29. Pohaku Mitchell, K.K.; Liberman, A.; Kummel, A.C.; Trogler, W.C. Iron(III)-Doped, Silica Nanoshells: A Biodegradable Form of Silica. *J. Am. Chem. Soc.* **2012**, *134*, 13997–14003. [CrossRef] [PubMed]
30. Alexis, F.; Pridgen, E.; Molnar, L.K.; Farokhzad, O.C. Factors Affecting the Clearance and Biodistribution of Polymeric Nanoparticles. *Mol. Pharm.* **2008**, *5*, 505–515. [CrossRef]
31. Patwardhan, S.V.; Emami, F.S.; Berry, R.J.; Jones, S.E.; Naik, R.R.; Deschaume, O.; Heinz, H.; Perry, C.C. Chemistry of Aqueous Silica Nanoparticle Surfaces and the Mechanism of Selective Peptide Adsorption. *J. Am. Chem. Soc.* **2012**, *134*, 6244–6256. [CrossRef] [PubMed]
32. Dechézelles, J.-F.; Ciotonea, C.; Catrinescu, C.; Ungureanu, A.; Royer, S.; Nardello-Rataj, V. Emulsions Stabilized with Alumina-Functionalized Mesoporous Silica Particles. *Langmuir ACS J. Surf. Colloids* **2020**, *36*, 3212–3220. [CrossRef]
33. Hu, Y.; Bai, S.; Wu, X.; Tan, S.; He, Y. Biodegradability of Mesoporous Silica Nanoparticles. *Ceram. Int.* **2021**, *47*, 31031–31041. [CrossRef]
34. Liu, C.-G.; Han, Y.-H.; Zhang, J.-T.; Kankala, R.K.; Wang, S.-B.; Chen, A.-Z. Rerouting Engineered Metal-Dependent Shapes of Mesoporous Silica Nanocontainers to Biodegradable Janus-Type (Sphero-Ellipsoid) Nanoreactors for Chemodynamic Therapy. *Chem. Eng. J.* **2019**, *370*, 1188–1199. [CrossRef]
35. Zhang, Y.; Lin, X.; Chen, X.; Fang, W.; Yu, K.; Gu, W.; Wei, Y.; Zheng, H.; Piao, J.; Li, F. Strategies to Regulate the Degradation and Clearance of Mesoporous Silica Nanoparticles: A Review. *Int. J. Nanomed.* **2024**, *19*, 5859–5878. [CrossRef]

36. Liu, C.; Tang, X.; Huang, G. Biodegradable Ca^{2+} Doped Mesoporous Silica Nanoparticles Promote Chemotherapy Synergism with Calcicoptosis and Activate Anti-Tumor Immunity. *Inorganics* **2024**, *12*, 152. [CrossRef]
37. Sarkar, R.; Biswas, S.; Ghosh, R.; Samanta, P.; Pakhira, S.; Mondal, M.; Dutta Gupta, Y.; Bhandary, S.; Saha, P.; Bhowmik, A.; et al. Exosome-Sheathed Porous Silica Nanoparticle-Mediated Co-Delivery of 3,3′-Diindolylmethane and Doxorubicin Attenuates Cancer Stem Cell-Driven EMT in Triple Negative Breast Cancer. *J. Nanobiotechnol.* **2024**, *22*, 285. [CrossRef]
38. Tang, L.; Chen, M.; Wang, D.; He, Y.; Ge, G.; Zeng, Z.; Shu, J.; Guo, W.; Wu, S.X.; Xiong, W. Doxorubicin and Iron-Doped Mesoporous Silica Nanoparticles for Chemodynamic Therapy and Chemotherapy of Breast Cancer. *New J. Chem.* **2024**, *48*, 17294–17309. [CrossRef]
39. He, Y.; Shao, L.; Hu, Y.; Zhao, F.; Tan, S.; He, D.; Pan, A. Redox and pH Dual-Responsive Biodegradable Mesoporous Silica Nanoparticle as a Potential Drug Carrier for Synergistic Cancer Therapy. *Ceram. Int.* **2021**, *47*, 4572–4578. [CrossRef]
40. Yu, L.; Chen, Y.; Lin, H.; Gao, S.; Chen, H.; Shi, J. Magnesium-Engineered Silica Framework for pH-Accelerated Biodegradation and DNAzyme-Triggered Chemotherapy. *Small* **2018**, *14*, e1800708. [CrossRef] [PubMed]
41. Wang, Y.; Chen, X.; He, D.; Zhou, Y.; Qin, L. Surface-Modified Nanoerythrocyte Loading DOX for Targeted Liver Cancer Chemotherapy. *Mol. Pharm.* **2018**, *15*, 5728–5740. [CrossRef] [PubMed]
42. Chen, Y.; Ma, H.; Wang, W.; Zhang, M. A Size-Tunable Nanoplatform: Enhanced MMP2-Activated Chemo-Photodynamic Immunotherapy Based on Biodegradable Mesoporous Silica Nanoparticles. *Biomater. Sci.* **2021**, *9*, 917–929. [CrossRef]
43. Esfahani, M.K.M.; Islam, N.; Cabot, P.J.; Izake, E.L. Development of Thiabendazole-Loaded Mesoporous Silica Nanoparticles for Cancer Therapy. *ACS Biomater. Sci. Eng.* **2022**, *8*, 4153–4162. [CrossRef] [PubMed]
44. Cheng, D.; Ji, Y.; Wang, B.; Wang, Y.; Tang, Y.; Fu, Y.; Xu, Y.; Qian, X.; Zhu, W. Dual-Responsive Nanohybrid Based on Degradable Silica-Coated Gold Nanorods for Triple-Combination Therapy for Breast Cancer. *Acta Biomater.* **2021**, *128*, 435–446. [CrossRef] [PubMed]
45. Choi, G.; Rejinold, N.S.; Piao, H.; Choy, J.-H. Inorganic-Inorganic Nanohybrids for Drug Delivery, Imaging and Photo-Therapy: Recent Developments and Future Scope. *Chem. Sci.* **2021**, *12*, 5044–5063. [CrossRef] [PubMed]
46. Lin, F.-C.; Xie, Y.; Deng, T.; Zink, J.I. Magnetism, Ultrasound, and Light-Stimulated Mesoporous Silica Nanocarriers for Theranostics and Beyond. *J. Am. Chem. Soc.* **2021**, *143*, 6025–6036. [CrossRef]
47. Kankala, R.K.; Han, Y.-H.; Na, J.; Lee, C.-H.; Sun, Z.; Wang, S.-B.; Kimura, T.; Ok, Y.S.; Yamauchi, Y.; Chen, A.-Z.; et al. Nanoarchitectured Structure and Surface Biofunctionality of Mesoporous Silica Nanoparticles. *Adv. Mater.* **2020**, *32*, e1907035. [CrossRef]
48. Cheng, Y.; Jiao, X.; Fan, W.; Yang, Z.; Wen, Y.; Chen, X. Controllable Synthesis of Versatile Mesoporous Organosilica Nanoparticles as Precision Cancer Theranostics. *Biomaterials* **2020**, *256*, 120191. [CrossRef]
49. Nguyen, T.L.; Choi, Y.; Kim, J. Mesoporous Silica as a Versatile Platform for Cancer Immunotherapy. *Adv. Mater.* **2019**, *31*, e1803953. [CrossRef] [PubMed]
50. Yamada, H.; Urata, C.; Aoyama, Y.; Osada, S.; Yamauchi, Y.; Kuroda, K. Preparation of Colloidal Mesoporous Silica Nanoparticles with Different Diameters and Their Unique Degradation Behavior in Static Aqueous Systems. *Chem. Mater.* **2012**, *24*, 1462–1471. [CrossRef]
51. Liu, T.; Li, L.; Teng, X.; Huang, X.; Liu, H.; Chen, D.; Ren, J.; He, J.; Tang, F. Single and Repeated Dose Toxicity of Mesoporous Hollow Silica Nanoparticles in Intravenously Exposed Mice. *Biomaterials* **2011**, *32*, 1657–1668. [CrossRef] [PubMed]
52. Yang, G.; Phua, S.Z.F.; Bindra, A.K.; Zhao, Y. Degradability and Clearance of Inorganic Nanoparticles for Biomedical Applications. *Adv. Mater.* **2019**, *31*, e1805730. [CrossRef] [PubMed]
53. Kempen, P.J.; Greasley, S.; Parker, K.A.; Campbell, J.L.; Chang, H.-Y.; Jones, J.R.; Sinclair, R.; Gambhir, S.S.; Jokerst, J.V. Theranostic Mesoporous Silica Nanoparticles Biodegrade after Pro-Survival Drug Delivery and Ultrasound/Magnetic Resonance Imaging of Stem Cells. *Theranostics* **2015**, *5*, 631–642. [CrossRef]
54. Wang, X.; Li, X.; Ito, A.; Sogo, Y.; Watanabe, Y.; Tsuji, N.M.; Ohno, T. Biodegradable Metal Ion-Doped Mesoporous Silica Nanospheres Stimulate Anticancer Th1 Immune Response in Vivo. *ACS Appl. Mater. Interfaces* **2017**, *9*, 43538–43544. [CrossRef]
55. Hao, X.; Hu, X.; Zhang, C.; Chen, S.; Li, Z.; Yang, X.; Liu, H.; Jia, G.; Liu, D.; Ge, K.; et al. Hybrid Mesoporous Silica-Based Drug Carrier Nanostructures with Improved Degradability by Hydroxyapatite. *ACS Nano* **2015**, *9*, 9614–9625. [CrossRef] [PubMed]
56. He, Y.; Zeng, B.; Liang, S.; Long, M.; Xu, H. Synthesis of pH-Responsive Biodegradable Mesoporous Silica-Calcium Phosphate Hybrid Nanoparticles as a High Potential Drug Carrier. *ACS Appl. Mater. Interfaces* **2017**, *9*, 44402–44409. [CrossRef] [PubMed]
57. Wang, X.; Li, X.; Ito, A.; Sogo, Y. Synthesis and Characterization of Hierarchically Macroporous and Mesoporous CaO–MO–SiO_2–P_2O_5 (M = Mg, Zn, Sr) Bioactive Glass Scaffolds. *Acta Biomater.* **2011**, *7*, 3638–3644. [CrossRef] [PubMed]
58. Pan, L.; He, Q.; Liu, J.; Chen, Y.; Ma, M.; Zhang, L.; Shi, J. Nuclear-Targeted Drug Delivery of TAT Peptide-Conjugated Monodisperse Mesoporous Silica Nanoparticles. *J. Am. Chem. Soc.* **2012**, *134*, 5722–5725. [CrossRef] [PubMed]
59. Sarkar, S.; Ekbal Kabir, M.; Kalita, J.; Manna, P. Mesoporous Silica Nanoparticles: Drug Delivery Vehicles for Antidiabetic Molecules. *ChemBioChem* **2023**, *24*, e202200672. [CrossRef]

60. Chen, W.-S.; Chen, H.-R.; Lee, C.-H. The Photocatalytic Performance of Ag-Decorated SiO_2 Nanoparticles (NPs) and Binding Ability between Ag NPs and Modifiers. *Coatings* **2022**, *12*, 146. [CrossRef]
61. Nguyen, T.N.; Nguyen, T.T.; Nghiem, T.H.L.; Nguyen, D.T.; Tran, T.T.H.; Vu, D.; Nguyen, T.B.N.; Nguyen, T.M.H.; Nguyen, V.T.; Nguyen, M.H. Optical Properties of Doxorubicin Hydrochloride Load and Release on Silica Nanoparticle Platform. *Molecules* **2021**, *26*, 3968. [CrossRef] [PubMed]
62. Yang, M.; Yang, W.; Chen, L.; Ding, M.; Li, C.; Shi, D. A Novel Synthesis of Fe_3O_4@SiO_2@Au@Porous SiO_2 Structure for NIR Irradiation-Induced DOX Release and Cancer Treatment. *Dose-Response Publ. Int. Hormesis Soc.* **2020**, *18*, 1559325820906662. [CrossRef]
63. Segall, M.D.; Lindan, P.J.D.; Probert, M.J.; Pickard, C.J.; Hasnip, P.J.; Clark, S.J.; Payne, M.C. First-Principles Simulation: Ideas, Illustrations and the CASTEP Code. *J. Phys. Condens. Matter* **2002**, *14*, 2717–2744. [CrossRef]
64. Grimme, S.; Antony, J.; Ehrlich, S.; Krieg, H. A Consistent and Accurate Ab Initio Parametrization of Density Functional Dispersion Correction (DFT-D) for the 94 Elements H-Pu. *J. Chem. Phys.* **2010**, *132*, 154104. [CrossRef] [PubMed]
65. Cai, W.; Guo, M.; Weng, X.; Zhang, W.; Owens, G.; Chen, Z. Modified Green Synthesis of Fe_3O_4@SiO_2 Nanoparticles for pH Responsive Drug Release. *Mater. Sci. Eng. C Mater. Biol. Appl.* **2020**, *112*, 110900. [CrossRef] [PubMed]
66. Mirzaei-Kalar, Z.; Kiani Nejad, Z.; Khandar, A.A. New $ZnFe_2O_4$@SiO_2@graphene Quantum Dots as an Effective Nanocarrier for Targeted DOX Delivery and CT-DNA Binder. *J. Mol. Liq.* **2022**, *363*, 119904. [CrossRef]
67. Trushina, D.B.; Sapach, A.Y.; Burachevskaia, O.A.; Medvedev, P.V.; Khmelenin, D.N.; Borodina, T.N.; Soldatov, M.A.; Butova, V.V. Doxorubicin-Loaded Core-Shell UiO-66@SiO_2 Metal-Organic Frameworks for Targeted Cellular Uptake and Cancer Treatment. *Pharmaceutics* **2022**, *14*, 1325. [CrossRef]
68. Li, J. Chidamide Enhances Cytotoxicity of Doxorubicin by Promoting Autophagy and Apoptosis in Breast Cancer. *BMC Cancer* **2023**, *23*, 353. [CrossRef]
69. Alioghli Ziaei, A.; Erfan-Niya, H.; Fathi, M.; Amiryaghoubi, N. In Situ Forming Alginate/Gelatin Hybrid Hydrogels Containing Doxorubicin Loaded Chitosan/AuNPs Nanogels for the Local Therapy of Breast Cancer. *Int. J. Biol. Macromol.* **2023**, *246*, 125640. [CrossRef]

inorganics

MDPI

Article

Electrical Features of Liquid Crystal Composition Doped with Cobalt Ferrite: Possible Sensing Applications

Yaroslav Barnash [1,2,3,4], Sonja Jovanović [5], Zoran Jovanović [5] and Natalia Kamanina [1,2,3,4,*]

1 Joint Stock Company Scientific and Production Corporation S.I. Vavilov State Optical Institute, Babushkona Str. 36/1, St. Petersburg 192171, Russia; barnash.yaroslaw@yandex.ru
2 Vavilov State Optical Institute, Kadetskaya Liniya V.O, 5/2, St. Petersburg 199053, Russia
3 Electronics Department, Kafedra Photonica, St. Petersburg Electrotechnical University ("LETI"), ul. Prof. Popova 5, St. Petersburg 197376, Russia
4 Petersburg Nuclear Physics Institute, National Research Center «Kurchatov Institute», 1 md. Orlova Roshcha, Gatchina 188300, Russia
5 Laboratory of Physics, Vinča Institute of Nuclear Sciences—National Institute of the Republic of Serbia, University of Belgrade, Vinca, 11351 Belgrade, Serbia; sonja.jovanovic@vin.bg.ac.rs (S.J.); zjovanovic@vin.bg.ac.rs (Z.J.)
* Correspondence: nvkamanina@mail.ru; Tel.: +7-(812)-327-00-95

Abstract: The effects of $CoFe_2O_4$ nanoparticles on the properties of an electro-optical liquid crystal cell based on the nematic composition of 4-Cyano-4′-pentylbiphenyl (5CB) under the influence of different forms of bias voltage were studied. Detailed results were established for the application of sinusoidal voltages with various frequencies and amplitudes. At the input signal, with a frequency of 500 kHz, a resonant current increase was obtained in the electrical circuit, followed by a decrease in the current with an increase in the frequency. This indicates the formation of a consistent oscillatory circuit. The quality factor of the nanoparticle system does not depend on the amplitude of the controlled voltage. Liquid crystal cells with constant quality can be used in a number of devices and technologies, including extended sensing devices, where stable electrical properties are required.

Keywords: liquid crystal; cobalt ferrite nanoparticles; nano-sensitization process; dynamic parameters; quality factor; specific electrical set-up

1. Introduction

Liquid crystal (LC) mesophase, especially nematic liquid crystals (NLCs) and the compositions based on them, is widely used in materials science and in different techniques. This is due to the rather easy process of LC mesophase control using thermal, mechanical, electrical and optical signals [1–5]. LC-based compositions are applied in the laser technique, in the set-ups for holographic recording and for information reading [6–10]. LC cells and modulators based on them can be used for the switching, modulation, conversion and limitation of radiation [11–15], in various photovoltaic and sensor devices [16–20] and in displays [21–25]. At the same time, the control of the electrical external action can be varied by modification of the shape of the electric pulse, its amplitude, its duration and the frequency of its repetition [26–40]. Thus, a wide range of parameters of the controlled electrical signal allows one to obtain unique optimized LC device properties. On the other hand, the recently used nanostructuring process for LC and polymers with different nanoparticles (NPs) makes it possible to change the polarization of the new organic composite and its order parameter, the refractive index, which also affects both the basic physical–chemical properties of conjugated organic composites, including LC ones, and, in particular, their performance [41–46].

In the present study, the effect of $CoFe_2O_4$ NPs on the properties of an LC cell based on the 5CB group under the influence of different forms of bias voltage were studied. Detailed results were established for the application of sinusoidal voltages with various frequencies and amplitudes. At the input signal, with a frequency of 500 kHz, a resonant current increase was obtained in the electrical circuit, followed by a decrease in the current with an increase in the frequency. Thus, if the resonant current is increased, it causes a decrease in LC cell impedance. This indicates the formation of a consistent oscillatory circuit. Under these conditions, the quality factor of the nanostructured LC system does not depend on the amplitude of the controlled voltage.

2. Results

The purpose of this study is to analyze the electrical response of an LC cell based on 5CB injected with $CoFe_2O_4$ nanoparticles when exposed to sinusoidal voltages of various amplitudes and frequencies. Special attention was paid to identifying resonant effects, changing the current properties of the system, and establishing dependencies that can serve as the basis for new types of the devices developed with controlled frequency properties.

In this study, two LC cells (pure and structured with $CoFe_2O_4$ NPs) based on the nematic composition 4-Cyano-4′-pentylbiphenyl (5CB) were manufactured. The structure of the manufactured cells is shown in Figure 1. The sandwich construction of an LC cell is presented, which is connected with two glass substrates (Crown glass K8), two conducting and orientation layers (where conducting ITO coating works as both the conducting and orienting layers) and LC layers inside this construction. The ITO layers are brought to the outer edges of the cell, where the contacts of the control voltage generator are connected to them. It should be noted once again that the ITO conducting layers also act as an orienting (alignment layer) for the LC molecules. The laser orienting deposition (LOD) technique has been used to achieve this aim, which is described in papers [47–49]. One cell was constructed by using pure LC, and the second cell was injected with $CoFe_2O_4$ nanoparticles.

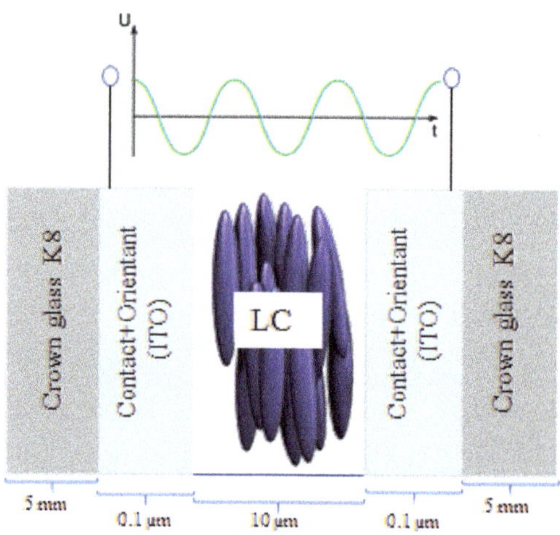

Figure 1. The structure of a liquid crystal cell doped with $CoFe_2O_4$ nanoparticles.

During the experiment, sinusoidal voltages of various amplitudes, including 1 V, 5 V and 10 V, were applied to both LC cells; the changes in the amplitude of the current in the circuit were recorded, with the quality factor measured based on the resonant properties. Figure 2 shows the resulting relationship. One can see an increase in the current with increasing frequency obtained at different applied voltage amplitudes for the pure and sensitized LC cells. It should be noted that the form of the applied voltage is shown in the middle part of Figure 2 as a sine wave.

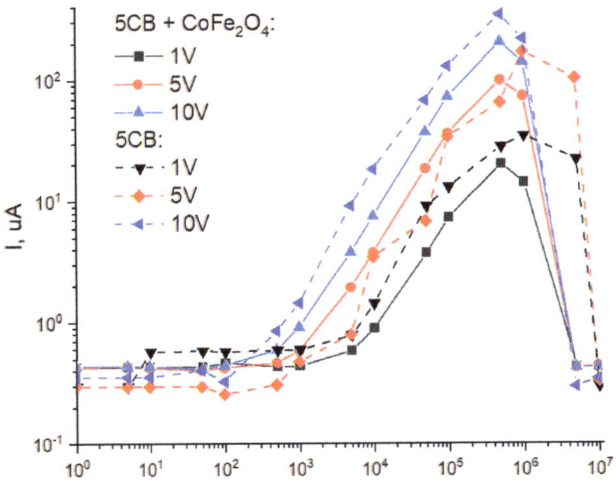

Figure 2. The dependence of the amplitude of the current strength on the frequency at different amplitudes of the applied sinusoidal voltage.

For a more detailed display of the changes, the quality factors of the systems were calculated. The calculated Q-factor values for a pure LC cell and a cell with the addition of magnetic nanoparticles at various amplitudes of the control voltage are shown in Table 1.

Table 1. The calculated Q-factor values of the systems at different values of control voltage amplitude.

Systems	The Amplitude of the Control Voltage, V		
	1	5	10
5CB	0.245	0.390	1.17
5CB + $CoFe_2O_4$	0.58	0.59	0.56

The Q-factor was calculated according to the above graph using the formula

$$Q = \frac{f_{res}}{\Delta f},$$

where f_{res} is the resonant frequency, Hz, and Δf is the frequency dependence at a current amplitude level of $\sqrt{2}$ times less than the resonant frequency, Hz.

Based on the obtained results the following can be concluded:

Resonance occurs for both pure 5CB and its sensitized version, while the resonant frequency of the system, measured when exposed to a sinusoidal voltage, remains unchanged at 500 kHz, as in the case of a pure LC cell and with the addition of $CoFe_2O_4$ nanoparticles. Although ferritic nanoparticles have magnetic properties, their concentration and effect on 5CB molecules are limited within local regions. Nanoparticles do not significantly affect the macroscopic inductance of the system, which could shift the resonant frequency.

Moreover, in a pure 5CB cell, the Q-factor increases with an increase in control voltage, since it lacks nanoparticles that could create local fields and additional magnetic losses. Cobalt ferrite nanoparticles injected into the cell create local fields and can interact with liquid crystal molecules, thus limiting their orientation in one direction. This leads to additional losses and a decrease in Q-factor, especially at high voltages, when the external electric field begins to compete with the local fields of the nanoparticles.

Furthermore, the quality factor of the nanoparticle system does not depend on the amplitude of the controlled voltage. The $CoFe_2O_4$ nanoparticles can partially shield an external electric field, creating local areas with a high field around themselves. These local fields stabilize the LC molecules and make the system less susceptible to changes in external voltage. Thus, regardless of the increase in the control voltage (in the studied range), the electrical response of the cell with nanoparticles remains at the same level, maintaining a constant quality factor.

We should note that the introduction of $CoFe_2O_4$ nanoparticles into the 5CB LC system potentially makes it possible to significantly change its electrical and optical properties, which opens up opportunities for creating more sensitive sensors and stable displays with improved properties [49]. In addition, the presence of ferrite nanoparticles can cause the effect of local electric and magnetic fields, which can affect the orientation and dynamics of the liquid crystal molecules, modifying both the electrical and optical responses of the LC cell. Indeed, this warrants the future study of these features.

3. Materials and Methods

We sensitized the composite materials based on the NLC structure with the addition of $CoFe_2O_4$ nanoparticles. LC cells were made based on the nematic composition 4-Cyano-4′-pentylbiphenyl (5CB), purchased from Alfa Aesar Co., Lancashire, England, LOT 10166530, No. B21856. To construct the LC cells, we used additional alignment layers. The conductive ITO coatings were treated with a CO_2 laser in order to make the gratings. These gratings permitted us to align the LC molecules with good accuracy.

$CoFe_2O_4$ nanoparticles were synthesized following the approach proposed in [50], providing non-agglomerated nanoparticles of a uniform shape, size and surface chemistry; $CoFe_2O_4$ nanoparticles were synthesized solvothermally. Using a typical synthesis method, NaOH (10 mmol) was dissolved in 2 mL of distilled water. After that, 10 mL of 1-pentanol was added, followed by the addition of oleic acid, which is used as a capping agent. Amounts of 2 mmol of iron nitrate and 1 mmol of cobalt nitrate were dissolved in 18 mL of distilled water, and aqueous solution was poured into the above solution and underwent vigorous stirring for 2 h. After that, the autoclave was placed into an oven at 180 °C for 8 h. Then, the obtained particles were separated from the liquid phase and washed three times by redispersing them in *n*-hexane and precipitating them with ethanol. Finally, the nanoparticles were redispersed in *n*-hexane, transferred to a watch glass and left to dry in air overnight. The prepared $CoFe_2O_4$ nanoparticles were non-agglomerated, uniform in size (5 ± 1 nm) and sphere-like in shape, covered with a monolayer of oleic acid [50].

Based on our previous experiments with different nanoparticles (the fullerenes C_{60} and C_{70}, carbon nanotubes, shungites, graphene oxides, WS_2 and MoS_2 nanotubes, quantum dots, etc.), we used approximately 0,1 wt.% of ferrite NPs in the LC matrix. This concentration permitted us to make a homogenous composition with increased polarizability. Thus, we predicted that the LC cells would have a high speed. Moreover, our previous study of this system showed that the introduction of cobalt ferrite NPs into the nematic LC has significant effects [46]. Based on the data presented in [46], it is possible to detect a tendency towards an increase in the refractive index of the LC medium when $CoFe_2O_4$ nanoparticles are introduced. This is of additional practical interest, since the possibility of the refractive index varying is an additional tool for the optical matching of LC media with orientation coatings and other functional layers in LC cells. Moreover, this material has a number of valuable properties, including a high level of the magnetic anisotropy, mechanical and chemical stability, and pronounced electrical properties.

In the present study, the 5CB and $CoFe_2O_4$ composite was examined under the conditions of different shapes, amplitudes, durations and repetition rates of the electric pulse.

It should be noted that the composition of LC 4′-Pentyl-4-biphenyl-carbonitrile was used as a classical LC matrix. $CoFe_2O_4$ nanoparticles were synthesized following the approach in Ref. [50], providing non-agglomerated nanoparticles of a uniform shape, size and surface chemistry. The concentration of nanoparticles injected into the LC composition was at the level of 0.1 wt.%. The thickness of the LC cells was 10 μm.

An experimental scheme for the current experiments is shown in Figure 3.

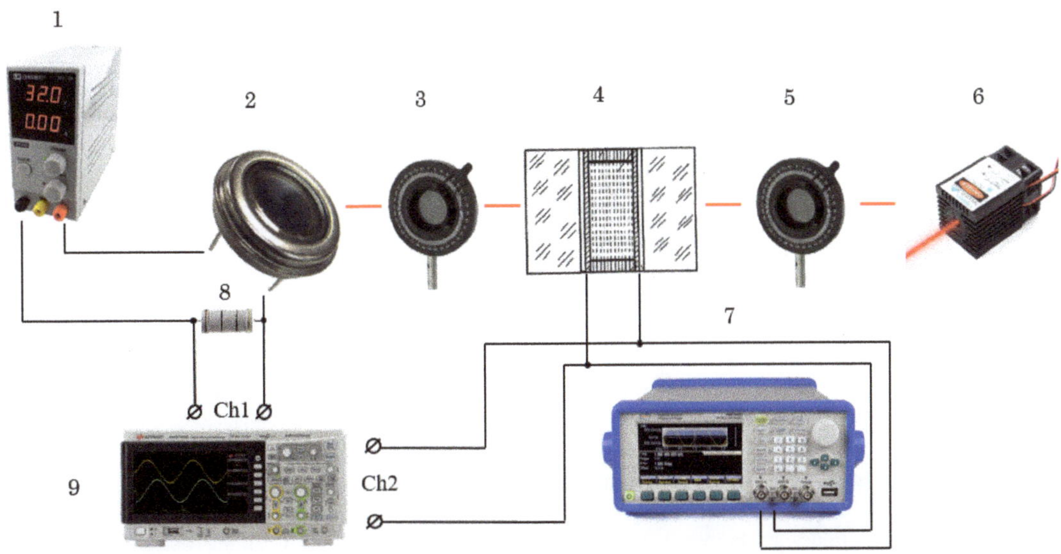

Figure 3. The set-up used: functional scheme. (1) Voltage source; (2) photodiode FD-26; (3) polarizer; (4) LC cell; (5) analyzer; (6) laser; (7) signal generator; (8) bias resistor; (9) oscilloscope.

The light source in the experimental set-up was a red semiconductor laser with a wavelength of λ = 633 nm and an output power of 200 mW (Model: LSM-SRA650-200; country of manufacture: China). The laser provided a stable and narrow beam, essential for passing through the liquid crystal cell. The laser was powered by a 12V constant voltage source, ensuring its stable operation. Neutral density filters were used to vary the laser intensity. The laser beam was linearly polarized using a polarizer placed before the liquid crystal cell. An analyzer was positioned after the LC cell at a perpendicular orientation relative to the polarizer. This configuration allowed for the detection of light intensity changes caused by the reorientation of the liquid crystal molecules under the influence of an electric field. It should be mentioned that both the thin-film polarizer (element 3 shown in the set-up) and analyzer (element 5 shown in the set-up) were developed in Prof. N. Kamanina's lab (Vavilov State Optical Institute, Saint-Petersburg, Russia) using their own patented technology [51].

The light transmitted through the analyzer was detected by a photodiode (model: FD-26; country of manufacture: Saint-Petersburg-Moscow, Russia, Key Specifications of Xibo XBI3005). The photodiode converted the optical signal into an electrical current proportional to the light intensity. A 12 V bias voltage, supplied by a Xibo XBI3005 power source, was applied through a load resistor. The resistor acted as a current-to-voltage converter, enabling the registration of the electrical signals corresponding to changes in the light intensity. These voltage signals were then analyzed using an oscilloscope. The control voltage was generated using a Tektronix AFG3011C signal generator (country of manufacture: Beaverton, OR, USA), which allowed for the selection of sinusoidal, triangular or rectangular waveforms. The generated signal was amplified using a non-inverting amplifier based on an operational amplifier. The amplifier ensured stable output signal amplitude, which was critical for ensuring the accuracy of the experiment. The system's response to the applied control voltage was recorded using a Rigol DS1102D oscilloscope (country of manufacture: Rigol Technologies, Inc. Headquarters, Suzhou, China) connected to the photodiode's load resistor. The oscilloscope captured changes in the electrical signal, reflecting variations in the light intensity at the cell's output. This set-up enabled real-time visualization and analysis of the system's dynamic response to the control voltage, including its temporal properties.

The developed set-up facilitates the determination of response times, including the device turn-on time and the relaxation time of the electro-optical response (device turn-off time).

Other instruments used: To study the spectral properties, a SF-26 spectrophotometer (USSR production) was used, operating in the wavelength range of 200–1200 nm. The calibrated filters were used to control the spectral measurements in the visible spectral range. The error in the measurements of the spectra was about 0.2%. The spectra in the IR range were diagnosed on an FSM-1211 device (country of manufacture: Infraspec Co., Saint-Petersburg, Russia).

4. Conclusions

It should be noted that constant-quality liquid crystal cells, including those with the addition of cobalt ferrite nanoparticles, can be used in a number of devices and technologies where stable electrical properties are critically important. For example, in radio frequency devices, such as filters, frequency stabilizers and modulators, the use of the constant-quality cells can ensure stable operation at different signal power levels. This is especially important in telecommunications, where the frequency properties must remain stable regardless of changes in the input signal power.

It is worth noting that these LC cells worked according to Frederick's scheme, taking into account the installation of the thin-film polarizers and analyzers based on a polyvinyl–alcohol matrix (PVA). It should be mentioned that Frederick's scheme has always been used by scientists working with LC materials. This scheme permits us to measure the time-on and time-off parameters, as well as the contact.

So, in terms of identifying the novelty of this work, it is worth saying that both the geometric factors and the physical processes of the functioning of the sensitized LC cell were taken into account to optimize its operation, but a different pulse shape of the control signal was also used. At the same time, this case was shown to yield the best result precisely when controlling the waveform as a sine wave.

Author Contributions: Software, investigation, visualization, Y.B.; nanoparticle synthesis, Z.J. and S.J.; conceptualization, formal analysis, writing—review and editing, supervision, N.K. All authors have read and agreed to the published version of the manuscript.

Funding: This research was partially supported (in terms of polarizer and analyzer synthesis and use) by the Russian Science Foundation, grant number 24-23-00021 (2024–2025), (https://rscf.ru/prjcard_int?24-23-00021, accessed on 13 March 2025). S.J. and Z.J. acknowledge the financial support from the Ministry of Science, Technological Development and Innovation (MoSTI) of the Republic of Serbia (Grant no. 451-03-66/2024-03/200017).

Data Availability Statement: The original data about the ITO conducting layer used as the orienting one in the study are openly available at: Russian Patent No. 2355001 (RU 2 355 001 C2). "Optical coating based on carbon nanotubes for optical instrumentation and nanoelectronics"; priority dated 9 January 2007; registered in the State Register of Inventions of the Russian Federation on 10 May 2009. Authors: N.V. Kamanina, P.Y. Vasiliev.

Acknowledgments: The authors would like to thank their colleagues from Vavilov State Optical Institute for the seminar discussions. S.J. and Z.J. acknowledge the financial support from the Ministry of Science, Technological Development and Innovation (MoSTI) of the Republic of Serbia (Grant no. 451-03-66/2024-03/200017). The current research was funded by the Russian Scientific Fund, project No. 24-23-00021, (2024–2025), (https://rscf.ru/prjcard_int?24-23-00021 (accessed on 28 November 2023)).

Conflicts of Interest: Authors Yaroslav Barnash and Natalia Kamanina are employed by Joint Stock Company Scientific and Production Corporation S.I. Vavilov State Optical Institute. The authors of the paper declare that the research was conducted in the absence of any commercial or financial relationships that could be construed as a potential conflict of interest. The affiliated company was not involved in the study design, collection, analysis, interpretation of data, the writing of this article or the decision to submit it for publication.

References

1. Vasil'ev, A.A.; Kasacent, D.; Kompanets, I.N.; Parfenov, A.V. *Spatial Light Modulators;* Radio & Svyaz': Moscow, Russia, 1987; 320p. (In Russian)
2. McEwen, R.S. Liquid crystals, displays and devices for optical processing. *J. Phys. B Sci. Instrum.* **1987**, *20*, 364–377. [CrossRef]
3. Simoni, F.; Cipparrone, G.; Umeton, C.; Arabia, G.; Chidichimo, G. Optical nonlinearities induced by thermal effects in polymer dispersed liquid crystals. *Appl. Phys. Lett.* **1989**, *54*, 896–897. [CrossRef]
4. Khoo, I.C.; Wood, M.V.; Shih, M.Y.; Chen, P.H. Extremely nonlinear photosensitive liquid crystals for image sensing and sensor protection. *Opt. Express* **1999**, *4*, 432–442. [CrossRef] [PubMed]
5. Prasad, P.N. *Nanophotonics;* John Wiley & Sons: Hoboken, NJ, USA, 2004; 432p.
6. Hur, S.-T.; Lee, B.R.; Gim, M.-J.; Park, K.-W.; Song, M.-H.; Choi, S.-W. Liquid-crystalline blue phase laser with widely tunable wavelength. *Adv. Mater.* **2013**, *25*, 3002–3006. [CrossRef]
7. Khoo, I.C. Holographic grating formation in dye- and fullerene C60-doped nematic liquid-crystal film. *Opt. Lett.* **1995**, *20*, 2137–2139. [CrossRef]
8. Lee, Y.-H.; Yin, K.; Wu, S.-T. Reflective polarization volume gratings for high efficiency waveguide-coupling augmented reality displays. *Opt. Express* **2017**, *25*, 27008–27014. [CrossRef]
9. Cho, S.Y.; Yoshida, H.; Ozaki, M. Tunable polarization volume gratings based on blue phase liquid crystals. *Opt. Express* **2022**, *30*, 1607–1614. [CrossRef]
10. Suzuki, M.; Ohara, S.; Sakamoto, M.; Noda, K.; Sasaki, T.; Kawatsuki, N.; Ono, H. Temperature dependence of the diffraction efficiency of circular polarization gratings made by liquid crystal molecules with anisotropic absorption. *Opt. Mater. Express* **2024**, *14*, 1857–1864. [CrossRef]
11. Wahle, M.; Kasdorf, O.; Kitzerow, H.-S.; Liang, Y.; Feng, X. Electrooptic Switching in Graphene-Based Liquid Crystal Cells. *Mol. Cryst. Liq. Cryst.* **2011**, *543*, 187–193. [CrossRef]
12. Zhang, Y.; Ren, Y.; Chen, J.; Hu, Z.; Bai, J.; Zhao, L.; Wang, K. Fast testing of partial camera lenses based on a liquid crystal spatial light modulator. *Appl. Opt.* **2022**, *61*, 6420–6429. Available online: https://opg.optica.org/ao/abstract.cfm?URI=ao-61-22-6420 (accessed on 13 March 2025). [CrossRef]
13. Hadjichristov, G.B. Control of Coherent Light Through Microperiodic Director Modulation in Nematic Films Under Low-Voltage DC Electric Field. *Materials* **2023**, *16*, 6014. [CrossRef] [PubMed]
14. Zhang, X.; Li, K. Phase-Only Liquid-Crystal-on-Silicon Spatial-Light-Modulator Uniformity Measurement with Improved Classical Polarimetric Method. *Crystals* **2023**, *13*, 958. [CrossRef]
15. Zhang, R.; Zhang, Z.; Han, J.; Yang, L.; Li, J.; Song, Z.; Wang, T.; Zhu, J. Advanced liquid crystal-based switchable optical devices for light protection applications: Principles and strategies. *Light Sci. Appl.* **2023**, *12*, 11. [CrossRef] [PubMed]
16. Wu, G.; Hsieh, L.-H.; Chien, H.-W. Enhanced Solar Cell Conversion Efficiency Using Birefringent Liquid Crystal Polymer Homeotropic Films from Reactive Mesogens. *Int. J. Mol. Sci.* **2013**, *14*, 21319–21327. [CrossRef]
17. Zhou, D.; Wang, Q.; Lan, Z.; Chen, Y.; Peng, Z.; Zhang, L.; Liu, Y. Liquid-crystal-based fiber laser sensor for non-invasive gas detection. *Opt. Lett.* **2023**, *48*, 4508–4511. Available online: https://opg.optica.org/ol/abstract.cfm?URI=ol-48-17-4508 (accessed on 13 March 2025). [CrossRef]
18. Shadkami, R.; Chan, P.K. A Numerical Study on the Performance of Liquid Crystal Biosensor Microdroplets. *Crystals* **2023**, *13*, 1237. [CrossRef]

19. Zhang, R.; Du, W.; Shao, F.; Li, S.; Kuai, Y.; Cao, Z.; Xu, F.; Liu, Y.; Luo, Y.; Peng, G.-D.; et al. Voltage, thermal and magnetic field fiber sensors based on magnetic nanoparticles-doped photonic liquid crystal fibers. *Opt. Express* **2023**, *31*, 25372–25384. [CrossRef]
20. Wu, G. Patterned Liquid Crystal Polymer Thin Films Improved Energy Conversion Efficiency at High Incident Angles for Photovoltaic Cells. *Polymers* **2024**, *16*, 1358. [CrossRef]
21. Kudreyko, A.; Chigrinov, V.; Neyts, K.; Chausov, D.; Perestoronina, A. Photonic Devices with Multi-Domain Liquid Crystal Structures. *Crystals* **2024**, *14*, 512. [CrossRef]
22. Chigrinov, V.G.; Kudreyko, A.A.; Podgornov, F. Optically rewritable liquid crystal displays: Properties and performance. *Crystals* **2021**, *11*, 1053. [CrossRef]
23. Jhong, G.C.; Chen, T.J.; Lin, Y.-T.; Wu, J.-J.; Yang, Y.J. Effect of chiral dopant on electro-optical properties of nematic liquid crystal cells under in-plane switching and non-uniform vertical electric fields. *Opt. Mater. Express* **2014**, *4*, 2468–2477.
24. Liu, J.; Weng, C. Tuning Electro-Optical Properties Through Polymerization Monomer Content in PSVA Liquid Crystal Displays: Simulation and Experimentation. *Polymers* **2024**, *16*, 1606. [CrossRef]
25. Sang, J.; Lai, Y.; Sun, J.; Shang, J.; Zhao, S. A Mesoporous Silica Nanoparticle-Doped Photo-Alignment Layer and Liquid Crystal Layer for Optimizing the Rewriting Speed and the Response Time of Optically Driving Liquid Crystal Displays. *Crystals* **2022**, *12*, 1088. [CrossRef]
26. Soref, R.A.; Rafuse, M.J. Electrically controlled birefringence of thin nematic films. *Appl. Phys.* **1972**, *43*, 2029–2036.
27. Casasent, D. Performance Evaluation of Spatial Light Modulators. *Appl. Opt.* **1979**, *18*, 2445–2453. [PubMed]
28. Montgomery, G.P. Optical properties of a liquid crystal image transducer at normal incidence: Mathematical analysis and applications to the off-state. *J. Opt. Soc. Am.* **1980**, *70*, 287–300.
29. Montgomery, G.P. Optical properties of a liquid crystal image transducer at normal incidence: Mathematical analysis and applications to the on-state. *J. Opt. Soc. Am.* **1980**, *70*, 843–856.
30. Macfaden, A.J.; Wilkinson, T.D. Characterization, design, and optimization of a two-pass twisted nematic liquid crystal spatial light modulator system for arbitrary complex modulation. *J. Opt. Soc. Am. A* **2017**, *34*, 161–170. [CrossRef]
31. Zhou, L.; Liu, S. Development and Prospect of Viewing Angle Switchable Liquid Crystal Devices. *Crystals* **2022**, *12*, 1347. [CrossRef]
32. Cai, W.; Cheng, M.; Kong, D.; Ma, Z.; Liu, Y.J. Switchable liquid crystal lenticular microlens arrays based on photopolymerization-induced phase separation for 2D/3D autostereoscopic displays. *Opt. Express* **2024**, *32*, 625–638. [CrossRef]
33. Kovalchuk, O.V.; Kovalchuk, T.M.; Garbovskiy, Y. Eliminating Ambiguities in Electrical Measurements of Advanced Liquid Crystal Materials. *Crystals* **2023**, *13*, 1093. [CrossRef]
34. Zid, M.; Cordoyiannis, G.; Kutnjak, Z.; Kralj, S. Criticality Controlling Mechanisms in Nematic Liquid Crystals. *Nanomaterials* **2024**, *14*, 320. [CrossRef] [PubMed]
35. An, C.-H.; Oh, S.-W. Review of Angular-Selective Windows with Guest–Host Liquid Crystals for Static Window Applications. *Crystals* **2024**, *14*, 931. [CrossRef]
36. Feng, W. Frequency-stable liquid crystal lenses using a circumferentially segmented concentric electrode. *Opt. Express* **2024**, *32*, 48315–48328. [CrossRef]
37. Bag, D.; Chakraborty, S.; Sinha, A. Nanoscale surface metrology with a liquid crystal-based phase-shifting angular shearing interferometer. *Opt. Lett.* **2024**, *49*, 1705–1708.
38. Gasparyan, P.K.; Margaryan, H.L.; Abrahamyan, V.K.; Sargsyan, T.K.; Hakobyan, N.H. Switchable anisotropic diffraction mode filter based on fork-like liquid crystal gratings. *Appl. Opt.* **2024**, *63*, 8712–8716. [CrossRef]
39. Voronov, A.A.; Bachiller, C.; Villacampa, B.; Boria, V.E. Development of Reconfigurable High-Frequency Devices Using Liquid Crystal in Substrate-Integrated Gap Waveguide Technology. *Crystals* **2024**, *14*, 735. [CrossRef]
40. Wang, Y.-M.; Fan, F.; Zhao, H.-J.; Liu, J.; Ji, Y.-Y.; Cheng, J.-R.; Chang, S.-J. Active broadband unidirectional focusing of terahertz surface plasmons based on a liquid-crystal-integrated on-chip metadevice. *Photonics Res.* **2024**, *12*, 2148–2157. [CrossRef]
41. Kamanina, N.V. Fullerene-dispersed liquid crystal structure: Dynamic properties and self-organization processes. *Physics-Uspekhi* **2005**, *48*, 419–427. [CrossRef]
42. Kamanina, N.V. New optical effects in liquid crystals: Self-organization and dynamic properties of fullerene-doped nematic mesophase. *Proc. SPIE* **2005**, *5947*, 594704. [CrossRef]
43. Kamanina, N.V.; Komolkin, A.V.; Yevlampieva, N.P. Variation of the Orientational Order Parameter in a Nematic Liquid Crystal–COANP–C_{70} Composite Structure. *Tech. Phys. Lett.* **2005**, *31*, 478–480. [CrossRef]
44. Kamanina, N.V.; Zubtsova, Y.A.; Toikka, A.S.; Likhomanova, S.V.; Zak, A.; Tenne, R. Temporal properties of liquid crystal cell with WS_2 nanoparticles: Mesophase sensitization and relief features. *Liq. Cryst. Their Appl.* **2020**, *20*, 34–40. Available online: http://nano.ivanovo.ac.ru/journal/ru/articles/article.php?year=2020&issue=1&first_page=34 (accessed on 31 March 2020). [CrossRef]
45. Kamanina, N. Refractive Properties of Conjugated Organic Materials Doped with Fullerenes and Other Carbon-Based Nano-Objects. *Polymers* **2023**, *15*, 2819. [CrossRef] [PubMed]

46. Kamanina, N.V.; Toikka, A.S.; Barnash, Y.V.; Redka, D.N.; Lihkomanova, S.V.; Zybtsova, Y.A.; Kyzhakov, P.V.; Jovanovic, Z.; Jovanovic, S. Functioning features of liquid crystalline cells doped with $CoFe_2O_4$ nanoparticles. *Liq. Cryst. Their Appl.* **2022**, *22*, 66–74. [CrossRef]
47. Kamanina, N.; Kukharchik, A.; Kuzhakov, P.; Vasilyev, P. Nano-structured conducting layers based on ITO and their modified properties. In Proceedings of the 14th Israeli Russian Bi-National Workshop "The Optimization of the Composition, Structure and Properties of Metals, Oxides, Composites, Nano and Amorphous Materials", Ariel, Israel, 12–16 July 2015; pp. 115–118.
48. Toikka, A.; Ilin, M.; Kamanina, N. Perspective Coatings Based on Structured Conducting ITO Thin Films for General Optoelectronic Applications. *Coatings* **2024**, *14*, 178. [CrossRef]
49. Toikka, A.S.; Kamanina, N.V. Surfaces modifications in functional layers for liquid crystal devices. *St. Petersburg Polytech. Univ. J. Phys. Math.* **2024**, *17*, 262–265. [CrossRef]
50. Jovanović, S.; Spreitzer, M.; Tramšek, M.; Trontelj, Z.; Suvorov, D. Effect of Oleic Acid Concentration on the Physicochemical Properties of Cobalt Ferrite Nanoparticles. *J. Phys. Chem. C* **2014**, *118*, 13844–13856. [CrossRef]
51. Kamanina, N.V.; Vasiliev, P.Y.; Studenov, V.I. Polarizing Films for the Visible Range of the Spectrum with a Nanostructured Surface Based on Carbon Nanoparticles. Russian Patent No. RU 2426157 C1, 10 August 2011.

inorganics

MDPI

Article

The Coumarin-Based Silver(I) Complex Showed Enhanced Antitumor and Antimicrobial Activity than Ligand Itself †

Jakub Kurjan [1], Zuzana Jendželovská [2], Viktória Dečmanová [2], Mária Vilková [3], Katarina Ćirković [4], Ivana Radojević [4], Miroslava Litecká [5], Rastislav Jendželovský [2] and Ivan Potočňák [1,*]

1 Department of Inorganic Chemistry, Faculty of Science, Pavol Jozef Šafárik University in Košice, Moyzesova 11, 041 54 Košice, Slovakia; jakub.kurjan@student.upjs.sk

2 Department of Cell Biology, Faculty of Science, Pavol Jozef Šafárik University in Košice, Šrobárova 2, 041 54 Košice, Slovakia; zuzana.jendzelovska@upjs.sk (Z.J.); viktoria.decmanova@upjs.sk (V.D.); rastislav.jendzelovsky@upjs.sk (R.J.)

3 NMR Laboratory, Faculty of Science, Pavol Jozef Šafárik University in Košice, Moyzesova 11, 041 54 Košice, Slovakia; maria.vilkova@upjs.sk

4 Department of Biology and Ecology, Faculty of Science, University of Kragujevac, Radoje Domanovića 12, 34000 Kragujevac, Serbia; katarina.cirkovic@pmf.kg.ac.rs (K.Ć.); ivana.radojevic@pmf.kg.ac.rs (I.R.)

5 Department of Materials Chemistry, Institute of Inorganic Chemistry of the Czech Academy of Sciences, Husinec-Řež 1001, 250 68 Řež, Czech Republic; litecka@iic.cas.cz

* Correspondence: ivan.potocnak@upjs.sk

† Published in the series "Low-dimensional compounds containing bioactive ligands" as Part XXIV.

Abstract: In this study, a novel silver(I) complex $[Ag(HL^1)_2]NO_3$ (**AgHL¹**) with coumarin derivative (3*E*)-3-(1-{[(pyridin-2-yl)methyl]amino}ethylidene)-3,4-dihydro-2*H*-benzopyran-2,4-dione (**HL¹**) was prepared. The compounds **HL¹** and **AgHL¹** were characterized by IR and NMR spectroscopy, elemental analysis, and single crystal X-ray structural analysis. Specifically, the single crystal X-ray analysis determined the structures of both compounds **HL¹** and **AgHL¹** in their solid state, while NMR spectroscopy was used for structural determination in a solution. The **HL¹** proved to be a monodentate ligand and is coordinated to the Ag(I) atom through a nitrogen atom from the 2-picolylamine fragment. In the complex **AgHL¹**, two molecules of neutral **HL¹** are coordinated forming a nearly linear N-Ag-N arrangement. An uncoordinated nitrate anion balances the positive charge of the complex cation. NMR spectroscopy also confirmed the stability of **AgHL¹** in DMSO-d_6 for 3 days. In vitro cytotoxicity of **HL¹** and **AgHL¹** was performed over two cancerous cell lines A549 and HT-29 and their selectivity was verified on a healthy CCD-18Co cell line. **AgHL¹** exhibited low anticancer nonselective activity while the ligand was inactive. Also, the complex shows better antimicrobial activity than the positive controls on the *Pseudomonas aeruginosa* standard and clinical strain as well as on the tested molds.

Keywords: coumarin; silver complex; anticancer activity; antimicrobial activity; crystal structure

1. Introduction

Metal-based drugs have been successfully used in the treatment of various diseases [1]. The discovery of cisplatin represented a milestone in the treatment of cancer and led to a deeper investigation of the therapeutic potential of metallodrugs. Nowadays, platinum compounds are effective in the treatment of various types of cancer [2]. Ferrochloroquine is used as an antimalarial agent [3], gold compounds treat rheumatoid arthritis, bismuth is known for antiulcer drugs [4], and compounds of gadolinium are used as MRI contrast agents [5]. Silver is one of the metal ions that are used clinically, too. The use of silver as an antimicrobial agent has been known since 1000 BC [6]. Silver nitrate and silver sulfadiazine were widely used in the treatment of wounds [7,8]. Nowadays, silver sulfadiazine is the most frequently used prophylactic agent in burn patients. It exhibits in vitro activity against microorganisms including *S. aureus*, *E. coli*, *Klebsiella* species, *Pseudomonas aeruginosa*, *Proteus* species, and *Candida albicans* [9]. However, there are some limitations in the usage of these compounds. The effect of these drugs is not long-term, so there is a risk of reinfection of the wound. Also, the development of resistance of some organisms is one of the limitations [10]. In addition to antimicrobial activity, silver complexes play an important role as antitumor agents with various mechanisms including DNA binding and cleavage, topoisomerase enzyme inhibition, generation of ROS, and induction of apoptosis [11].

To overcome problems like resistance, poor solubility, and severe side effects, a new strategy, seems to be, designing metal complexes with ligands that already have some biological activity. Therefore, N-heterocyclic ligands have been used in preparing anticancer agents [12]. Another widely used ligand is coumarin and its derivatives. Many derivatives of coumarin are being investigated as possible drugs with strong pharmacological activity, low toxicity, and side effects or with reduced resistance to these substances [13]. Out of the many studied effects, there can be mentioned anticoagulant [14], antimicrobial [15], anticancer [16], anti-inflammatory [17], or anti-HIV effect [18].

Thati et al. [19] investigated the antiproliferative effect of 4-hydroxy-3-nitro-coumarin and the mixed-ligand silver(I) complex of 4-oxy-3-nitro-coumarin-bis(phenanthroline) on four human-derived cell lines. The results showed a decrease in the proliferation of neoplastic cell lines of renal cancer A-498 and hepatic carcinoma HepG2, while non-neoplastic cell line Chang appeared to be less sensitive to the effect of the complex. Furthermore, regarding the results of IC_{50} values, the investigated complex was almost four times more potent than cisplatin, using HepG2 cell line. In another study, Mujahid et al. [20] prepared a series of silver(I) complexes with coumarin-based ligands and one of their phenanthroline adducts. The cytotoxicity of compounds was also studied against HepG2 and A-498 cancer cell lines. Tests showed that complexes were more cytotoxic than clinically used chemotherapeutic mitoxantrone. Also, it seems that complexes have quite different mechanisms of action to many metal-based therapeutics, due to excellent superoxide dismutase activity instead of interaction with DNA.

Considering the above-stated information, the main objective of the present manuscript is to investigate the antimicrobial and anticancer activity of a novel silver(I) complex $[Ag(HL^1)_2]NO_3$ (**AgHL^1**) with the ligand (3*E*)-3-(1-{[(pyridin-2-yl)methyl]amino}ethylidene)-3,4-dihydro-2*H*-benzopyran-2,4-dione (**HL^1**). The compound **HL^1** has been previously synthesized by Brawley et al. [21] who studied its myosin inhibiting activity; however, its antimicrobial and anticancer properties as well as its crystal structure remain unexplored. Furthermore, our study compares the biological activities of the ligand **HL^1** itself with that of the complex **AgHL^1**.

Given that the ligand has been synthesized before, our subsequent research phases focused on preparing halogen-modified **HL^1** (F, Cl, and Br) and the formation of complexes with different metal ions with the aim of enabling an exploration of the correlation between structural modifications and the biological activity of the modified ligand **HL^1** and diverse metal-containing complexes. However, all our attempts to synthesize Ag complexes with these ligands failed.

Therefore, in this manuscript, we only expound on the synthesis of the complex **AgHL^1** and its characterization (structure and stability). The structure of **HL^1** and **AgHL^1** in the solid state is investigated by single crystal X-ray structure analysis. The NMR spectroscopy is used to determine the structure of the ligand **HL^1** and complex **AgHL^1** in the solution and in the investigation of the stability of the complex **AgHL^1**.

Additionally, their anticancer activity was assessed using two cancerous cell lines: A549 (lung carcinoma) and HT-29 (colorectal carcinoma), and their selectivity was confirmed on a healthy cell line, CCD-18Co (normal colon tissue cell line). According to statistical data from 2020, colorectal and lung carcinoma were the predominant cancer types diagnosed in males in Slovakia. In females, colorectal carcinoma was the second most prevalent, superseded only by breast cancer. Consequently, the selection of cell lines for testing was informed by these statistical insights [22]. Moreover, it is widely acknowledged that silver(I) and its compounds possess antimicrobial activity [23]. Therefore, the antimicrobial potency of both compounds was tested against 15 microorganisms, encompassing bacteria and fungi, with results benchmarked against the standards doxycycline and fluconazole, respectively, and $AgNO_3$.

2. Results and Discussion

2.1. Synthesis

In this work we prepared the novel silver(I) complex $[Ag(HL^1)_2]NO_3$ (**AgHL^1**) with coumarin derivative (3*E*)-3-(1-{[(pyridin-2-yl)methyl]amino}ethylidene)-3,4-dihydro-2*H*-benzopyran-2,4-dione (**HL^1**). The ligand was synthesized through a two-step process. Initially, acetylation of 4-hydroxy-2*H*-chromen-2-one was carried out using acetic acid in the presence of phosphorus oxychloride [24]. Subsequently, a condensation reaction between 3-acetyl-4-hydroxy-2*H*-chromen-2-one and 2-picolylamine was taken in methanol to yield the final product, in accordance with the established literature [21,25]. The reaction was performed under reflux conditions, following which the reaction mixture was cooled in a refrigerator for approximately 5 h. A white solid product was then isolated by filtration and air dried. To obtain crystals for X-ray analysis, the ligand was recrystallized from methanol at 24 °C and the identity of crystalline and powder was proved by comparison of their IR spectra. The ligand was then fully characterized by NMR spectroscopy, elemental analysis, and X-ray analysis. The purity of the ligand (from the reflux) was checked by NMR spectroscopy, so the ligand was used in the synthesis of the complex without another purification.

The complex **AgHL^1** was synthesized by mixing an aqueous solution of $AgNO_3$ and an ethanolic solution of **HL^1** in a molar ratio of 1:2. As the ligand is not completely soluble in ethanol at room temperature, the solution of the ligand was heated up to about 50 °C and stirred on a magnetic stirrer until complete dissolution of the ligand. Then, aqueous solution of the $AgNO_3$ was added. The beaker was wrapped in aluminum foil to prevent light from entering. The reaction mixture was cooled and allowed to crystallize in the dark at a temperature of 24 °C for approximately 2 weeks to give the beige **AgHL^1** crystals. Crystals were stored in the dark at room temperature.

As mentioned in the introduction, halogen derivatives of the ligand **HL^1** were prepared [26]; however, the attempt to synthesize Ag complexes with these ligands failed.

2.2. IR and NMR Spectra

The IR spectra (Figure S5) of the ligand and the complex are similar and comprise stretching vibrations at 3000–2800 (C-H aromatic and aliphatic), 1610–1560 (C=C aromatic), and 1400–1300 cm^{-1} (C-O). However, there are some differences in the spectra. There is a slight shift up of the lactone C=O (coumarin) band from 1689 (**HL^1**) to 1698 cm^{-1} (**AgHL^1**). C-N vibrations of pyridine ring change from 1335 and 1301 cm^{-1} (**HL^1**) to 1337, 1359 and 1287 cm^{-1} (**AgHL^1**), indicating coordination through nitrogen atom originating in the pyridine ring of the ligand, which is confirmed by X-ray analysis. Also, pyridine ring deformation bands are slightly shifted from 779 and 753 cm^{-1} (**HL^1**) to 776 and 751 cm^{-1}

(**AgHL¹**). In addition, in the IR spectrum of the complex there is a new band at 431 cm^{-1}, which could represent Ag-N interaction.

The ligand **HL¹** and the complex **AgHL¹** were fully characterized by NMR spectroscopy. The ^{1}H and ^{13}C NMR (Figure S1) assignments were determined using splitting patterns, chemical shifts, and 2D NMR data from COSY, HSQC, and HMBC (Figure S2) spectra. The ^{1}H and ^{13}C NMR data for the ligand **HL¹** and the complex **AgHL¹** in DMSO-d_6 are summarized in Table 1. DMSO-d_6 was chosen as the solvent due to the poor solubility of **HL¹** and **AgHL¹** in non-coordinating solvents. The analysis of NMR data for both samples demonstrates a minimal effect of the coordinated Ag atom on the ligand's proton and carbon chemical shifts [27] (Table 1). The interaction between the silver atom and the ligand may not have been sufficiently strong to induce significant alterations in the ligand's electronic environment, as evidenced by cases where the NMR chemical shifts in ligands in complexes closely approximate those observed in their unbound state [28,29]. However, the X-ray crystal structure analysis confirmed the structure of the complex **AgHL¹**. Furthermore, the ^{15}N-HMBC NMR verified the coordination of N1′ atom to the central atom. The chemical shift for the N1′ atom is −71.8 ppm in the unbound ligand and −72.9 ppm in the **AgHL¹** complex (Figures S3 and S4).

Table 1. ^{1}H and ^{13}C NMR chemical shifts (δ, ppm) of the ligand **HL¹** and the complex **AgHL¹** together with structure of **HL¹** showing atom numbering used for NMR assignment.

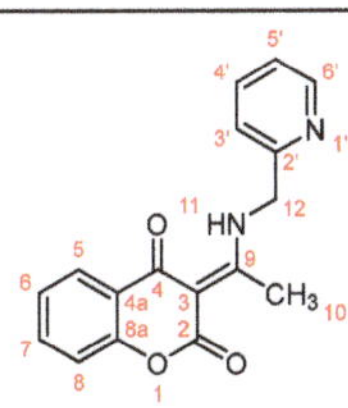

Atom	HL¹ δ$_H$, mult. (*J*, Hz)	HL¹ δ$_C$	AgHL¹ δ$_H$, mult. (*J*, Hz)	AgHL¹ δ$_C$
2	-	162.0	-	162.0
3	-	96.4	-	96.4
4	-	179.5	-	179.5
4a	-	120.3	-	120.3
5	7.96 (dd, 7.8, 1.7)	125.7	7.96 (dd, 7.8, 1.5)	125.7
6	7.29 (ddd, 8.1, 7.3, 1.0)	123.7	7.29 (ddd, 8.1, 7.2, 1.1)	123.6
7	7.63 (ddd, 8.2, 7.2, 1.7)	134.0	7.63 (ddd, 8.2, 7.2, 1.7)	134.0
8	7.26 (dd, 8.2, 1.1)	116.2	7.26 (dd, 8.2, 1.0)	116.2
8a	-	153.1	-	153.1
9	-	176.2	-	176.2
10	2.75 (s)	19.0	2.75 (s)	19.0
11	14.14 (s)	-	14.14 (s)	-
12	5.04 (d, 5.1)	48.7	5.04 (d, 5.2)	48.8
2′	-	154.6	-	154.6
3′	7.47 (d, 7.9)	122.2	7.48 (dt, 7.8, 1.0)	122.2
4′	7.87 (td, 7.7, 1.8)	137.3	7.87 (td, 7.8, 1.8)	137.4
5′	7.39 (ddd, 7.5, 4.8, 1.1)	123.1	7.40 (ddd, 7.6, 4.8, 1.0)	123.1
6′	8.64 (ddd, 4.8, 1.8, 0.9)	149.3	8.64 (ddd, 4.9, 1.8, 1.0)	149.4

Notably, in ^{1}H NMR spectra of both the free ligand and the complex show a downfield shift for the proton of the NH11 group upon formation of an H-bond with the oxygen atom of the carbonyl group C4=O. This shift to higher frequency, indicative of H-bond formation, results from a decrease in diamagnetic shielding due to a reduced electron density near the hydrogen [30]. The formation of the H-bond was further confirmed by X-ray crystal structure analysis.

The stability of the complex was studied by measuring ^{1}H NMR at different time intervals (Figure 1). The ^{1}H NMR spectra exhibited sharp and well-resolved signals, indicating stable and consistent coordination of ligand donor atoms to the silver center at 298 K.

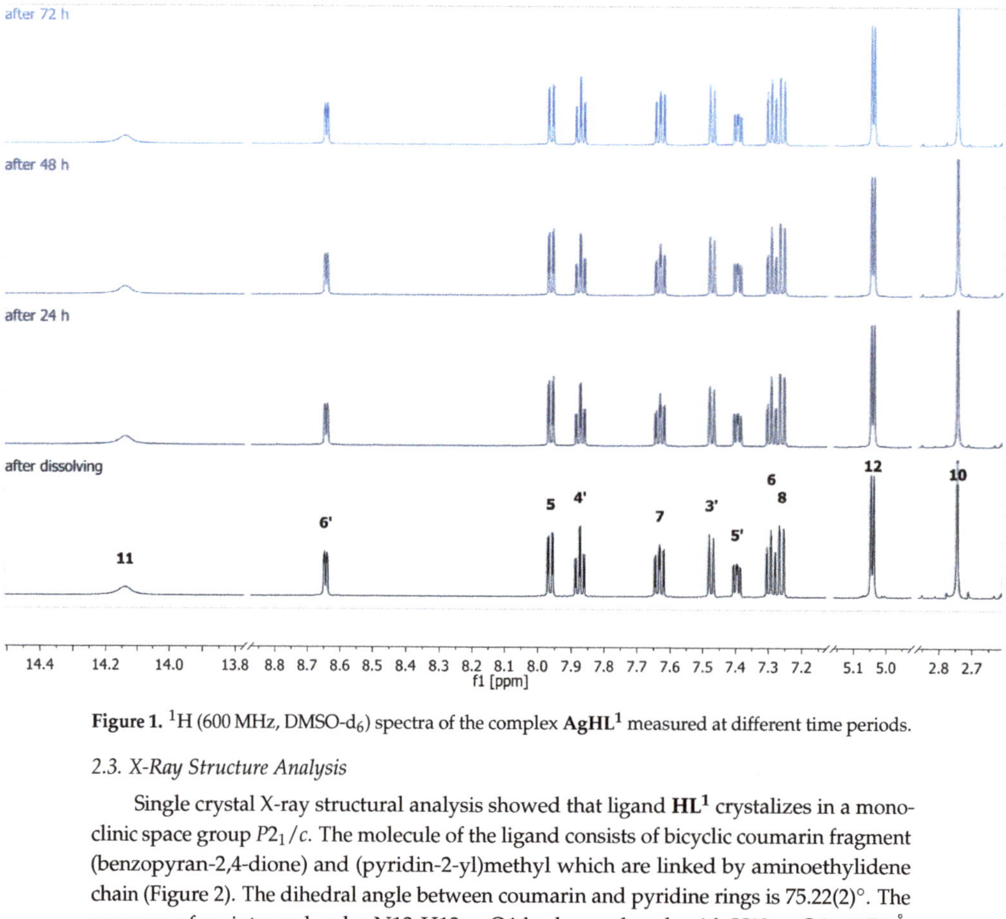

Figure 1. ^{1}H (600 MHz, DMSO-d_6) spectra of the complex **AgHL^1** measured at different time periods.

2.3. X-Ray Structure Analysis

Single crystal X-ray structural analysis showed that ligand **HL^1** crystalizes in a monoclinic space group $P2_1/c$. The molecule of the ligand consists of bicyclic coumarin fragment (benzopyran-2,4-dione) and (pyridin-2-yl)methyl which are linked by aminoethylidene chain (Figure 2). The dihedral angle between coumarin and pyridine rings is 75.22(2)°. The presence of an intramolecular N13-H13···O4 hydrogen bond, with H13···O4 = 1.74 Å, N13···O4 = 2.5372(13) Å and N13-H13···O4 = 147°, causes the formation of a six-membered ring and the molecule is in keto-amine tautomeric form. Due to the formation of O4=C4-C3=C11-N13-H13 conjugated bond ring system, the flattening of the C4-C3 (1.4392(17) Å) and the C3=C11 (1.4413(17) Å) (Table 2) bonds is observed, although the bonds are formally single and double, respectively. The hydrogen bond mentioned above can be classified as a resonance-assisted hydrogen bond caused by π-electron delocalization [31]. Benzopyran-2,4-dione rings are connected by π-π interactions while a 3D structure is formed (Figure 3). The distance between centroids of the benzopyran-2,4-dione ring is 3.9407(7) Å and the angle γ between the vector joining the two centroids and the normal to the ring is 26.4°.

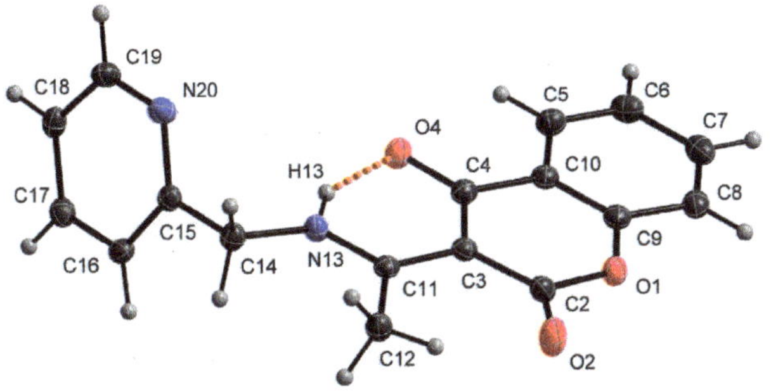

Figure 2. Molecular structure of $\mathbf{HL^1}$, displacement ellipsoids are drawn at 50% probability. The orange dashed line describes a hydrogen bond.

Table 2. Selected bond lengths and angles in $\mathbf{HL^1}$ [Å, °].

Bonds		Angles	
O4-C4	1.2539(16)	O4-C4-C3	124.01(11)
C4-C3	1.4392(17)	C4-C3-C11	119.94(11)
C3-C11	1.4413(17)	C3-C11-N13	118.09(11)
C11-N13	1.3168(16)		

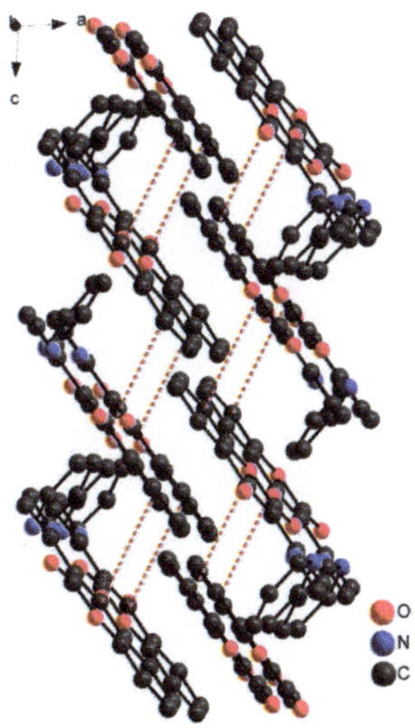

Figure 3. Part of the 3D structure of $\mathbf{HL^1}$ viewed along the *ac* plane with π-π interactions (red dashed lines). Hydrogen atoms are omitted because of clarity.

At the outset, it was hypothesized that the ligand $\mathbf{HL^1}$ would demonstrate a tridentate bonding mode, involving donor atoms N13, N20, and O4, analogous to complexes with similar coumarin derivatives as previously documented [32,33]. However, the single crystal X-ray analysis of the $\mathbf{AgHL^1}$ complex revealed that this assumption was incorrect. In subsequent discussion, we elaborate on the binding mode of the ligand $\mathbf{HL^1}$ to the central atom and the structure of the complex $\mathbf{AgHL^1}$. The complex crystallizes in a triclinic space group *P*-1. Two molecules of $\mathbf{HL^1}$ are bound to the central Ag1 atom only via N20 (and N20′) atoms, with an average Ag1-N20 bond distance of 2.170(4) Å and the N20-Ag1-N20′ bond angle being 176.69(17)° (Table 3), thus the ligand is monodentate with the donor N20 atom originating from the (pyridin-2-yl)methyl fragment. This aligns with the usual range for Ag-N(pyridyl) bonds, spanning from 1.869 to 2.987 Å [34]. Additionally, Wang et al. [35] reported a complex where the central silver atom is bonded to two nitrogen atoms derived from pyridyl groups. In that complex, the Ag-N bond length is 2.158(4) Å, closely paralleling our findings. However, the N-Ag-N angle in their case is perfectly linear at 180°. As a consequence, atom N13 in **AgHL¹** remains protonated and, thus, the ligand is electroneutral. The positive charge of the formed complex cation is balanced by the uncoordinated nitrate anion. The shape of the coordination polyhedron is linear (Figure 4). Selected bond lengths and angles are given in Table 3. Due to intermolecular hydrogen bonds and weak interactions (Table 4), the **AgHL¹** complex creates a 3D structure shown in Figure 5.

Table 3. Selected bonds and angles in **AgHL¹** [Å, °].

Bonds		Angles	
Ag1-N20	2.170(4)	N20-Ag1-N20′	176.69(17)
O4-C4	1.250(7)	O4-C4-C3	124.4(5)
C4-C3	1.450(8)	C4-C3-C11	120.2(5)
C3-C11	1.437(8)	C3-C11-N13	117.6(5)
C11-N13	1.319(7)	O4′-C4′-C3′	124.6(5)
Ag1-N20′	2.170(4)	C4′-C3′-C11′	120.2(5)
O4′-C4′	1.254(7)	C3′-C11′-N13′	119.2(5)
C4′-C3′	1.450(7)		
C3′-C11′	1.453(8)		
C11′-N13′	1.315(7)		
Ag1-O31	2.806(4)		

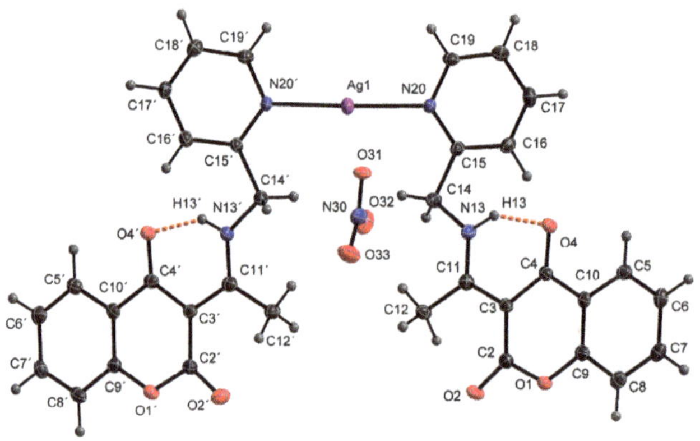

Figure 4. Molecular structure of **AgHL¹**, displacement ellipsoids are drawn at 50% probability. Orange dashed lines describe hydrogen bonds.

Table 4. Hydrogen bonds in **AgHL¹** [Å, °].

D-H···A	*d*(D-H)	*d*(H···A)	*d*(D···A)	<(DHA)
N13-H13···O4	0.88	1.79	2.548(6)	142.4
N13′-H13′···O4′	0.88	1.78	2.545(6)	143.8
C18-H18···O4′ [i]	0.95	2.40	3.190(7)	140.0
C18′-H18′···O4 [ii]	0.95	2.47	3.220(7)	135.6
C19′-H19′···O32 [i]	0.95	2.51	3.371(8)	151.3
C14′-H14D···O32 [iii]	0.99	2.56	3.461(7)	150.7
C19-H19···O32 [i]	0.95	2.48	3.350(7)	152.2
C19-H19···N30 [i]	0.95	2.57	3.303(7)	134.6
C14-H14A···O31	0.99	2.49	3.046(7)	115.1
C14-H14A···O33	0.99	2.53	3.325(8)	137.5

i = $-x + 1, -y + 2, -z + 1$, ii = $-x, -y + 2, -z + 1$, iii = $x - 1, y, z$.

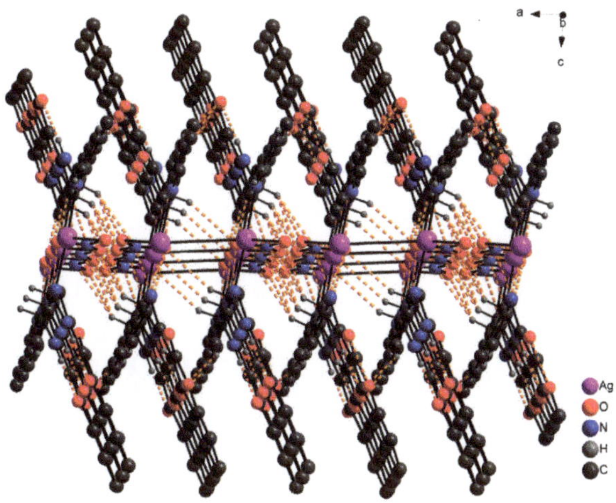

Figure 5. Part of the 3D structure of **AgHL¹** viewed along the *ac* plane with hydrogen bond interactions (orange dashed lines). Hydrogen atoms not involved in hydrogen bond interactions are omitted because of clarity.

2.4. In Vitro Antitumor Activity

2.4.1. Effect of HL^1 and $AgHL^1$ on Metabolic Activity of A549, HT-29 and CCD-18Co Cell Lines and Determination of IC_{50} Values

To determine the potential inhibitory effect of $\mathbf{HL^1}$ and $\mathbf{AgHL^1}$ on the metabolic activity of A549 lung and HT-29 colorectal carcinoma cell lines, an MTT assay was performed 24 and 48 h after the exposure of the cells to tested compounds (1–100 μM) and 1% DMSO. Moreover, the metabolic activity of CCD-18Co healthy colon fibroblasts was also analyzed to verify the selectivity of tested complexes to cancer cells.

$\mathbf{HL^1}$ did not inhibit the metabolic activity of A549 (Figure 6A) and HT-29 cell lines (Figure 6C). A slight and significant decrease in metabolic activity was evident in the case of higher concentrations of $\mathbf{HL^1}$ (100 μM in A549 cells; 50, 75, and 100 μM in HT-29 cells); however, it was more likely caused by DMSO as a diluent of tested compounds. The drop of metabolic activity caused by 100 μM $\mathbf{HL^1}$ in A549 (83.77 ± 3.36% after 48 h) (Figure 6A) and HT-29 cell lines (78.12 ± 6.47% after 24 h; 73.43 ± 5.43% after 48 h) (Figure 6C) corresponds to the metabolic activity reduced by 1% DMSO (82.39 ± 5.53% in A549 cell line after 48 h; 79.71 ± 8.90% in HT-29 cell line after 24 h; 72.05 ± 7.38% in HT-29 cell line after 48 h). $\mathbf{HL^1}$ did not even inhibit the metabolic activity of CCD-18Co fibroblasts (Figure 6E).

On the other hand, $\mathbf{AgHL^1}$ showed a time- and dose-dependent inhibitory effect on the metabolic activity of both cancer cell lines (Figure 6B,D). In the case of A549 cell line, 1 and 2.5 μM $\mathbf{AgHL^1}$ slightly but significantly increased the metabolic activity after 24 h above the untreated control levels (>100%). However, significant inhibition in metabolic activity was observed in the experimental groups treated with 50 μM (83.59 ± 7.29% after 24 h; 68.88 ± 11.60% after 48 h), 75 μM (58.33 ± 11.32% after 24 h; 26.58 ± 8.36% after 48 h) and 100 μM $\mathbf{AgHL^1}$ (9.83 ± 1.90% after 24 h; 3.60 ± 0.59% after 48 h) (Figure 6B). Moreover, the inhibitory effect of $\mathbf{AgHL^1}$ against HT-29 cell line was even more pronounced (50 μM group: 68.89 ± 7.28% after 24 h and 57.45 ± 11.93% after 48 h; 75 μM group: 5.11 ± 2.83% after 24 h and 3.34 ± 1.72% after 48 h; 100 μM group: 2.43 ± 1.12% after 24 h and 2.45 ± 1.02% after 48 h) (Figure 6D). However, a strong inhibitory effect on the metabolic activity was also observed in the case of CCD-18Co fibroblasts (25 μM group: 81.24 ± 8.65% after 24 h and 74.56 ± 3.39% after 48 h; 50 μM group: 6.69 ± 2.32% after 24 h and 4.39 ± 1.12% after 48 h; 75 μM group: 5.80 ± 2.97% after 24 h and 4.82 ± 2.46% after

48 h; 100 μM group: 6.47 ± 1.86% after 24 h and 6.81 ± 2.35% after 48 h) (Figure 6F). The obtained results suggest no selectivity of **AgHL[1]** towards cancer cell lines.

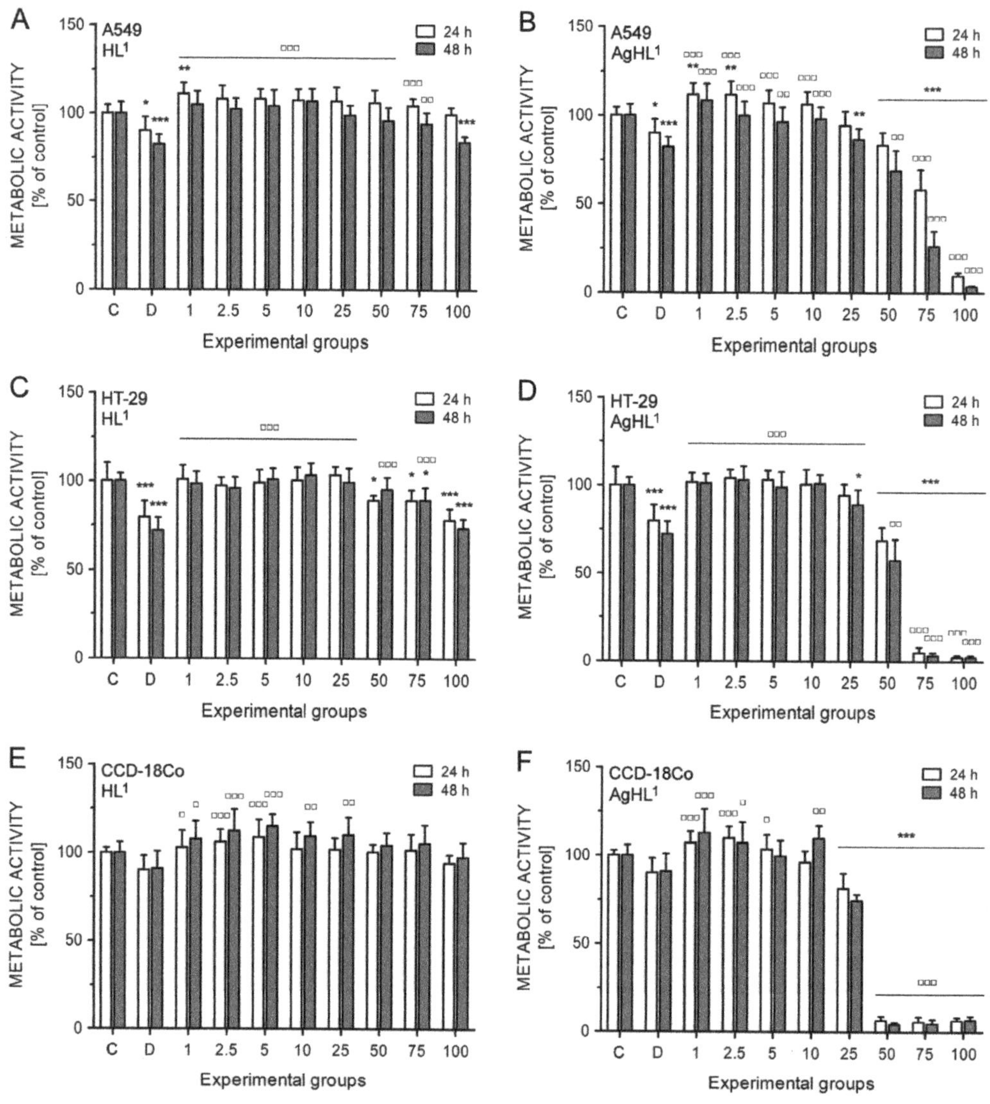

Figure 6. The effect of **HL[1]** (**left** (**A**,**C**,**E**)) and **AgHL[1]** (**right** (**B**,**D**,**F**)) on metabolic activity of A549 (top (**A**,**B**)), HT-29 (**middle** (**C**,**D**)), and CCD-18Co cell lines (**bottom** (**E**,**F**)). The metabolic activity of the cells was estimated by MTT assay. The analyses were performed 24 and 48 h after the treatment of the cells with **HL[1]** and **AgHL[1]** at concentrations of 1–100 μM (all tested concentrations are shown on the x-axis in μM) and with 1% DMSO (marked as D on the x-axis). The results were evaluated as percentages of the untreated control (marked as C on the x-axis) and are expressed as the mean values ± SD of at least three independent experiments. The groups treated with 1% DMSO, **HL[1]**, or **AgHL[1]** were compared with the untreated control (* $p < 0.05$, ** $p < 0.01$, *** $p < 0.001$) and the groups treated with **HL[1]** or **AgHL[1]** were also compared with the group treated with 1% DMSO (□ $p < 0.05$, □□ $p < 0.01$, □□□ $p < 0.001$).

Based on the results of the MTT assay, IC_{50} values for **HL¹** and **AgHL¹** for both carcinoma cell lines were evaluated (Table 5). As **HL¹** was not effective against A549 and HT-29 cell lines, its IC_{50} values could not be defined from the tested range of concentrations (>100 μM). In the case of **AgHL¹**, lower IC_{50} values were obtained for HT-29 cell line indicating slightly higher sensitivity of HT-29 colorectal carcinoma cell line to **AgHL¹** compared to A549 lung carcinoma cell line. Moreover, IC_{50} values for **HL¹** and **AgHL¹** were also determined in CCD-18Co healthy fibroblasts (Table 5) to verify the selective anticancer effect of tested complexes. **HL¹** had no inhibitory effect on CCD-18Co fibroblasts (IC_{50} > 100 μM); however, IC_{50} values for **AgHL¹** were the lowest in this cell line (32.44 ± 0.60 μM after 24 h and 29.09 ± 1.14 μM after 48 h). These results suggest stronger inhibitory effects of **AgHL¹** on CCD-18Co healthy fibroblasts than on HT-29 and A549 carcinoma cell lines.

Table 5. Half maximal inhibitory concentrations (IC_{50}) of **HL¹** and **AgHL¹** in A549, HT-29, and CCD-18Co cell lines. Estimated IC_{50} values (μM) were derived from metabolic activity values of at least three independent experiments by OriginPro 8.5.0 SR1.

	A549		HT-29		CCD-18Co	
	24 h	48 h	24 h	48 h	24 h	48 h
HL¹	>100	>100	>100	>100	>100	>100
AgHL¹	79 ± 5	60 ± 8	56 ± 2	53 ± 2	32 ± 1	29 ± 1
$AgNO_3$	61 ± 3	62 ± 2	35 ± 2	36 ± 1	-	-

The obtained results indicate that, unlike **HL¹**, **AgHL¹** has inhibitory potential towards the metabolic activity of cancer cell lines. However, the decrease in metabolic activity of the cells may either be the consequence of the cytotoxic and/or cytostatic action of the tested compound on the cancer cells or may just reflect the inhibitory potential of the compound towards cellular metabolism. Thus, more specific analyses will be needed to elucidate the effect of **AgHL¹** on A549 and HT-29 cell lines. On the other hand, **AgHL¹** also showed an inhibitory effect on CCD-18Co healthy fibroblasts, indicating no selectivity of this complex towards cancer cell lines.

Additionally, IC_{50} values for $AgNO_3$ were assessed. As shown in Table 5, the complex **AgHL¹** exhibited slightly higher activity than $AgNO_3$ on the A549 cell line. Conversely, on the HT-29 cell line, $AgNO_3$ demonstrated greater cytotoxicity compared to the complex **AgHL¹**. This increased effectiveness can be attributed to the rapid and nonspecific release of Ag^+ ions from $AgNO_3$. Owing to its high aqueous solubility, silver nitrate undergoes rapid dissociation, leading to the immediate release of Ag^+ ions. These cations exhibit broad reactivity, engaging in non-specific interactions with diverse biomolecular targets, thereby contributing to pronounced cytotoxicity [36]. On the other hand, the formation of Ag–N bonds in complexes results in a more controlled and gradual release of Ag^+, enhancing their interaction with cellular targets such as membranes, enzymes, and nucleic acids [37].

2.4.2. Effect of **HL¹** and **AgHL¹** on Proliferation of A549 and HT-29 Carcinoma Cell Lines

To verify the potential cytostatic action of **AgHL¹**, a real-time analysis of cell proliferation was performed. The impact of **HL¹** and **AgHL¹** on the proliferation of A549 and HT-29 cell lines was determined by the monitoring of the confluence of seeded and treated cells using the IncuCyte™ ZOOM system (IncuCyte™ ZOOM 2016B) every 2 h for a total duration of 72 h.

The obtained results are in agreement with the results of the MTT assay. **HL¹** did not inhibit the proliferation of the HT-29 cell line, as the confluence was continuously increasing in all treated experimental groups equally to untreated control throughout the whole duration of monitoring (from approx. 24% at 0 h to approx. 80% in all groups at 72 h) (Figure 7C). In the case of the A549 cell line, 75 μM and 100 μM **HL[1]** slightly but significantly delayed the rate of cell growth, as the percentage of confluence was lower compared to the untreated control group (80.65 ± 9.14% in 75 μM group and 49.18 ± 15.25% in 100 μM group vs. 99.59 ± 0.51% in control) (Figure 7A). However, an increasing percentage of confluence in time suggests that A549 cell line maintained their proliferative activity even after **HL[1]** treatment.

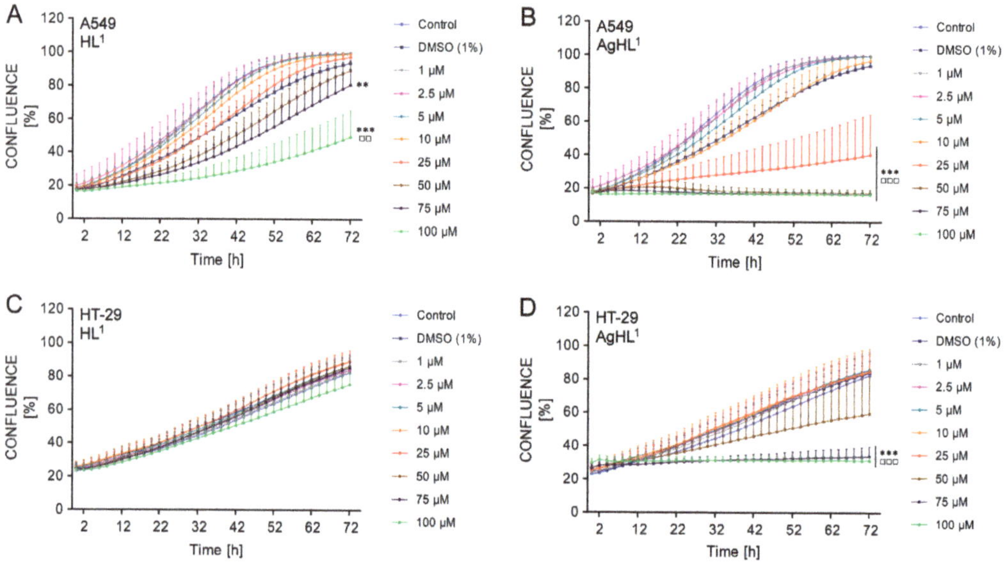

Figure 7. The effect of **HL[1]** (**left** (**A**,**C**)) and **AgHL[1]** (**right** (**B**,**D**)) on proliferation of A549 (**top** (**A**,**B**)) and HT-29 cell lines (**bottom** (**C**,**D**)). The proliferation of the cells is presented as a percentage of confluence (the percentage of the cell-occupied area of the culture well surface) and was monitored by IncuCyte™ ZOOM (Essen Bioscience) every 2 h after the treatment of the cells with **HL[1]** and **AgHL[1]** (1–100 μM) and with 1% DMSO for a total duration of 72 h. The results are expressed as the mean values ± SD of three independent experiments. The groups treated with 1% DMSO, **HL[1]** or **AgHL[1]** were compared with the untreated control (** $p < 0.01$, *** $p < 0.001$) and the groups treated with **HL[1]** or **AgHL[1]** were also compared with the group treated with 1% DMSO (□□ $p < 0.01$; □□□ $p < 0.001$).

On the other side, higher concentrations of **AgHL[1]** inhibited the proliferation of A549 and HT-29 carcinoma cell lines, as the confluence of both cell lines was not increasing in time. The confluence of HT-29 cell line treated with 75 μM and 100 μM **AgHL[1]** remained almost constant (26.52 ± 2.68% at 0 h vs. 33.80 ± 5.43% at 72 h in 75 μM group; 29.61 ± 2.40% at 0 h vs. 31.09 ± 2.69% at 72 h in 100 μM group) (Figure 7D). In the case of the A549 cell line, the confluence even slightly decreased after the treatment with 50, 75, and 100 μM **AgHL[1]** below the percentage of confluence at 0 h (17.44 ± 1.71% at 0 h vs. 17.21 ± 2.00% at 72 h in 50 μM group; 17.10 ± 0.95% at 0 h vs. 16.32 ± 1.05% at 72 h in 75 μM group; 17.21 ± 1.59% at 0 h vs. 16.27 ± 1.87% at 72 h in 100 μM group) (Figure 7B). Moreover, 25 μM **AgHL[1]** in the A549 cell line (Figure 7B) and 50 μM **AgHL[1]** in the HT-29 cell line (Figure 7D) also significantly delayed the rate of cell proliferation, as the percentage of confluence was lower compared to the untreated control group.

The obtained results suggest that **AgHL[1]** was more effective against A549 and HT-29 carcinoma cell lines than **HL[1]**. Unlike **HL[1]**, **AgHL[1]** inhibited not only the metabolic activity but also the proliferation of A549 and HT-29 cell lines.

So far, the antitumor properties of Ag(I) complexes of coumarin derivatives have been poorly studied. However, some evidence of the antitumor effects of other metal complexes, especially palladium complexes with a coumarin-derived ligand, may indicate their potential use in the treatment of tumors. A palladium(II) complex of 3-(1-(2-hydroxyethylamino)ethylidene)chroman-2,4-dione was shown to be effective against L929 mouse fibrosarcoma, B16 mouse melanoma, and U251 human glioma cell lines [25]. Similarly, in another study, palladium(II) complex of 3-(1-(3-hydroxypropylamino)ethylidene)chroman-2,4-dione decreased the viability of B16 and U251 cell lines, while the ligand showed only moderate cytotoxic activity [32]. In both cases, the IC_{50} values of these complexes were lower than the corresponding IC_{50} values for cisplatin treatment, suggesting that the complexes could be therapeutically effective at low concentrations with overall lower systemic toxicity after administration to the patient. The cytotoxic activity of palladium(II) complex of 3-(1-aminoethylidene)-2*H*-chromene-2,4(3*H*)-dione prepared by Budzisz et al. [38] was also more potent than carboplatin treatment in HeLa (cervix epithelial carcinoma), K562 (chronic myelogenous leukemia), and A549 cell lines. However, none of these studies verified the selectivity of the complexes towards cancer cells. In the current study, the Ag(I) complex of coumarin derivative, **AgHL[1]**, inhibited not only the metabolic activity but also the proliferation of A549 and HT-29 cell lines. Compared with the above-mentioned complexes, the relatively high IC_{50} values for **AgHL[1]** in A549 cells (79.34 μM at 24 h and 59.66 μM at 48 h) appear to be insufficient to compete the cytotoxic effect of the chemotherapeutic agent, mitoxantrone (37.2 μM at 24 h and 4.46 μM at 48 h), demonstrated in the same cancer cell line [39]. Conversely, in the case of HT-29 cell line, **AgHL[1]** (IC_{50} values 55,81 μM at 24 h and 52,86 μM at 48 h) could be more potent than documented cisplatin treatment (IC_{50} values 181,1 μM at 24 h and 104,5 μM at 48 h) [40]. Despite the demonstrated cytotoxic and even cytostatic activity, **AgHL[1]** also showed an inhibitory effect on CCD-18Co healthy fibroblasts, indicating no selectivity of this complex towards cancer cells.

2.5. *In Vitro Antimicrobial Activity*

The results of in vitro testing of antimicrobial activities of the synthesized ligand **HL[1]**, corresponding complex **AgHL[1]**, positive controls doxycycline and fluconazole are shown in Table 6. Since Ag^+ ions are antimicrobially active, as stated below, experiments were also conducted using the $AgNO_3$ control to assess **AgHL[1]** efficacy. The intensity of antimicrobial action varied depending on the type of substances.

The complex has significantly better activity than the ligand, nevertheless, the activity of $AgNO_3$ is comparable to the activity of **AgHL[1]**. MIC values for the ligand were obtained from 250 to >1000 $\mu g \cdot mL^{-1}$ and for the complex from 7.81 to 500 $\mu g \cdot mL^{-1}$. There is no difference in the effect of the tested substances on Gram-positive and Gram-negative bacteria and fungi. The complex shows better activity than the positive controls doxycycline on the *Pseudomonas aeruginosa* standard and clinical strain as well as on the tested molds.

Similar results were also obtained by the authors of other studies where the Ag complexes were significantly more antimicrobial active in relation to ligands derived from coumarin [41–47].

Sahoo et al. gave an overview of the current knowledge about coumarin derivatives as promising antibacterial agents, including complexes with silver [48]. Silver complexes with phenathroline adducts exhibited good antibacterial activity [42]. The compound with coumarin-bearing naphthalene-benzimidazole was a good antibacterial agent against

S. aureus and *E. faecalis* in comparison to positive controls [43]. Similarly, silver salts of butyl, pentyl, and hexyl substituted imidazolium of 6-methyl coumarin had shown good antibacterial action against *E. coli* and *P. aeruginosa* in comparison to the standard [44–47]. Also, a sol–gel matrix with Ag-coumarin complexes may provide non-toxic surfaces with antibacterial properties [49].

Table 6. Antimicrobial activity of $\mathbf{HL^1}$ and $\mathbf{AgHL^1}$, and positive controls (MIC and MMC values for substances are given as $\mu g \cdot mL^{-1}$).

Species/Substances	HL^1		$AgHL^1$		$AgNO_3$		Doxycycline/ Fluconazole	
Bacillus cereus	>1000	>1000	62.5	125	15.62	31.25	0.98	7.81
Bacillus subtilis ATCC 6633	>1000	>1000	7.81	15.78	7.81	15.62	1.953	31.25
Staphylococcus aureus ATCC 25923	>1000	>1000	62.5	125	15.62	31.25	0.224	3.75
Staphylococcus aureus	>1000	>1000	125	250	15.62	31.25	0.45	7.81
Proteus mirabilis ATCC 12453	>1000	>1000	31.25	62.5	15.62	31.25	7.81	15.63
Pseudomonas aeruginosa ATCC 27853	>1000	>1000	31.25	62.5	31.25	62.50	62.5	125
Pseudomonas aeruginosa	1000	>1000	15.78	31.25	15.62	15.62	250	1000
Escherichia coli ATCC 25922	1000	>1000	62.5	125	15.62	62.50	15.63	31.25
Escherichia coli	>1000	>1000	62.5	125	15.62	31.25	15.63	62.5
Salmonella enterica	>1000	>1000	62.5	125	15.62	31.25	15.63	31.25
Candida albicans ATCC 10231	1000	>1000	250	500	62.50	125	7.81	31.25
Aspergillus fumigatus	1000	>1000	31.25	62.5	15.62	15.62	1000	1000
Trichoderma viridae	1000	1000	62.5	125	31.25	31.25	500	1000
Mucor mucedo	250	250	15.78	31.25	31.25	31.25	250	250
Aspergillus niger ATCC 16404	1000	1000	31.25	62.5	31.25	31.25	1000	1000

The minimum inhibitory concentrations (MICs) of silver ions were 0.5 $\mu g \cdot mL^{-1}$ against *E. coli*, and 1 $\mu g \cdot mL^{-1}$ against *P. aeruginosa*, and *S. aureus* [50]. In the case [51] MICs for silver ions were 1.69 $\mu g \cdot mL^{-1}$ against *S. aureus*, and 0.84 $\mu g \cdot mL^{-1}$ against *P. aeruginosa*. In contrast to these values the MIC and MBC of silver nanoparticles against *S. aureus* was found to be 625 $\mu g \cdot mL^{-1}$ [52].

It is believed that the antibacterial activity of silver-based complexes is strictly related to their solubility and stability in water, lipophilicity, redox ability, and release rate of silver ions [53]. Silver ions (Ag^+) preferentially target the cell membrane, inducing bacteriolysis by inhibiting peptidoglycan elongation in *Staphylococcus aureus* and compromising the integrity of the outer membrane in *Escherichia coli* [54]. This process culminates in DNA damage and the subsequent loss of replication capabilities [55]. *Staphylococci* exhibit universal sensitivity to Ag^+ and cannot create resistant strains, whereas *E. coli* rapidly acquires resistance through mutations that alter Ag^+ efflux mechanisms and outer membrane porins [56]. The antimicrobial efficacy of Ag^+ is influenced by interactions with other ions; magnesium ions (Mg^{2+}) diminish its effectiveness by inhibiting cellular uptake of Ag^+, whereas zinc ions (Zn^{2+}) marginally enhance its activity by synergistically augmenting cell death and reducing ferrihydrite reduction [57]. Upon entry into a bacterial cell, silver accumulates, resulting in cell death. Such bacteria serve as an effective reservoir for prolonged release of silver and prolonged antimicrobial activity that can last up to 40 days [58], which further encourages its practical use.

3. Materials and Methods

3.1. Materials and Chemicals

Reagents were purchased from the following commercial sources and used as obtained: phosphorus oxychloride (Merck, Darmstadt, Germany, 99%), 4-hydroxycoumarin (4-hydroxy-2H-chromen-2-one) (Sigma Aldrich, Co., St. Louis, MO, USA, 98%), 2-picolylamine (Sigma Aldrich, Co., St. Louis, MO, USA, 99%), methanol (Centralchem, Bratislava, Slovakia, 99.5%), acetone (Centralchem, Bratislava, Slovakia, 99.5%), silver(I) nitrate (Sigma Aldrich, Co., St. Louis, MO, USA, 99%), and acetic acid (Honeywell, Seelze, Germany, 99.8%).

3.2. Syntheses

3.2.1. Synthesis of **HL^1**

Starting compound 3-acetyl-4-hydroxy-2*H*-chromen-2-one (**1**) was prepared by acetylation of 4-hydroxy-2*H*-chromen-2-one (Scheme 1) [24]. To a solution of 4-hydroxy-2*H*-chromen-2-one (3 g, 18.6 mmol) in acetic acid (16 mL), phosphorus oxychloride $POCl_3$ (5.6 mL) was added dropwise. The mixture was then refluxed for 45 min. After cooling in the refrigerator, a slightly yellow precipitate of **1** was filtered under suction, dried on air, and used without another purification. Yield: 54%.

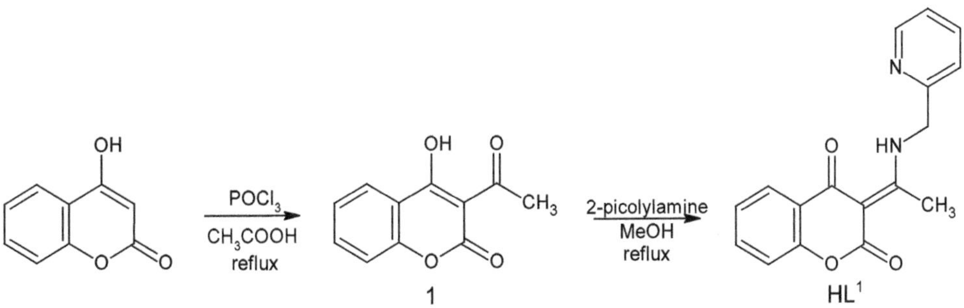

Scheme 1. Reaction scheme of the **HL^1** synthesis.

Ligand **HL^1** was prepared by the condensation reaction of **1** with 2-picolylamine (Scheme 1) based on reactions described previously [21]. The mixture of **1** (0.205 g, 1 mmol) and 2-picolylamine (0.108 g, 1 mmol) in 25 mL methanol was refluxed for about 90 min. The reaction was monitored by TLC, using mobile phase toluene: acetone = 7:3 (v/v). After the reaction was complete, the clear solution was placed in a freezer to give a white product, which was filtered under suction and dried on air. Yield: 60%.

HL^1—Calc. for $C_{17}H_{14}N_2O_3$ (294.30 g·mol^{-1}): C, 69.38; H, 4.79; N, 9.52%. Found: C, 69.44; H, 5.13; N, 9.53%.

IR (ATR, cm^{-1}): 1689(m), 1606(m), 1586(m), 1567(m), 1460(m), 1434(w), 1335(m), 1302(m), 1225(w), 951(w), 898(m), 779(w), 753(s), 438(m) (Figure S5).

^{1}H NMR (600 MHz, DMSO-d$_6$): δ 14.14 (s, 1H, H11), 8.64 (ddd, J = 4.8, 1.8, 0.9 Hz, 1H, H6′), 7.96 (dd, J = 7.8, 1.7 Hz, 1H, H5), 7.87 (td, J = 7.7, 1.8 Hz, 1H, H4′), 7.63 (ddd, J = 8.2, 7.2, 1.7 Hz, 1H, H7), 7.47 (d, J = 7.9 Hz, 1H, H3′), 7.39 (ddd, J = 7.5, 4.8, 1.1 Hz, 1H, H5′), 7.29 (ddd, J = 8.1, 7.3, 1.0 Hz, 1H, H6), 7.26 (dd, J = 8.2, 1.1 Hz, 1H, H8) 5.04 (d, J = 5.1 Hz, 2H, H12), 2.75 (s, 3H, H10) ppm.

^{13}C NMR (150 MHz, DMSO-d$_6$): δ 179.5 (C4), 176.2 (C9), 162.0 (C2), 154.6 (C2′), 153.1 (C8a), 149.3 (C6′), 137.3 (C4′), 134.0 (C7), 125.7 (C5), 123.7 (C6), 123.1 (C5′), 122.2 (C3′), 120.3 (C4a), 116.2 (C8), 96.4 (C3), 48.7 (C12), 19.0 (C10) ppm.

3.2.2. Synthesis of [Ag(HL1)$_2$]NO_3 (**AgHL^1**)

The complex [Ag(HL^1)$_2$]NO_3 (**AgHL^1**) was prepared by mixing 30 mL of warm (50 °C) ethanolic solution of the ligand **HL^1** (88.2 mg, 0.3 mmol) and 10 mL aqueous solution of $AgNO_3$ (25.4 mg, 0.15 mmol) under strong stirring (molar ratio of $AgNO_3$: **HL^1** = 1:2) in the beaker wrapped in aluminum foil. The formed solution was slowly cooled down and crystalized in the dark at temperature of 24 °C for about 2 weeks. Beige crystals of **AgHL^1** were then filtered off and dried on air. Yield: 73%.

[Ag(HL^1)$_2$]NO_3 (**AgHL^1**)—Calc. for $C_{34}H_{26}AgN_5O_9$ (758.48 g·mol^{-1}): C, 53.84; H, 3.72; N, 9.23%. Found: C, 53.71; H, 3.79; N, 9.13%.

IR (ATR, cm^{-1}): 1698(s), 1607(m), 1568(m), 1465(m), 1337(m), 1287(s), 1231(w), 900(w), 751(s), 431(m) (Figure S5).

^{1}H NMR (600 MHz, DMSO-d$_6$): δ 14.14 (s, 1H, H11), 8.64 (ddd, J = 4.9, 1.8, 1.0 Hz, 1H, H6′), 7.96 (dd, J = 7.8, 1.5 Hz, 1H, H5), 7.87 (td, J = 7.8, 1.8 Hz, 1H, H4′), 7.63 (ddd, J = 8.2, 7.2, 1.7 Hz, 1H, H7), 7.48 (dt, J = 7.8, 1.0 Hz, 1H, H3′), 7.40 (ddd, J = 7.6, 4.8, 1.0 Hz, 1H, H5′), 7.29 (ddd, J = 8.1, 7.2, 1.1 Hz, 1H, H6), 7.26 (dd, J = 8.2, 1.0 Hz, 1H, C8), 5.04 (d, J = 5.2 Hz, 2H, H12), 2.75 (s, 3H, H10) ppm (Figure S1).

^{13}C NMR (150 MHz, DMSO-d$_6$): δ 179.5 (C4), 176.2 (C9), 162.0 (C2), 154.6 (C2′), 153.1 (C8a), 149.4 (C6′), 137.4 (C4′), 134.0 (C7), 125.7 (C5), 123.6 (C6), 123.1 (C5′), 122.2 (C3′), 120.3 (C4a), 116.2 (C8), 96.4 (C3), 48.8 (C12), 19.0 (C10) ppm (Figure S1).

3.3. *Physical Measurements*

Elemental analyses of C, H, and N were obtained on CHNOS Elemental Analyzer vario MICRO from Elementar Analysensysteme GmbH, Langenselbold, Germany.

The infrared spectra of the prepared compounds were recorded on a Nicolet 6700 FT-IR spectrophotometer from Thermo Scientific equipped with a diamond crystal Smart Orbit™ in the range 4000–400 cm^{-1}. To process the results, the program OMNIC™ from Thermo Scientific, Waltham, MA, USA, was used.

NMR spectra were recorded on a Varian VNMRS (599.87 MHz for ^{1}H, 150.84 MHz for ^{13}C, and 60.80 MHz for ^{15}N) spectrometer with a 5 mm inverse-detection H-X probe equipped with a z-gradient coil at 298.15 K. All the pulse programs were taken from the Varian sequence library. Chemical shifts (δ in ppm) are given from internal solvent and the partially deuterated residual DMSO-d$_6$ 39.5 ppm for ^{13}C; DMSO-d$_5$ 2.5 ppm for ^{1}H. External nitromethane (0.0 ppm) was used for ^{15}N references. NMR spectra were processed and analyzed in MestReNova v. 15.0.1 (Mestrelab Research, Santiago de Compostela, Spain).

3.4. *X-Ray Data Collection and Structure Refinement*

The data collections for **HL^1** and **AgHL^1** were carried out on an XtaLAB Synergy diffractometer equipped with a HyPix detector. CrysAlisPro software, Version 1.0.43 [59] was used for data collection and cell refinement, data reduction, and absorption correction. The structures were solved by SHELXT [60] and refined by subsequent Fourier syntheses using SHELXL-2018 [61], implemented in WinGX program suit [62]. A geometric analysis was performed using SHELXL-2018, PLATON [63] was used to analyze π-π interaction, while DIAMOND [64] was used for molecular graphics. Carbon-bonded hydrogen atoms were placed in calculated positions and refined riding on their parent C atoms. Nitrogen-bonded hydrogen atoms were found in the different Fourier maps and then refined riding on their parent N atoms. A summary of crystal data and structure refinement for the ligand **HL^1** and the complex **AgHL^1** are presented in Table S1.

3.5. *In Vitro Antitumor Activity*

3.5.1. Cell Lines and Growth Conditions

A human lung adenocarcinoma cell line A549, human colorectal adenocarcinoma cell line HT-29, and human colon fibroblasts CCD-18Co were used in experiments and were purchased from the American Type Culture Collection (ATCC, Rockville, MD, USA). CCD-18Co cell line was derived from normal colon tissue; thus, it is used as a normal non-transformed control in cancer studies. A549 and HT-29 cells were grown in complete RPMI-1640 medium (Sigma-Aldrich, St. Louis, MO, USA) and CCD-18Co cells were grown in MEM medium (Biosera, Nuaillé, France) in an incubator at 37 °C, 95% humidity and 5% CO_2 atmosphere. All cultivation media were supplemented with 10% fetal bovine serum (FBS; Biosera), 7.5% $NaHCO_3$, and antibiotics (1% antibiotic–antimycotic 100× and 50 $\mu g \cdot mL^{-1}$ gentamicin, Biosera).

3.5.2. Reagents and Experimental Design

$\mathbf{HL^1}$ and $\mathbf{AgHL^1}$ were dissolved in dimethyl sulfoxide (DMSO) at a final concentration of 10 mM and stored at room temperature in the dark. For the experiments, the working solutions of $\mathbf{HL^1}$ and $\mathbf{AgHL^1}$ were always freshly prepared immediately before the addition to the cell culture.

For the MTT assay and cell proliferation assay, the cells were seeded in 96-well plates (TPP, Trasadingen, Switzerland) in a density of 6.700 cells per well (A549), 10.000 cells per well (HT-29) or 5000 cells per well (CCD-18Co), and were left to settle for 24 h before treatment. Afterward, the cells were treated with different concentrations of $\mathbf{HL^1}$ and $\mathbf{AgHL^1}$ ranging from 1 µM to 100 µM and with 1% DMSO (the highest final concentration of the diluent corresponding to DMSO concentration in the experimental groups treated with 100 µM $\mathbf{HL^1}$ or $\mathbf{AgHL^1}$).

3.5.3. MTT Assay and IC_{50} Values Evaluation

MTT assay was performed to determine the changes in the metabolic activity of A549, HT-29, and CCD-18Co cell lines treated with $\mathbf{HL^1}$ and $\mathbf{AgHL^1}$ and to calculate the half-maximal inhibitory concentrations (IC_{50}) of $\mathbf{HL^1}$ and $\mathbf{AgHL^1}$.

MTT (3-(4,5-dimethylthiazol-2-yl)-2,5-diphenyltetrazolium bromide; Sigma-Aldrich) was added to the cells (final concentration 0.5 $mg \cdot mL^{-1}$) 24 and 48 h after the treatment with $\mathbf{HL^1}$, $\mathbf{AgHL^1}$, and 1% DMSO. MTT is yellow water-soluble tetrazolium salt that is reduced in metabolically active cells by mitochondrial dehydrogenases to purple water-insoluble formazan crystals. This metabolic conversion was stopped after 4 h incubation at 37 °C and the formazan crystals were dissolved by the addition of sodium dodecyl sulfate (SDS; SERVA Electrophoresis GmbH, Heidelberg, Germany) at a final concentration of 3.3%. The absorbance (λ = 584 nm) was measured using a BMG FLUOstar Optima (BMG Labtechnologies GmbH, Offenburg, Germany). Results were evaluated as percentages of the absorbance of the untreated control.

The IC_{50} values of $\mathbf{HL^1}$ and $\mathbf{AgHL^1}$ were extrapolated from a dose–response fit to the metabolic activity data using OriginPro 8.5.0 SR1 (OriginLab Corp., Northampton, MA, USA) and are expressed as mean IC_{50} values ± standard deviation (SD) from at least three independent experiments.

3.5.4. Cell Proliferation Assay

To determine the potential antiproliferative (cytostatic) effect of $\mathbf{HL^1}$ and $\mathbf{AgHL^1}$ on A549 and HT-29 cell lines, a real-time label-free proliferation assay was performed using the IncuCyte™ ZOOM live cell imaging system (Essen BioScience, Ann Arbor, MI, USA).

The 96-well plates were placed into an IncuCyte^{TM} ZOOM system within a regular cell culture incubator (37 °C, 95% humidity, 5% CO_2) immediately after the treatment of the cells with **HL^1**, **AgHL^1**, and 1% DMSO. Proliferation of the cells was monitored by analyzing the percentage of confluence (the cell occupied area—the percentage of the culture well surface area that is covered by a layer of cells) using the IncuCyte^{TM} ZOOM automated software every 2 h for a total duration of 72 h. Results were evaluated as percentages of the confluence for untreated control and each experimental group.

3.5.5. Statistical Analysis

The results are expressed as the mean values ± standard deviation (SD) of at least three independent experiments. The data were analyzed using one-way ANOVA with Tukey's post-test and the significance levels are indicated in the legend for each particular figure.

3.6. *In Vitro Antimicrobial Activity*

3.6.1. Test Substances, Microorganisms, and Microbial Suspension

DMSO was purchased from Acros Organics (Geel, Belgium). Resazurin was obtained from Alfa Aesar GmbH & Co. (KG, Karlsruhe, Germany). An antibiotic, doxycycline (Galenika A.D., Belgrade, Serbia), was dissolved in a nutrient liquid medium, a Mueller–Hinton broth (Torlak, Belgrade, Serbia), while an antimycotic, fluconazole (Pfizer Inc., New York, NY, USA) was dissolved in Tryptone soya broth (Torlak, Belgrade, Serbia). The tested compounds were dissolved in DMSO and then diluted into a nutrient liquid medium to achieve a concentration of 10%.

The list of tested microorganisms is presented in Table 6. All clinical isolates were a generous gift from the Institute of Public Health, Kragujevac. The other microorganisms were provided from a collection held by the Microbiology Laboratory, Faculty of Science, University of Kragujevac.

Bacterial and yeast suspensions were prepared by the direct colony method. The turbidity of the initial suspension was adjusted using a densitometer (DEN-1, BioSan, Latvia). Initial bacterial suspensions contain about 10^8 colony forming units (CFU)$\cdot mL^{-1}$ for bacteria and the suspension of yeast contains 10^6 CFU per mL. A total of 1:100 dilutions of initial suspension were additionally prepared into sterile 0.85% saline [65]. The suspensions of fungal spores were prepared by gentle stripping of spores from agar slants with growing aspergilli. The resulting suspensions were 1:1000 diluted in a sterile 0.85% saline.

3.6.2. Microdilution Method

Antimicrobial activity was tested by determining the minimum inhibitory concentration (MIC) and the minimum microbicidal concentration (MMC) using the microdilution method with resazurin [66]. The 96-well plates were prepared by dispensing 100 μL of nutrient broth, Mueller–Hinton broth for bacteria, and Tryptone soya broth for yeasts and molds, into each well. A 100 μL aliquot from the stock solution of the tested compound (with a concentration of 2000 $\mu g\cdot mL^{-1}$) was added to the first row of the plate. Then, twofold serial dilutions were performed by using a multichannel pipette. The obtained concentration range was from 1000 to 7.8 $\mu g\cdot mL^{-1}$. The method is described in detail in the reported paper [67].

The growth of bacteria and yeasts was monitored by adding resazurin, an indicator of microbial growth. Resazurin is a blue non-fluorescent dye that becomes pink and fluorescent when reduced to resorufin by oxidoreductases within viable cells. The inoculated plates were incubated at 37 °C for 24 h for bacteria, 28 °C for 48 h for yeasts, and 28 °C for 72 h for molds. MIC was defined as the lowest concentration of the tested substance that prevented resazurin color change from blue to pink. For molds, MIC values of the tested substance were determined as the lowest concentration that visibly inhibited mycelia growth. Minimum microbicidal concentration was determined by plating 10 mL of samples from wells, where no indicator color change and mycelia growth were recorded, on a nutrient agar medium. At the end of the incubation period, the lowest concentration with no growth (no colony) was defined as the minimum microbicidal concentration.

Doxycycline and fluconazole, dissolved in nutrient liquid medium, were used as positive controls. It was observed that 10% DMSO did not inhibit the growth of microorganisms. Each test included growth control and sterility control. All tests were performed in duplicate and MICs were constant.

4. Conclusions

A novel silver(I) complex $[Ag(HL^1)_2]NO_3$ (**AgHL^1**) with coumarin derivative (3*E*)-3-(1-{[(pyridin-2-yl)methyl]amino}ethylidene)-3,4-dihydro-2*H*-benzopyran-2,4-dione (**HL^1**), has been synthesized and subjected to characterization. The molecular structure of **AgHL^1** and **HL^1** has been elucidated through single-crystal X-ray crystallography. In vitro antitumor efficacy assessments of the ligand, its complex and $AgNO_3$ were conducted on A549 lung cancer and HT-29 colorectal carcinoma cell lines utilizing the MTT assay and cell proliferation analysis. Remarkably, while the ligand exhibited moderate activity, the **AgHL^1** complex displayed a pronounced, time- and dose-dependent inhibitory effect on the metabolic activities of both cancerous cell lines. Nevertheless, **AgHL^1** also induced inhibitory effects on CCD-18Co healthy fibroblasts, revealing an absence of selectivity for cancerous cells. Corroborating the MTT assay findings, **AgHL^1** impeded both the metabolic activity and proliferation in A549 and HT-29 cell lines. Moreover, an in vitro examination of antimicrobial efficacy was performed for both the ligand and **AgHL^1**, evaluated against established positive controls doxycycline and fluconazole, and $AgNO_3$. The complex and $AgNO_3$ demonstrated a similar activity which was substantially higher compared to the ligand's activity. There was no discrepancy observed in the effect of the substances tested against Gram-positive and Gram-negative bacteria as well as fungi. The complex exhibited enhanced efficacy compared to the positive controls against the *Pseudomonas aeruginosa* standard and clinical strains, in addition to the molds tested.

Supplementary Materials: The following supporting information can be downloaded at: https://www.mdpi.com/article/10.3390/inorganics13050164/s1, Figure S1. ^{1}H (600 MHz, DMSO-d_6) and ^{13}C (150 MHz, DMSO-d_6) spectra of the complex **AgHL^1**; Figure S2. ^{1}H,^{13}C-HSQC and ^{1}H,^{13}C-HMBC spectra of the complex **AgHL^1**; Figure S3. ^{15}N-HMBCspectra of **HL^1**; Figure S4. ^{15}N-HMBCspectra of **AgHL^1**; Figure S5. Comparison of the IR spectra of **HL^1** and **AgHL^1**; Table S1. Crystal data and structure refinement of **HL^1** and **AgHL^1**.

Author Contributions: Conceptualization, I.P. and J.K.; investigation, J.K., Z.J., V.D., M.V., M.L., K.Ć., I.R., R.J. and I.P.; resources, I.P., R.J. and I.R.; writing—original draft preparation, J.K., Z.J., M.V., I.R. and I.P.; writing—review and editing, J.K., Z.J., M.V., I.R. and I.P.; visualization, J.K., Z.J. and V.D.; supervision, I.P., I.R. and R.J.; project administration, I.P., I.R. and R.J.; funding acquisition, I.P., I.R. and R.J. All authors have read and agreed to the published version of the manuscript.

Funding: The financial support of Slovak grant agency VEGA 1/0126/23 is gratefully acknowledged. This work was further supported by the project Open Scientific Community for Modern Interdisciplinary Research in Medicine (OPENMED) under Grant ITMS2014+: 313011V455. This work was supported by the Serbian Ministry of Science, Technological development, and Innovations (Agreement No. 451-03-137/2025-03/200122 and 451-03-136/2025-03/200122).

Institutional Review Board Statement: Not applicable.

Informed Consent Statement: Not applicable.

Data Availability Statement: The supplementary crystallographic data for **HL^1** and **AgHL^1** presented in the study are openly available at "http://www.ccdc.cam.ac.uk/structures" with deposition Numbers CCDC 2384413 and 2384414, provided by the Cambridge Crystallographic Data Centre.

Conflicts of Interest: The authors declare no conflicts of interest. The funders had no role in the design of the study; in the collection, analyses, or interpretation of data; in the writing of the manuscript; or in the decision to publish the results.

References

1. Mcquitty, R.J. Metal-Based Drugs. *Sci. Prog.* **2014**, *97*, 1–19. [CrossRef] [PubMed]
2. Peng, K.; Liang, B.-B.; Liu, W.; Mao, Z.-W. What Blocks More Anticancer Platinum Complexes from Experiment to Clinic: Major Problems and Potential Strategies from Drug Design Perspectives. *Coord. Chem. Rev.* **2021**, *449*, 214210. [CrossRef]
3. Biot, C.; Delhaes, L.; N'Diaye, C.M.; Maciejewski, L.A.; Camus, D.; Dive, D.; Brocard, J.S. Synthesis and Antimalarial Activity in Vitro of Potential Metabolites of Ferrochloroquine and Related Compounds. *Bioorg. Med. Chem.* **1999**, *7*, 2843–2847. [CrossRef] [PubMed]
4. Guo, Z.; Sadler, P.J. Metals in Medicine. *Angew. Chem. Int. Ed.* **1999**, *38*, 1512–1531. [CrossRef]
5. Yan, G.-P.; Robinson, L.; Hogg, P. Magnetic Resonance Imaging Contrast Agents: Overview and Perspectives. *Radiography* **2007**, *13*, e5–e19. [CrossRef]
6. Fraise, A.P. Historical Introduction. In *Principles and Practice of Disinfection, Preservation & Sterilization*; Fraise, A.P., Lambert, P.A., Maillard, J.-Y., Eds.; Blackwell Publishing Ltd.: Hoboken, NJ, USA, 2004; ISBN 1-4051-0199-7.
7. Moyer, C.A. Treatment of Large Human Burns with 0.5% Silver Nitrate Solution. *Arch. Surg.* **1965**, *90*, 812. [CrossRef]
8. Fox, C.L. Silver Sulfadiazine—A New Topical. *Arch. Surg.* **1968**, 184–188. [CrossRef]
9. Bessey, P.Q. Wound Care. In *Total Burn Care*; Herndon, D.N., Ed.; W.B. Saunders: Edinburgh, Scotland, 2007; pp. 127–135, ISBN 978-1-4160-3274-8.
10. Dai, T.; Huang, Y.-Y.; Sharma, S.K.; Hashmi, J.T.; Kurup, D.B.; Hamblin, M.R. Topical Antimicrobials for Burn Wound Infections. *Recent Pat. Anti-Infect. Drug Discov.* **2010**, *5*, 124–151. [CrossRef]
11. Raju, S.K.; Shridharshini, K.; Praveen, S.; Maruthamuthu, M.; Mohanapriya, K. Silver Complexes as Anticancer Agents: A Perspective Review. *Ger. J. Pharm. Biomater.* **2022**, *1*, 06–28. [CrossRef]
12. Balcıoğlu, S.; Olgun Karataş, M.; Ateş, B.; Alıcı, B.; Özdemir, İ. Therapeutic Potential of Coumarin Bearing Metal Complexes: Where Are We Headed? *Bioorg. Med. Chem. Lett.* **2020**, *30*, 126805. [CrossRef]
13. Matos, M.J.; Santana, L.; Uriarte, E.; Abreu, O.A.; Molina, E.; Yordi, E.G. Coumarins—An Important Class of Phytochemicals. In *Phytochemicals-Isolation, Characterisation and Role in Human Health*; Rao, A.V., Rao, L.G., Eds.; InTech: Tokyo, Japan, 2015; ISBN 978-953-51-2170-1.
14. Rettie, A.E. The Pharmocogenomics of Warfarin: Closing in on Personalized Medicine. *Mol. Interv.* **2006**, *6*, 223–227. [CrossRef]
15. Matos, M.; Vazquez-Rodriguez, S.; Santana, L.; Uriarte, E.; Fuentes-Edfuf, C.; Santos, Y.; Muñoz-Crego, A. Synthesis and Structure-Activity Relationships of Novel Amino/Nitro Substituted 3-Arylcoumarins as Antibacterial Agents. *Molecules* **2013**, *18*, 1394–1404. [CrossRef]
16. Kapoor, S. The Anti-Neoplastic Effects of Coumarin: An Emerging Concept. *Cytotechnology* **2013**, *65*, 787–788. [CrossRef] [PubMed]
17. Bansal, Y.; Sethi, P.; Bansal, G. Coumarin: A Potential Nucleus for Anti-Inflammatory Molecules. *Med. Chem. Res.* **2013**, *22*, 3049–3060. [CrossRef]
18. Xu, Z.; Chen, Q.; Zhang, Y.; Liang, C. Coumarin-Based Derivatives with Potential Anti-HIV Activity. *Fitoterapia* **2021**, *150*, 104863. [CrossRef]
19. Thati, B.; Noble, A.; Creaven, B.S.; Walsh, M.; McCann, M.; Devereux, M.; Kavanagh, K.; Egan, D.A. Role of Cell Cycle Events and Apoptosis in Mediating the Anti-Cancer Activity of a Silver(I) Complex of 4-Hydroxy-3-Nitro-Coumarin-Bis(Phenanthroline) in Human Malignant Cancer Cells. *Eur. J. Pharmacol.* **2009**, *602*, 203–214. [CrossRef] [PubMed]
20. Mujahid, M.; Kia, A.F.-A.; Duff, B.; Egan, D.A.; Devereux, M.; McClean, S.; Walsh, M.; Trendafilova, N.; Georgieva, I.; Creaven, B.S. Spectroscopic Studies, DFT Calculations, and Cytotoxic Activity of Novel Silver(I) Complexes of Hydroxy Ortho-Substituted-Nitro-2H-Chromen-2-One Ligands and a Phenanthroline Adduct. *J. Inorg. Biochem.* **2015**, *153*, 103–113. [CrossRef] [PubMed]
21. Brawley, J.; Etter, E.; Heredia, D.; Intasiri, A.; Nennecker, K.; Smith, J.; Welcome, B.M.; Brizendine, R.K.; Gould, T.W.; Bell, T.W.; et al. Synthesis and Evaluation of 4-Hydroxycoumarin Imines as Inhibitors of Class II Myosins. *J. Med. Chem.* **2020**, *63*, 11131–11148. [CrossRef]
22. OECD. Onkologický Profil Krajiny: Slovenská Republika 2023. In *EU Country Cancer Profiles*; OECD Publishing: Paris, France, 2023; ISBN 9789264710344.

23. Fromm, K.M. Silver Coordination Compounds with Antimicrobial Properties. *Appl. Organomet. Chem.* **2013**, *27*, 683–687. [CrossRef]
24. Sukdolak, S.; Solujić, S.; Manojlović, N.; Vuković, N.; Krstić, L.J. Hantzsch Reaction of 3-(2-Bromoacetyl)-4-Hydroxy-Chromen-2-One. Synthesis of 3-(Thiazol-4-Yl)-4-Hydroxy Coumarines. *J. Heterocycl. Chem.* **2004**, *41*, 593–596. [CrossRef]
25. Ilić, D.R.; Jevtić, V.V.; Radić, G.P.; Arsikin, K.; Ristić, B.; Harhaji-Trajković, L.; Vuković, N.; Sukdolak, S.; Klisurić, O.; Trajković, V.; et al. Synthesis, Characterization and Cytotoxicity of a New Palladium(II) Complex with a Coumarine-Derived Ligand. *Eur. J. Med. Chem.* **2014**, *74*, 502–508. [CrossRef] [PubMed]
26. Kurjan, J.; Jendželovská, Z.; Smolková, R.; Volarevic, V.; Smolko, L.; Jendželovský, R.; Potočňák, I. *Manuscript in Preparation*; Pavol Jozef Šafárik University: Košice, Slovakia, 2025.
27. Butera, V.; D'Anna, L.; Rubino, S.; Bonsignore, R.; Spinello, A.; Terenzi, A.; Barone, G. How the Metal Ion Affects the ^{1}H NMR Chemical Shift Values of Schiff Base Metal Complexes: Rationalization by DFT Calculations. *J. Phys. Chem. A* **2023**, *127*, 9283–9290. [CrossRef]
28. El-Naggar, M.A.; Sharaf, M.M.; Albering, J.H.; Abu-Youssef, M.A.M.; Kassem, T.S.; Soliman, S.M.; Badr, A.M.A. One Pot Synthesis of Two Potent Ag(I) Complexes with Quinoxaline Ligand, X-Ray Structure, Hirshfeld Analysis, Antimicrobial, and Antitumor Investigations. *Sci. Rep.* **2022**, *12*, 20881. [CrossRef] [PubMed]
29. Sakharov, S.G.; Kovalev, V.V.; Gorbunova, Y.E.; Tokmakov, G.P.; Skabitskii, I.V.; Kokunov, Y.V. Structures of Silver Nitrate Complexes with Quinolines According to NMR Data. *Russ. J. Coord. Chem.* **2017**, *43*, 75–81. [CrossRef]
30. Dalvit, C.; Veronesi, M.; Vulpetti, A. ^{1}H and ^{19}F NMR Chemical Shifts for Hydrogen Bond Strength Determination: Correlations between Experimental and Computed Values. *J. Magn. Reson.* **2022**, *12–13*, 100070. [CrossRef]
31. Gilli, G.; Bellucci, F.; Ferretti, V.; Bertolasi, V. Evidence for Resonance-Assisted Hydrogen Bonding from Crystal-Structure Correlations on the Enol Form of the Beta-Diketone Fragment. *J. Am. Chem. Soc.* **1989**, *111*, 1023–1028. [CrossRef]
32. Avdović, E.H.; Stojković, D.L.J.; Jevtić, V.V.; Kosić, M.; Ristić, B.; Harhaji-Trajković, L.; Vukić, M.; Vuković, N.; Marković, Z.S.; Potočňák, I.; et al. Synthesis, Characterization and Cytotoxicity of a New Palladium(II) Complex with a Coumarin-Derived Ligand 3-(1-(3-Hydroxypropylamino)Ethylidene)Chroman-2,4-Dione. Crystal Structure of the 3-(1-(3-Hydroxypropylamino)Ethylidene)-Chroman-2,4-Dione. *Inorg. Chim. Acta* **2017**, *466*, 188–196. [CrossRef]
33. Usman, M.; Zaki, M.; Khan, R.A.; Alsalme, A.; Ahmad, M.; Tabassum, S. Coumarin Centered Copper(II) Complex with Appended-Imidazole as Cancer Chemotherapeutic Agents against Lung Cancer: Molecular Insight via DFT-Based Vibrational Analysis. *RSC Adv.* **2017**, *7*, 36056–36071. [CrossRef]
34. Bruno, I.J.; Cole, J.C.; Edgington, P.R.; Kessler, M.; Macrae, C.F.; McCabe, P.; Pearson, J.; Taylor, R. New Software for Searching the Cambridge Structural Database and Visualizing Crystal Structures. *Acta Crystallogr. Sect. B Struct. Sci.* **2002**, *58*, 389–397. [CrossRef]
35. Wang, P.-N.; Yeh, C.-W.; Tsou, C.-H.; Ho, Y.-W.; Lin, S.-C.; Suen, M.-C. Structural Diversity in the Self-Assembly of Ag(i) Complexes Containing 2,6-Dimethyl-3,5-Dicyano-4-(3-Pyridyl)-1,4-Dihydropyridine. *New J. Chem.* **2014**, *38*, 1079. [CrossRef]
36. Ratte, H.T. Bioaccumulation and Toxicity of Silver Compounds: A Review. *Environ. Toxicol. Chem.* **1999**, *18*, 89–108. [CrossRef]
37. Żyro, D.; Radko, L.; Śliwińska, A.; Chęcińska, L.; Kusz, J.; Korona-Głowniak, I.; Przekora, A.; Wójcik, M.; Posyniak, A.; Ochocki, J. Multifunctional Silver(I) Complexes with Metronidazole Drug Reveal Antimicrobial Properties and Antitumor Activity against Human Hepatoma and Colorectal Adenocarcinoma Cells. *Cancers* **2022**, *14*, 900. [CrossRef]
38. Budzisz, E.; Keppler, B.K.; Giester, G.; Wozniczka, M.; Kufelnicki, A.; Nawrot, B. Synthesis, Crystal Structure and Biological Characterization of a Novel Palladium(II) Complex with a Coumarin-Derived Ligand. *Eur. J. Inorg. Chem.* **2004**, *2004*, 4412–4419. [CrossRef]
39. Bul'ková, V.; Vargová, J.; Babinčák, M.; Jendželovský, R.; Zdráhal, Z.; Roudnický, P.; Košuth, J.; Fedoročko, P. New Findings on the Action of Hypericin in Hypoxic Cancer Cells with a Focus on the Modulation of Side Population Cells. *Biomed. Pharmacother.* **2023**, *163*, 114829. [CrossRef]
40. Leong, K.H.; Looi, C.Y.; Loong, X.-M.; Cheah, F.K.; Supratman, U.; Litaudon, M.; Mustafa, M.R.; Awang, K. Cycloart-24-Ene-26-Ol-3-One, a New Cycloartane Isolated from Leaves of Aglaia Exima Triggers Tumour Necrosis Factor-Receptor 1-Mediated Caspase-Dependent Apoptosis in Colon Cancer Cell Line. *PLoS ONE* **2016**, *11*, e0152652. [CrossRef]
41. Creaven, B.S.; Egan, D.A.; Kavanagh, K.; McCann, M.; Noble, A.; Thati, B.; Walsh, M. Synthesis, Characterization and Antimicrobial Activity of a Series of Substituted Coumarin-3-Carboxylatosilver(I) Complexes. *Inorg. Chim. Acta* **2006**, *359*, 3976–3984. [CrossRef]
42. Mujahid, M.; Trendafilova, N.; Arfa-Kia, A.F.; Rosair, G.; Kavanagh, K.; Devereux, M.; Walsh, M.; McClean, S.; Creaven, B.S.; Georgieva, I. Novel Silver(I) Complexes of Coumarin Oxyacetate Ligands and Their Phenanthroline Adducts: Biological Activity, Structural and Spectroscopic Characterisation. *J. Inorg. Biochem.* **2016**, *163*, 53–67. [CrossRef] [PubMed]
43. Karataş, M.O.; Olgundeniz, B.; Günal, S.; Özdemir, İ.; Alıcı, B.; Çetinkaya, E. Synthesis, Characterization and Antimicrobial Activities of Novel Silver(I) Complexes with Coumarin Substituted N-Heterocyclic Carbene Ligands. *Bioorg. Med. Chem.* **2016**, *24*, 643–650. [CrossRef]

44. Achar, G.; Ramya, V.C.; Upendranath, K.; Budagumpi, S. Coumarin-tethered (Benz)Imidazolium Salts and Their Silver(I) N-heterocyclic Carbene Complexes: Synthesis, Characterization, Crystal Structure and Antibacterial Studies. *Appl. Organomet. Chem.* **2017**, *31*, e3770. [CrossRef]
45. Achar, G.; Shahini, C.R.; Patil, S.A.; Budagumpi, S. Synthesis, Structural Characterization, Crystal Structures and Antibacterial Potentials of Coumarin–Tethered N–Heterocyclic Carbene Silver(I) Complexes. *J. Organomet. Chem.* **2017**, *833*, 28–42. [CrossRef]
46. Achar, G.; Uppendranath, K.; Ramya, V.C.; Biffis, A.; Keri, R.S.; Budagumpi, S. Synthesis, Characterization, Crystal Structure and Biological Studies of Silver(I) Complexes Derived from Coumarin-Tethered N-Heterocyclic Carbene Ligands. *Polyhedron* **2017**, *123*, 470–479. [CrossRef]
47. Achar, G.; Agarwal, P.; Brinda, K.N.; Małecki, J.G.; Keri, R.S.; Budagumpi, S. Ether and Coumarin–Functionalized (Benz)Imidazolium Salts and Their Silver(I)–N–Heterocyclic Carbene Complexes: Synthesis, Characterization, Crystal Structures and Antimicrobial Studies. *J. Organomet. Chem.* **2018**, *854*, 64–75. [CrossRef]
48. Ranjan Sahoo, C.; Sahoo, J.; Mahapatra, M.; Lenka, D.; Kumar Sahu, P.; Dehury, B.; Nath Padhy, R.; Kumar Paidesetty, S. Coumarin Derivatives as Promising Antibacterial Agent(s). *Arab. J. Chem.* **2021**, *14*, 102922. [CrossRef]
49. Jaiswal, S.; Bhattacharya, K.; Sullivan, M.; Walsh, M.; Creaven, B.S.; Laffir, F.; Duffy, B.; McHale, P. Non-Cytotoxic Antibacterial Silver–Coumarin Complex Doped Sol–Gel Coatings. *Colloids Surf. B* **2013**, *102*, 412–419. [CrossRef]
50. Li, W.-R.; Sun, T.-L.; Zhou, S.-L.; Ma, Y.-K.; Shi, Q.-S.; Xie, X.-B.; Huang, X.-M. A Comparative Analysis of Antibacterial Activity, Dynamics, and Effects of Silver Ions and Silver Nanoparticles against Four Bacterial Strains. *Int. Biodeterior. Biodegrad.* **2017**, *123*, 304–310. [CrossRef]
51. Kvitek, L.; Panacek, A.; Prucek, R.; Soukupova, J.; Vanickova, M.; Kolar, M.; Zboril, R. Antibacterial Activity and Toxicity of Silver – Nanosilver versus Ionic Silver. *J. Phys. Conf. Ser.* **2011**, *304*, 012029. [CrossRef]
52. Parvekar, P.; Palaskar, J.; Metgud, S.; Maria, R.; Dutta, S. The Minimum Inhibitory Concentration (MIC) and Minimum Bactericidal Concentration (MBC) of Silver Nanoparticles against *Staphylococcus aureus. Biomater. Investig. Dent.* **2020**, *7*, 105–109. [CrossRef]
53. Claudel, M.; Schwarte, J.V.; Fromm, K.M. New Antimicrobial Strategies Based on Metal Complexes. *Chemistry* **2020**, *2*, 849–899. [CrossRef]
54. Ishida, T. Antibacterial Mechanism of Ag^+ Ions for Bacteriolyses of Bacterial Cell Walls via Peptidoglycan Autolysins, and DNA Damages. *MOJ Toxicol.* **2018**, *4*, 345–350. [CrossRef]
55. Feng, Q.L.; Wu, J.; Chen, G.Q.; Cui, F.Z.; Kim, T.N.; Kim, J.O. A Mechanistic Study of the Antibacterial Effect of Silver Ions onEscherichia Coli andStaphylococcus Aureus. *J. Biomed. Mater. Res.* **2000**, *52*, 662–668. [CrossRef]
56. Randall, C.P.; Oyama, L.B.; Bostock, J.M.; Chopra, I.; O'Neill, A.J. The Silver Cation (Ag^+): Antistaphylococcal Activity, Mode of Action and Resistance Studies. *J. Antimicrob. Chemother.* **2013**, *68*, 131–138. [CrossRef]
57. Ge, C.; Huang, M.; Huang, D.; Dang, F.; Huang, Y.; Ahmad, H.A.; Zhu, C.; Chen, N.; Wu, S.; Zhou, D. Effect of Metal Cations on Antimicrobial Activity and Compartmentalization of Silver in Shewanella Oneidensis MR-1 upon Exposure to Silver Ions. *Sci. Total Environ.* **2022**, *838*, 156401. [CrossRef] [PubMed]
58. Mohamed, D.S.; Abd El-Baky, R.M.; Sandle, T.; Mandour, S.A.; Ahmed, E.F. Antimicrobial Activity of Silver-Treated Bacteria against Other Multi-Drug Resistant Pathogens in Their Environment. *Antibiotics* **2020**, *9*, 181. [CrossRef]
59. *CrysAlisPRO*, Version 1.0.43, Oxford Diffraction/Agilent Technologies UK Ltd.: Yarnton, UK, 2020.
60. Sheldrick, G.M. *SHELXT*–Integrated Space-Group and Crystal-Structure Determination. *Acta Crystallogr. Sect. A Found. Adv.* **2015**, *71*, 3–8. [CrossRef] [PubMed]
61. Sheldrick, G.M. Crystal Structure Refinement with *SHELXL*. *Acta Crystallogr. Sect. C Struct. Chem.* **2015**, *71*, 3–8. [CrossRef]
62. Farrugia, L.J. *WinGX* and *ORTEP for Windows*: An Update. *J. Appl. Crystallogr.* **2012**, *45*, 849–854. [CrossRef]
63. Spek, A.L. Structure Validation in Chemical Crystallography. *Acta Crystallogr. Sect. D Biol. Crystallogr.* **2009**, *65*, 148–155. [CrossRef]
64. Brandenburg, K. *DIAMOND*, Version 3.2k ed; Crystal Impact GbR: Bonn, Germany, 2014.
65. Andrews, J.M. Determination of Minimum Inhibitory Concentrations. *J. Antimicrob. Chemother.* **2001**, *48* (Suppl. 1), 5–16. [CrossRef] [PubMed]
66. Sarker, S.D.; Nahar, L.; Kumarasamy, Y. Microtitre Plate-Based Antibacterial Assay Incorporating Resazurin as an Indicator of Cell Growth, and Its Application in the in Vitro Antibacterial Screening of Phytochemicals. *Methods* **2007**, *42*, 321–324. [CrossRef]
67. Radić, G.P.; Glođović, V.V.; Radojević, I.D.; Stefanović, O.D.; Čomić, L.R.; Ratković, Z.R.; Valkonen, A.; Rissanen, K.; Trifunović, S.R. Synthesis, Characterization and Antimicrobial Activity of Palladium(II) Complexes with Some Alkyl Derivates of Thiosalicylic Acids: Crystal Structure of the Bis(S-Benzyl-Thiosalicylate)–Palladium(II) Complex, [Pd(S-Bz-Thiosal)2]. *Polyhedron* **2012**, *31*, 69–76. [CrossRef]

inorganics

MDPI

Article

Ca-, Li-, and Cu-Salicylatoborates for Potential Applications in Neutron Capture Therapy: A Computational Method for the Preliminary Discrimination of the More Promising Compounds

Domenica Marabello [1,2,*], Paola Benzi [1,2], Carlo Canepa [1] and Alma Cioci [1]

[1] Dipartimento di Chimica, Università degli Studi di Torino, 10125 Torino, Italy; paola.benzi@unito.it (P.B.); carlo.canepa@unito.it (C.C.); alma.cioci@unito.it (A.C.)

[2] Crisdi—Interdepartmental Center for Crystallography, University of Torino, 10125 Torino, Italy

* Correspondence: domenica.marabello@unito.it

Abstract: Boron Neutron Capture Therapy is a re-emerging therapy for the treatment of cancer, and the development of new neutron-reactive nuclei carriers with enhanced efficiency is of great importance. In this work we propose three new boron-based solid compounds, of formulas $[Ca(H_2O)_6](C_{14}H_8O_6B)_2$ (CaSB), $[Cu(C_{14}H_8O_6B)]$ (CuSB), and $[Li(C_{14}H_8O_6B)(H_2O)]$ (LiSB), usable as nanoparticles for the carriage of the ^{10}B isotope. The copper atom in CuSB was introduced because it is known that its presence magnifies the effect of the radiation on cells. Furthermore, the lithium atom in LiSB also allows us to include the ^6Li isotope, which can take part in the nuclear reactions, enhancing the efficiency of the anti-cancer treatment. The compounds were characterized with single-crystal X-ray diffraction to compare the densities of the reactive isotopes in the materials, a key parameter related to the efficiency of the materials. In this work, we used a computational method to calculate the dose absorbed by a tumor mass treated with nanoparticles of the compounds in order to select the most efficient one for the therapy. The results reported in this work are encouraging.

Keywords: salicylatoborate compounds; Neutron Capture Therapy; nanoparticles; absorbed dose; X-ray diffraction; anti-cancer treatment

1. Introduction

Among the radiation techniques used to treat cancer, Boron Neutron Capture Therapy (BNCT) is a re-emerging therapy due to its advantages of target therapy and low toxicity. In fact, it can overcome radio-resistance in different types of cancers, inducing minimal damage to the adjacent normal tissues [1]. It consists in the irradiation of cancer cells carrying the boron-10 isotope with thermal neutrons. When a thermal neutron is captured by the boron-10 isotope, a nuclear (n,α) reaction occurs: the unstable boron-11 isotope forms and decays into a lithium-7 and an α particle, which release a large amount of energy along their short pathway.

The efficacy of this technique requires a suitable selective concentration of the boron-10 isotope into cancer cells rather than the healthy cells, and for this reason, a lot of efforts are being made to develop more selective boron delivery agents [2]. Several boron-10-containing molecules were prepared and studied, but the low difference in boron concentration between tumor tissues and healthy tissues is still not satisfactory to justify the application of BNCT on a large scale.

Recently, boron-10-enriched nanoparticles were proposed as boron delivery agents, since in each nanoparticle hundreds of thousands of boron atoms were concentrated, and furthermore, the surface of nanoparticles can be easily modified with specific biomarkers to systematically penetrate inside the target tumor cells [3–6].

Boron-10 is not the only isotope that can produce a nuclear reaction with thermal neutrons; lithium-6 nuclei can capture neutrons to yield tritium and α particles that release energy into the cells. The combination of the nuclear reaction of both isotopes could greatly improve the efficacy of the Neutron Capture Therapy (NCT).

In our previous work, two compounds containing both boron and lithium, of formulas $Li[(C_6H_{12}O_6)_2B]\cdot 2H_2O$ (LiMB) and $Li_5[(C_4H_4O_6)_2B]\cdot 5.5H_2O$ (LiTB) (M = mannitol, T = L-(+)-tartrate), were synthesized with ^6Li and ^{10}B isotopes, and their behavior under thermal neutron radiation was tested [7]. The preliminary tests were encouraging, although not quantitatively conclusive, and the necessity to synthesize more efficient compounds was reported.

In this work, we propose three new solid salicylatoborate complexes, of formulas $[Ca(H_2O)_6](C_{14}H_8O_6B)_2\cdot 4H_2O$ (CaSB), $[Cu(C_{14}H_8O_6B)_2]$ (CuSB), and $[Li(C_{14}H_8O_6B)(H_2O)]$ (LiSB), as new boron and/or lithium delivery agents for the NCT. In general, salicylatoborate complexes are considered inexpensive, stable materials, usually non-toxic [8]. In our previous work [9], we developed, for analogue compounds, a method to both obtain particles of nanoscale dimension and to suitably cover them in order to permit their encapsulation in biological tumor cells. Thus, we are confident that we will also be able to reduce to nanoscale the dimension of the particles of the present compounds, and, due to the presence of many OH groups in the salicylate fragments, to exploit the same method to suitably functionalize the nanoparticle surface, which allows them to be concentrated into tumor tissues.

CuSB was synthesized since Cu^{2+} ions were found to enhance the radiosensitivity of cells towards gamma rays [10].

In this work, we employed a computational method to calculate the dose (D) absorbed by a tumor mass. To perform these calculations, the three compounds investigated in this work were characterized with the single-crystal X-ray diffraction technique (XRD), in order to calculate the concentration of the neutron-reactive isotopes in the nanoparticles. The calculations were also performed on the LiMB and LiTB compounds for comparison. The calculation method reported in this work allows for the preliminary prediction of the compound with the best therapeutic efficacy for the NCT before the testing at the neutron source.

2. Results and Discussion

2.1. Syntheses

The salicylatoborate compounds were synthesized by mixing salicylic acid, boric acid, and an inorganic base (a metal hydroxide or carbonate), and dissolving the reagents in warm water. An esterification reaction occurs, where the boric acid hydroxyl groups react with those of the salicylate, and the metal atom is coordinated by the oxygen atoms of either water molecules or the salicylate fragment. In Scheme 1, the molecular schemes of the three products obtained are reported.

The diesters of boric acid, based on aromatic or aliphatic diols, are known to be thermodynamically stable, almost indissociable in water, non-toxic, and inexpensive [8]. The products obtained precipitated directly in crystals suitable for the XRD characterization or quickly spontaneously recrystallized.

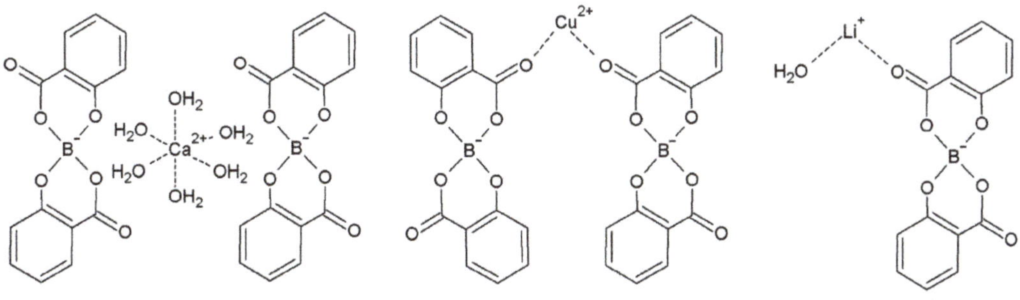

Scheme 1. Molecular schemes of the salicylatoborate compounds.

2.2. X-Ray Crystal Structures

2.2.1. Crystal Structure of CaSB

The asymmetric unit of compound CaSB consists of two salicylatoborate units (one boron atom covalently bonded to two salicylate fragments) connected through hydrogen bonds to one disordered $[Ca(H_2O)_6]^{2+}$ complex, and to four water molecules embedded in the crystal structure (Figure 1).

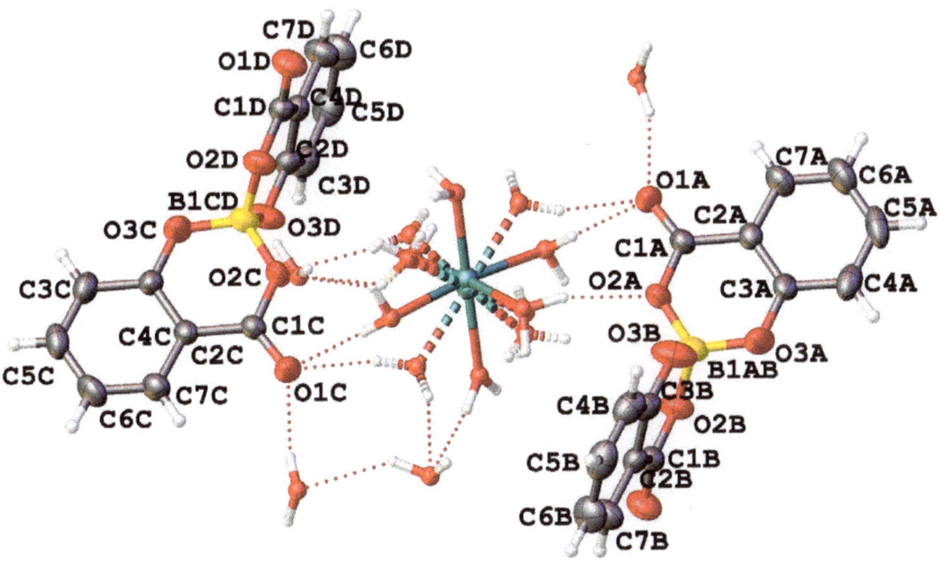

Figure 1. View of the asymmetric unit of $[Ca(H_2O)_6](C_{28}H_{16}O_{12}B)_2 \cdot 4H_2O$ (CaSB) with atom labelling.

In the crystal packing, the salicylatoborate units are intercalated by the calcium exahydrated complexes and by the free water molecules in a three-dimensional lattice. The relevant bond distances are reported in Table 1.

In each unit cell, there are eight boron atoms, and the density of the boron atoms in the crystals is 2.256×10^{27} atoms/m^3.

Table 1. Relevant bond distances (Å) from X-ray data for the LiSB compound.

CaSB				CuSB		LiSB	
						Li1–Li1 [2]	2.84(2)
Ca1A–O1X	2.078(5)	Ca1B–O1Y	2.105(9)	Cu1–O1A [1]	1.947(1)	Li1–O1A [3]	1.92(1)
Ca1A–O2X	2.106(5)	Ca1B–O2Y	2.096(9)	Cu1–O1A [2]	1.947(1)	Li1–O1B	1.88(1)
Ca1A–O3X	2.004(4)	Ca1B–O3Y	2.046(9)	Cu1–O1B [3]	1.939(1)	Li1–O1W	1.97(1)
Ca1A–O4X	2.100(4)	Ca1B–O4Y	1.943(8)	Cu1–O1B	1.939(1)	Li1–O1W [2]	2.08(1)
Ca1A–O5X	2.093(4)	Ca1B–O5Y	2.070(9)				
Ca1A–O6X	2.016(4)	Ca1B–O6Y	2.201(8)				
B1AB–O2A	1.467(6)	B1CD–O2C	1.484(6)	B1–O2A	1.486(2)	B1–O2A	1.496(8)
B1AB–O3A	1.446(7)	B1CD–O3C	1.449(7)	B1–O2B	1.483(3)	B1–O3A	1.437(7)
B1AB–O2B	1.499(6)	B1CD–O2D	1.480(6)	B1–O3B	1.450(3)	B1–O2B	1.471(8)
B1AB–O3B	1.432(7)	B1CD–O3D	1.442(6)	B1–O3A	1.443(3)	B1–O3B	1.436(8)
C1A–O1A	1.218(5)	C1C–O1C	1.224(5)	C1A–O1A	1.247(2)	C1A–O1A	1.256(7)
C1A–O2A	1.324(5)	C1C–O2C	1.311(5)	C1A–O2A	1.301(2)	C1A–O2A	1.316(7)
C1B–O1B	1.233(5)	C1D–O1D	1.224(5)	C1B–O1B	1.241(2)	C1B–O1B	1.218(7)
C1B–O2B	1.301(5)	C1D–O2D	1.310(5)	C1B–O2B	1.302(2)	C1B–O2B	1.308(7)
C3A–O3A	1.350(6)	C3C–O3C	1.346(5)	C3A–O3A	1.342(3)	C3A–O3A	1.360(7)
C3B–O3B	1.345(6)	C3D–O3D	1.347(5)	C3B–O3B	1.346(2)	C3B–O3B	1.348(7)
C1A–C2A	1.473(6)	C1C–C2C	1.469(6)	C1A–C2A	1.447(3)	C1A–C2A	1.448(8)
C2A–C3A	1.380(6)	C2C–C3C	1.391(6)	C2A–C3A	1.395(3)	C2A–C3A	1.389(8)
C3A–C4A	1.407(7)	C3C–C4C	1.402(6)	C3A–C4A	1.390(3)	C3A–C4A	1.395(8)
C4A–C5A	1.363(8)	C4C–C5C	1.366(7)	C4A–C5A	1.376(5)	C4A–C5A	1.379(9)
C5A–C6A	1.394(9)	C5C–C6C	1.386(8)	C5A–C6A	1.389(5)	C5A–C6A	1.375(9)
C6A–C7A	1.358(8)	C6C–C7C	1.373(7)	C6A–C7A	1.362(4)	C6A–C7A	1.382(8)
C2A–C7A	1.396(6)	C2C–C7C	1.406(6)	C2A–C7A	1.397(3)	C2A–C7A	1.382(8)
C1B–C2B	1.470(6)	C1D–C2D	1.473(6)	C1B–C2B	1.455(2)	C1B–C2B	1.457(9)
C2B–C3B	1.389(6)	C2D–C3D	1.382(6)	C2B–C3B	1.397(3)	C2B–C3B	1.391(8)
C3B–C4B	1.402(7)	C3D–C4D	1.401(6)	C3B–C4B	1.394(3)	C3B–C4B	1.373(8)
C4B–C5B	1.354(8)	C4D–C5D	1.353(7)	C4B–C5B	1.374(3)	C4B–C5B	1.379(9)
C5B–C6B	1.396(8)	C5D–C6D	1.404(8)	C5B–C6B	1.382(3)	C5B–C6B	1.357(11)
C6B–C7B	1.367(7)	C6D–C7D	1.382(7)	C6B–C7B	1.377(3)	C6B–C7B	1.389(10)
C2B–C7B	1.397(7)	C2D–C7D	1.393(6)	C2B–C7B	1.392(3)	C2B–C7B	1.388(8)
				[1] +X, −1 + Y, +Z [2] 3/2 − X, −1 + Y, 1/2 − Z [3] 3/2 − X, +Y, 1/2 − Z		[2] 1 − X, 2 − Y, 1 − Z [3] +X, 1 + Y, +Z	

2.2.2. Crystal Structure of CuSB

The asymmetric unit of CuSB contains half of the copper ion and one salicylatoborate ligand (Figure 2).

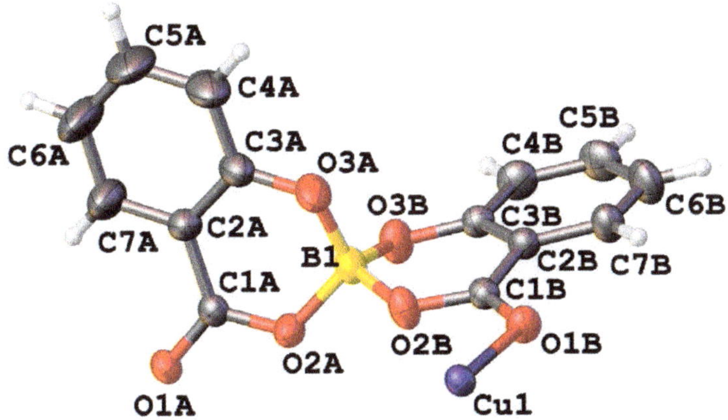

Figure 2. View of the asymmetric unit of $[Cu(C_{14}H_8O_6B)_2]$ (CuSB) with atom labelling.

In the crystal packing, each copper ion is coordinated by four salicylatoborate ligands, and each salicylatoborate ligand bridges two copper ions, forming an infinite row along the b axis, i.e., a one-dimensional polymeric structure (1D Metal Organic Framework, Figure 3). The rows are joined through strong hydrogen bonds that connect the hydrogens of the aromatic ring of the salicylate to the carboxylic oxygen bonded to the metal. Relevant bond distances are reported in Table 1.

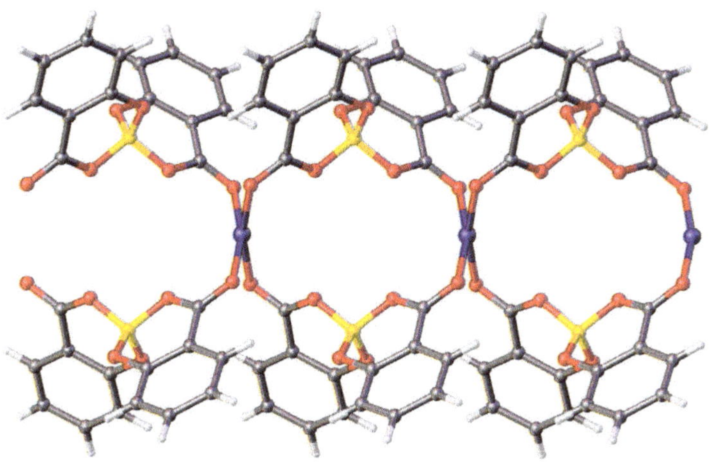

Figure 3. View of the packing motif of CuSB.

In each unit cell, there are two boron atoms, and the density of the boron atoms is 1.533×10^{27} atoms/m^3.

2.2.3. Crystal Structure of LiSB

The asymmetric unit of compound LiSB consists of one boron atom covalently bonded to two salicylate molecules, one Li^+ ion, and one water molecule coordinated to the lithium atom. (Figure 4).

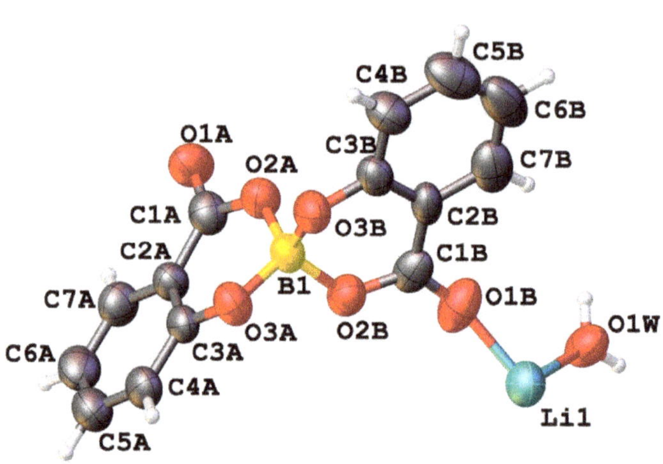

Figure 4. View of the asymmetric unit of $[Li(C_{14}H_8O_6B)(H_2O)]$ (LiSB) with atom labelling.

Along the b-axis direction, the water molecules bridge two lithium ions, and each salicylatoborate ligand also bridges two lithium ions, forming an infinite row, i.e., a one-dimensional polymeric structure (1D Metal Organic Framework, Figure 5). The rows are joined through strong hydrogen bonds that connect the hydrogens of the water molecules to the oxygen atoms of the carboxylic groups of the salicylate fragments. The relevant bond distances are reported in Table 1.

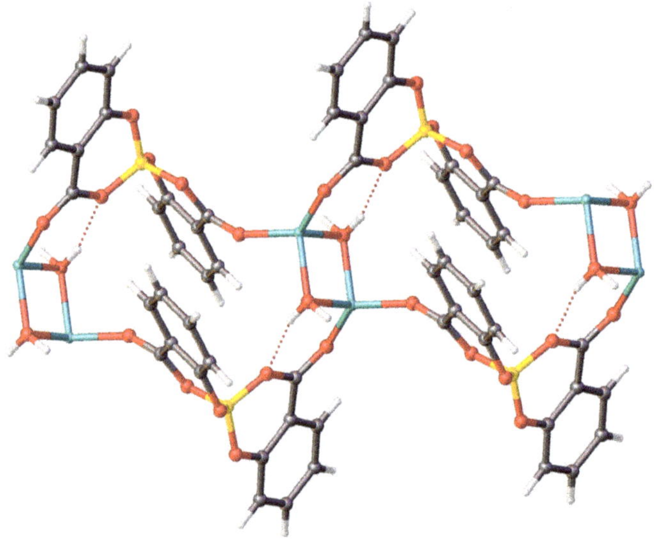

Figure 5. View of the packing motif of LiSB.

In each unit cell, there are four boron and four lithium atoms (cell volume = 1304.51 Å^3), and the density of the boron and lithium atoms is 2.954×10^{27} atoms/m^3.

2.3. *Calculation of the Dose (D) Absorbed by a Tumor Mass*

The dose absorbed by a tumor mass depends on the energy deposited in the tumor tissue by the particles along their paths. This, in turn, depends on the particle's kinetic energy, its charge, mass, and range. The calculations performed evaluate a realistic value of

D. However, given the uncertainty on the tumor dimension and the density of the neutron-reactive isotopes inside each tumor cell, in this work, we provide only a comparison of the therapeutic efficiency of the different compounds. For these reasons, the calculations were performed considering, for all the compounds, the same spherical tumor diameter (5 mm), number of nanoparticles inside each tumor cell (100), dimension of the nanoparticles (diameter 175 nm, the medium size between the smaller nanoparticles obtained with the grinding method proposed in ref. [9]), and a homogeneous distribution of ^{10}B and/or ^{6}Li in the tumor mass. These values were chosen considering that the nanoparticles should be able to concentrate inside a tumor cell. The model of tumor proposed is a little distant from real tumors, which are usually heterogeneous, and therefore, the drug can be unevenly distributed. However, the calculations cannot take into account this aspect, since in this case, the number of calculations to do would increase exponentially and would become quite prohibitive. Since the aim of the theoretical method is not the simulation of a real tumor situation but the discrimination of the more promising compound for the radiotherapy, the model used is a compromise between a possible real situation and the time consumed for calculations.

The calculation of the dose rate $\dot{D}$ (the dose absorbed by the tumor in the unit of time) is based on the following nuclear reactions, taking into account the different densities of the neutron-reactive isotopes in the three materials:

$$^{10}\mathrm{B} + \mathrm{n} \rightarrow {}^{7}\mathrm{Li} + \alpha + 2.31\ \mathrm{MeV} \quad (1)$$

$$^{6}\mathrm{Li} + \mathrm{n} \rightarrow {}^{3}\mathrm{H} + \alpha + 4.78\ \mathrm{MeV} \quad (2)$$

The $\dot{D}$ values depend on the total energy developed by the reaction (that is divided between the two particles produced), on the cross section of the reaction, on the individual energy of the particles produced, and on the value of the Linear Energy Transfer of each particle.

In Figure 6, the contributions of the products of the above-reported reactions to the $\dot{D}$ values (Gy h^{-1}) in function of the reduced radius r/R (r = point in which the contribution was calculated; R = radius of the tumor mass) are reported. For CaSB (a) and CuSB (b), only the products of reaction (1) are considered, while for LiSB (c), the contributions of the products of reaction (2) are also considered. Since the range of the products is very short in comparison to the tumor dimension, the contributions of the products to the $\dot{D}$ values drastically decrease near the surface of the tumor.

Comparing the three pictures in Figure 6, for all the particles produced, the value of $\dot{D}$ is quite constant while varying r/R. As expected, the products of reaction (1) show the highest $\dot{D}$ values, since most of the parameters that influence the energy transfer are more favorable compared to reaction (2). In fact, the major contribution to the transfer of energy is the cross section of ^{10}B with respect to neutron capture (neutron flux 2.35×10^{13} m^{-2} s^{-1}), which is more than four times higher compared to ^{6}Li (4210.71 vs. 1030.19 b, respectively).

In Figure 7, the sums of the contributions to the $\dot{D}$ values of the products of reactions (1) and (2) for CaSB (a), CuSB (b), and LiSB (c) are reported. The green lines represent the $\dot{D}$ values averaged on the sphere of the tumor mass, and their values are reported in Table 2. For all compounds, the $\dot{D}$ values, being different for each compound, slightly decrease from the center to the periphery of the tumor mass.

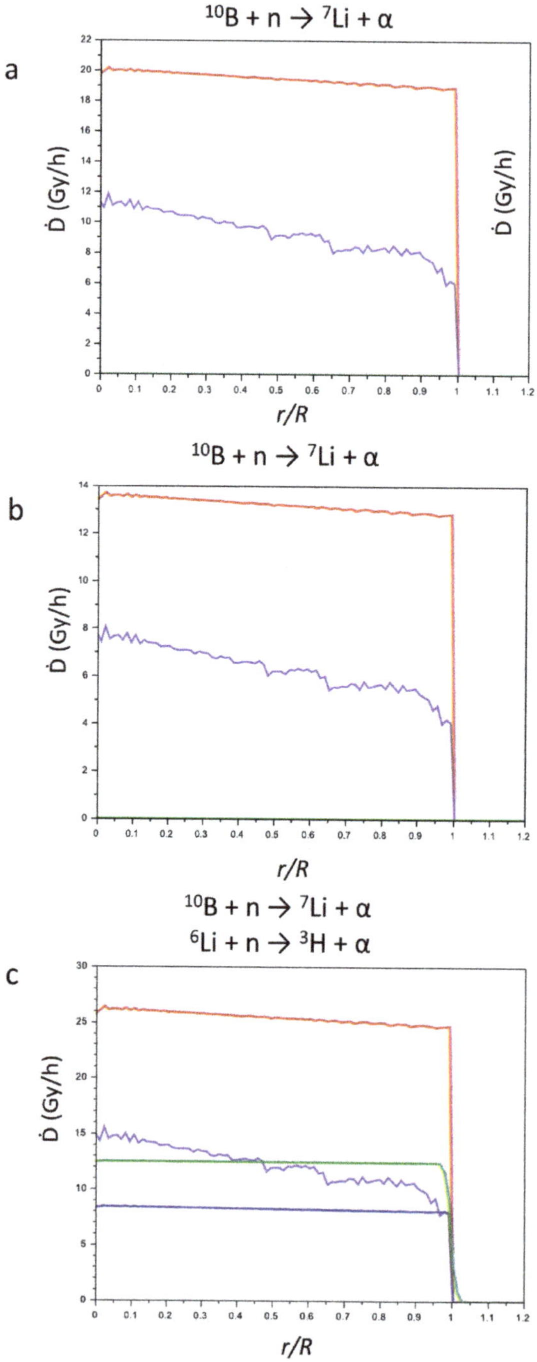

Figure 6. Contributions of the products of the nuclear reactions (1) and (2) to the Ḋ values vs. r/R for CaSB (**a**), CuSB (**b**), and LiSB (**c**). ^{7}Li, violet line; α from Equation (1), red line; ^{3}H, green line; α from Equation (2), blue line.

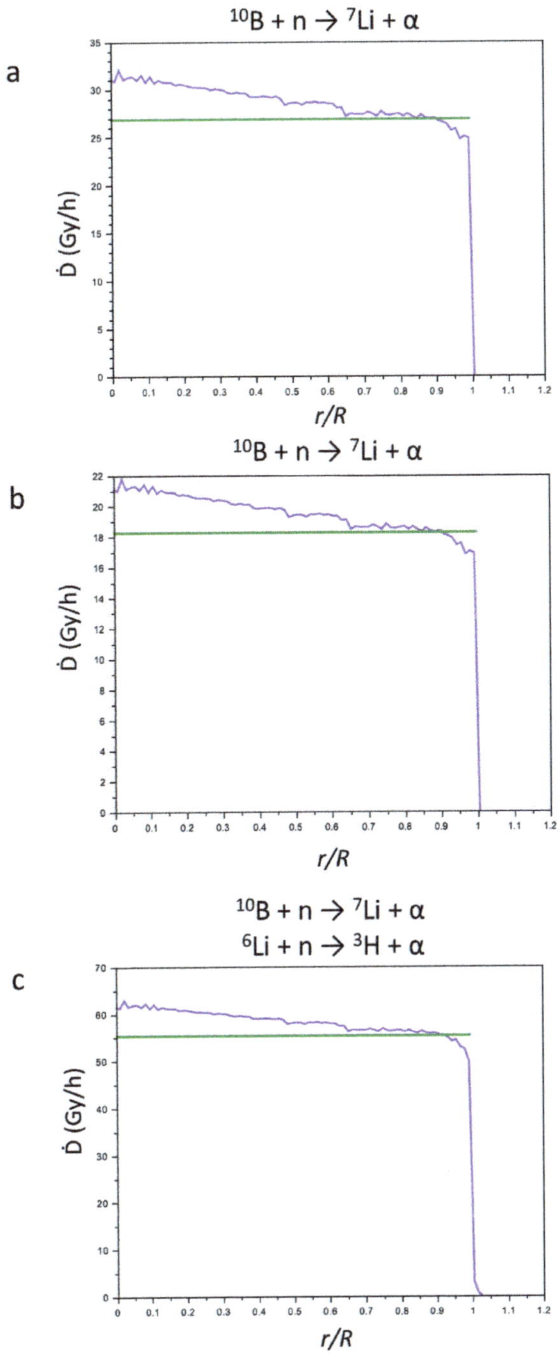

Figure 7. Sum (blue line) of the contribution of the $\dot{D}$ values of the products of the nuclear reactions for CaSB (**a**), CuSB (**b**), and LiSB (**c**). The green line represents the $\dot{D}$ values averaged on the sphere of the tumor mass.

Table 2. Densities of the isotopes ^{10}B and ^{6}Li and total $\dot{D}$ values for all the compounds.

	V_{cell} (Å^3)	N° B/cell, N° Li/cell	^{10}B Density (N° ^{10}B/m^3)	^{6}Li Density (N° ^{6}Li/m^3)	^{10}B + ^{6}Li Density	Total $\dot{D}$ (Gy h^{-1})
CaSB	3545.38	8, -	2.234×10^{27}	-	2.234×10^{27}	26.9
CuSB	1304.51	2, -	1.518×10^{27}	-	1.518×10^{27}	18.3
LiSB	1354.3	4, 4	2.924×10^{27}	2.806×10^{27}	5.555×10^{27}	55.4
LiMB	996.06	1, 1	9.939×10^{26}	9.538×10^{26}	1.948×10^{27}	18.8
LiTB	3578.61	4, 20	1.107×10^{27}	5.309×10^{27}	6.416×10^{27}	51.6

Table 2 also reports the density of the reactive isotopes in the materials, which was calculated from the density of the boron and lithium atoms in the crystal structure, taking into account the percentage of each isotope in the commercial enriched reagents (99% for $H_3{}^{10}BO_3$ and 95% for $^{6}LiCO_3$).

The total $\dot{D}$ values of CaSB, CuSB, and LiSB vary according to the densities of the reactive isotopes. The addition of ^{6}Li to the compounds enhances the efficiency of the LiSB for the NCT.

In Scheme 2, the molecular schemes of LiMB and LiTB, whose crystal structure was published in our previous work [7], are reported. In Figure 8, for these two compounds, the same graphs as in Figures 6 and 7 are reported.

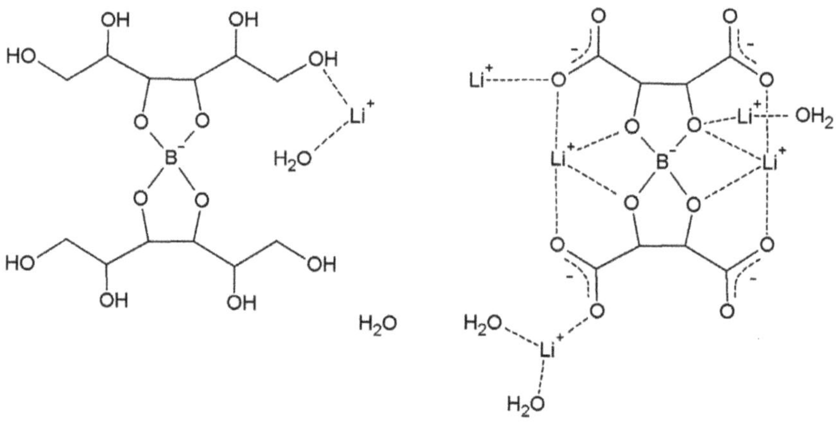

LiMB LiTB

Scheme 2. Molecular schemes of LiMB and LiTB.

In LiMB and LiTB, both boron and lithium are present, as in LiSB, but the averaged $\dot{D}$ values are lower since their ^{10}B + ^{6}Li density is also lower (Table 2). However, the averaged $\dot{D}$ values for LiTB are slightly lower than for LiSB, despite the fact that the ^{10}B + ^{6}Li density is greater. This indicates that the largest contribution to the averaged $\dot{D}$ values is from the products of the boron reaction rather than the lithium ones, since the cross section of Equation (1) is higher than that of Equation (2), and the higher presence of the ^{6}Li in the LiTB is not able to counterbalance the contribution of the ^{10}B. Thus, to design more efficient compounds for the NCT, it is more convenient to enhance the density of ^{10}B rather than the density of ^{6}Li. However, in a chemical synthesis, the lithium cation is easier to introduce in a compound compared to the boron atom, and thus, ^{6}Li can be an easy expedient to enhance the performance of the compounds for the NCT.

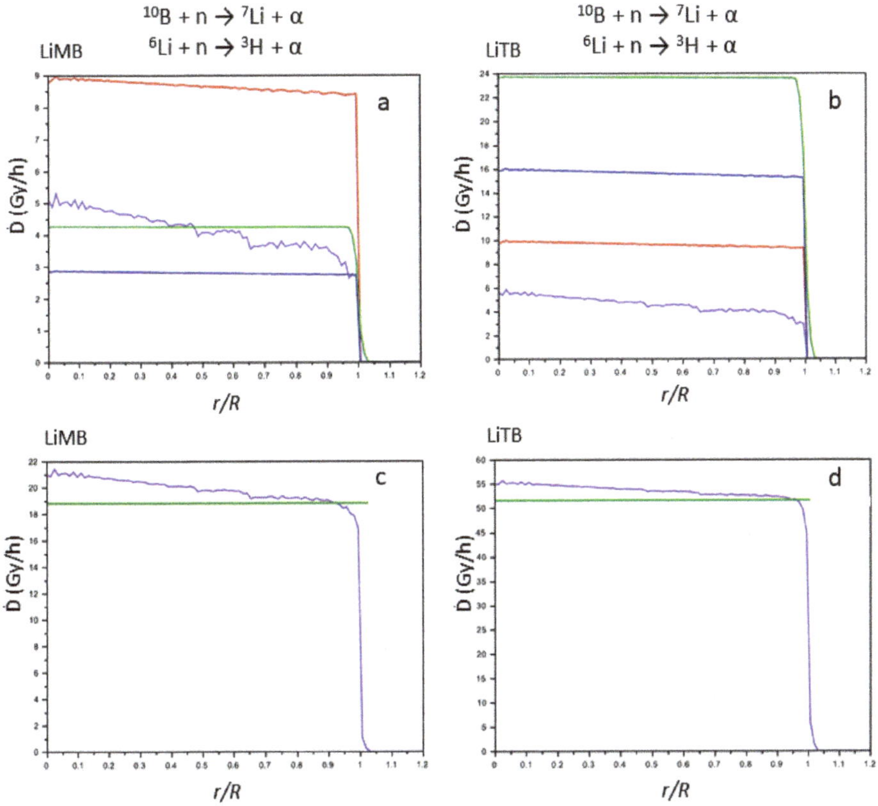

Figure 8. Contributions of the products of the nuclear reactions to the Ḋ values (Gy h^{-1}) vs. (r/R) for LiMB (**a**) and LiTB (**b**) (^{7}Li, violet line; α from Equation (1), red line; ^{3}H, green line; α from Equation (2), blue line). Sum (blue line) of the contributions of the products of the nuclear reactions for LiMB (**c**) and LiTB (**d**) (the green line represents the Ḋ values averaged on the sphere of the tumor mass).

3. Materials and Methods

3.1. Syntheses

All the reagents were purchased from Sigma-Aldrich (Darmstadt, Germany) and used as received.

3.1.1. Synthesis of $[Ca(H_2O)_6](C_{14}H_8O_6B)_2 \cdot 4H_2O$ (CaSB)

A total of 5 mmol of salicylic acid, 1.25 mmol of calcium hydroxide, and 2.50 mmol of boric acid were dissolved in 10 mL of warm water (323 K). The solution was evaporated at room temperature for a few days, and during the evaporation, faint green crystals of CaSB precipitated. The crystals were dried at room temperature.

3.1.2. Synthesis of $[Cu(C_{14}H_8O_6B)_2]$ (CuSB)

A total of 5 mmol of salicylic acid, 1.25 mmol of copper hydroxide, and 2.50 mmol of boric acid were dissolved in 10 mL of warm water (323 K) to obtain a light green solution with some undissolved residue. After keeping steel the solution at 288 K, crystalline needles formed in a few days. In about two more days, some of the crystals grew to fairly large green transparent crystals, which were filtered and dried in air. In a closed vial, the crystals slowly lost water and, after a few months, yielded dark green crystals of CuSB.

3.1.3. Synthesis of $[Li(C_{14}H_8O_6B)(H_2O)]$ (LiSB)

A total of 5 mmol of salicylic acid (5 mmol), 1.5 mmol of lithium carbonate, and 2.5 mmol of boric acid were added to about 5 mL of ethanol, and the mixture was warmed at 323 K until all the solids were dissolved. The solution was slowly evaporated at 290 K, and after a few hours, a white crystalline powder of LiSB precipitated. The crystals were dried at room temperature.

3.2. Single Crystal X-Ray Diffraction (XRD)

X-ray diffraction data were collected at room temperature using a Xcalibur AtlasS2 Gemini R Ultra diffractometer (Rigaku Oxford Diffraction, Wrocław, Poland). Data were collected with mirror monochromatized Cu-Kα (1.5418 Å) radiation. The CrysAlisPro (Version 1.171.37.31) [11] package was used for data collection and integration, SHELXT [12] for solving, SHELXL [13] for refinement, and Olex2 [14] for graphics. All atoms but hydrogen were refined with anisotropic thermal factors. Even if the hydrogen peaks were observed in the difference Fourier maps, hydrogen atoms were calculated and refined riding with the U_{iso} = 1.2 or 1.5 U_{eq} of the connected atom.

3.2.1. Crystal Data for CaSB

Monoclinic, space group $P2_1/c$, Z = 4, a = 9.8996(3), b = 12.4244(3) Å, c = 28.8806(8), β = 93.556(2)°, V = 3545,4(2) $Å^3$, 49769 reflections collected of which 6194 unique (R_{int} = 0.0617). R1 = 0.1031 (I > 2σ(I)), wR2 = 0.3803 (all data).

3.2.2. Crystal Data for CuSB

Monoclinic, space group $P2/n$, Z = 2, a = 12.6250(2), b = 6.5216(1) Å, c = 16.7556(2), β = 108.988(1)°, V = 1304.51(3) $Å^3$, 12674 reflections collected of which 2317 unique (R_{int} = 0.0235). R1 = 0.0300 (I > 2σ(I)), wR2 = 0.0815 (all data).

3.2.3. Crystal Data for LiSB

Monoclinic, space group $P2_1/c$, Z = 4, a = 10.720(5), b = 8.347(4) Å, c = 15.446(6), β = 101.52(4)°, V = 1354.3(10) $Å^3$, 6161 reflections collected of which 1417 unique (R_{int} = 0.1003). R1 = 0.0604 (I > 2σ(I)), wR2 = 0.1576 (all data).

The interested reader can find further details on crystal data, data collection, least-squares refinements, and bond angles in the Supporting Information (Tables S1–S4, Section S1) and CIF files (CCDC 2408779-2408781).

3.3. Calculation of the Dose (D) Absorbed by a Tumor Mass

The algorithm begins by calculating the loss of energy in each point along a radius of the spherical tumor (in our simulation, 100 points) from the flux of particles emitted by each element of volume in the tumor. This flux is, in turn, evaluated by calculating the nuclear reaction rate in each element of volume and integrating it over the sphere, taking into account the distance between the element of volume emitting the radiation and the point where we want to obtain the dose rate. Details are reported in the Supporting Information (Section S2).

The reaction cross sections were evaluated by averaging the experimental $\sigma(\epsilon)$ over the Boltzmann distribution at 311 K.

The thermal neutron flux used is $2.35 \times 10^{13}\ m^{-2}s^{-1}$. As suggested by the International Atomic Energy Agency for BNCT applications [15], a thermal distribution was used for the neutron flux. In a spherical tumor of arbitrary diameter 5 mm, 100 nanoparticles of diameter 175 nm of each compound were considered inside each tumor cell.

4. Conclusions

In this work, we reported three new solid salicylatoborate complexes for application as nanoparticles for the NCT. These nanoparticles are inexpensive and non-toxic materials, easily functionalizable on the surface in order to improve their concentration in tumor tissues. The XRD structures were of fundamental importance for the calculation of the densities of the reactive boron and lithium isotopes, which allowed us to employ a computational method to calculate the dose absorbed by the spherical tumor mass in order to preliminarily test their therapeutic efficacy for the NCT. This method was also applied to two other similar compounds containing boron and lithium with mannitol or tartaric acid, published in our previous work.

The calculated dose rate is in good agreement with the values of the ^{10}B + ^{6}Li densities in the compounds, except for the comparison of the $\dot{D}$ values for LiTB and LiSB, due to the higher cross section of the ^{10}B reaction compared to the ^{6}Li.

Among all the compounds studied, the LiSB gives the highest dose absorbed by the tumor mass, and thus it results in being the most efficient compound according to our model calculations, demonstrating that the synergy between the boron-10 and the lithium-6 isotopes enhances the efficacy of the NCT.

The density of the reactive nuclides in a compound is the only parameter that we can partially control. It can be precisely calculated from the XRD structure and thus allows both for the calculation of the dose absorbed by a tumor and the quantification of the amount of drug needed for a given therapy. Therefore, the method used in this work can be useful to both determine the therapeutic efficacy of a new compound and help design new ones with higher efficiency for the NCT. As expected, along with the density of the neutron-reactive isotopes in the material, among the most important parameters influencing the absorbed dose by a tumor mass are the cross sections for neutron capture of the nuclides involved in the nuclear reactions.

In conclusion, in this work, we propose a simple and relatively fast method for the preliminary discrimination of the most promising compound for therapeutic NCT applications. The assumptions made in the tumor-nanoparticle model used for the calculations do not take into account any type of tumor cells or other biological parameters, but once the most promising compound has been chosen, we can focus future experimental efforts on it, limiting the time and costs of the research.

Supplementary Materials: The following supporting information can be downloaded at: https://www.mdpi.com/article/10.3390/inorganics13050136/s1, Table S1: Crystal data and structure refinements for CaSB, CuSB, and LiSB. Table S2: Bond Angles for CaSB. Table S3: Bond Angles for CuSB. Table S4: Bond Angles for LiSB.

Author Contributions: Conceptualization, D.M. and C.C.; data curation, C.C.; formal analysis, A.C.; funding acquisition, P.B.; investigation, D.M. and A.C.; methodology, D.M. and C.C.; validation, P.B. and A.C.; writing—original draft, D.M.; writing—review and editing, P.B. and C.C. All authors have read and agreed to the published version of the manuscript.

Funding: This work was supported by MUR (Ministero dell'Università e della Ricerca), funded by the Italian Government.

Institutional Review Board Statement: Not applicable.

Informed Consent Statement: Not applicable.

Data Availability Statement: Data are contained within the article and Supplementary Materials.

Conflicts of Interest: The authors declare no conflicts of interest.

References

1. Wang, S.; Zhang, Z.; Miao, L.; Li, Y. Boron Neutron Capture Therapy: Current status and challenges. *Front. Oncol.* **2022**, *12*, 788770. [CrossRef] [PubMed]
2. Nedunchezhian, K.; Aswath, N.; Thiruppathy, M.; Thirugnanamurthy, S. Boron Neutron Capture Therapy–A Literature Review. *J. Clin. Diag. Res.* **2016**, *10*, ZE01–ZE04. [CrossRef] [PubMed]
3. Moss, R.L. Critical review, with an optimistic outlook, on Boron Neutron Capture Therapy (BNCT). *Appl. Rad. Isot.* **2014**, *88*, 2. [CrossRef] [PubMed]
4. Hwang, K.C.; Lai, P.D.; Chiang, C.-S.; Wang, P.-J.; Yuan, C.-J. Neutron capture nuclei-containing carbon nanoparticles for destruction of cancer cells. *Biomaterials* **2010**, *31*, 8419. [CrossRef] [PubMed]
5. Kuthala, N.; Vankayala, R.; Li, Y.-N.; Chiang, C.-S.; Hwang, K.C. Engineering novel targeted boron-10-enriched theranostic nanomedicine to combat against murine brain tumors via MR imaging-guided boron neutron capture therapy. *Adv. Mater.* **2017**, *29*, 1700850. [CrossRef] [PubMed]
6. Kuthala, N.; Shanmugam, M.; Yao, C.-L.; Chiang, C.-S.; Hwang, K.C. One step synthesis of ^{10}B-enriched $^{10}BPO_4$ nanoparticles for effective boron neutron capture therapeutic treatment of recurrent head-and-neck tumor. *Biomaterials* **2022**, *290*, 121861. [CrossRef] [PubMed]
7. Marabello, D.; Benzi, P.; Beccari, F.; Canepa, C.; Cariati, E.; Cioci, A.; Costa, M.; Durisi, E.A.; Monti, V.; Sans Planell, O.; et al. Synthesis and characterization of new lithium and boron based Metal Organic Frameworks with NLO properties for application in Neutron Capture Therapy. *Processes* **2020**, *8*, 558. [CrossRef]
8. Köse, D.A.; Zümreoglu-Karan, B.; Hökelek, T.; Şahin, E. Boric acid complexes with organic biomolecules: Mono-chelate complexes with salicylic and glucuronic acids. *Inorg. Chim. Acta* **2010**, *363*, 4031. [CrossRef]
9. Marabello, D.; Antoniotti, P.; Benzi, P.; Beccari, F.; Canepa, C.; Barge, A.; Boscaro, V.; Gallicchio, M.; Peira, E. Synthesis, characterization and cell uptake of nanoparticles for a novel approach to radionuclide therapy: A feasibility study. *Int. J. Res. Pharm. Nano Sci.* **2019**, *8*, 230.
10. Selvaraj, S.; Chabita (Saha), K.; Chakrabortya, A.; Bhattacharyya, S.N.; Saha, A. Influence of copper (II) ions and its derivatives on radiosensitivity of *Escherichia coli*. *Rad. Phys. Chem.* **2007**, *76*, 1160. [CrossRef]
11. *CrysAlisPro*, Version 1.171.37.31; Release 14-01-2014 CrysAlis171.NET; Agilent Technologies UK Ltd.: Oxford, UK, 2014.
12. Sheldrick, G.M. SHELXT-Integrated Space-Group and Crystal-Structure Determination. *Acta Cryst. A* **2015**, *71*, 3–8. [CrossRef] [PubMed]
13. Sheldrick, G.M. Crystal Structure Refinement with SHELXL. *Acta Cryst. C* **2015**, *71*, 3–8. [CrossRef] [PubMed]
14. Dolomanov, O.V.; Bourhis, L.J.; Gildea, R.J.; Howard, J.A.K.; Puschmann, H. OLEX2: A Complete Structure Solution, Refinement and Analysis Program. *J. Appl. Cryst.* **2009**, *42*, 339. [CrossRef]
15. International Atomic Energy Agency. Physical Dosimetry and Determination of Neutron Field Parameters. In *Advances in Boron Neutron Capture Therapy*; International Atomic Energy Agency: Vienna, Austria, 2023; Chapter 4; p. 29. Available online: https://www-pub.iaea.org/MTCD/Publications/PDF/CRCP-BOR-002_web.pdf (accessed on 18 March 2025).

inorganics

MDPI

Article

Significant Enhancement of Strength and Ductility in Bioresorbable Zn–0.1Mg Alloy via ECAP Processing

Iryna Cuperová [1,*], Martin Fujda [1], Róbert Kočiško [1], Patrik Petroušek [1], Zuzana Molčanová [2], Miloš Matvija [1], Róbert Džunda [2], Beáta Ballóková [2], Dávid Csík [2], Katarína Gáborová [1] and Karel Saksl [1,2]

1 Institute of Materials, Faculty of Materials, Metallurgy and Recycling, Technical University of Košice, Letná 1/9, 042 00 Košice, Slovakia; martin.fujda@tuke.sk (M.F.); robert.kocisko@tuke.sk (R.K.); patrik.petrousek@tuke.sk (P.P.); milos.matvija@tuke.sk (M.M.); katarina.gaborova@tuke.sk (K.G.); ksaksl@saske.sk (K.S.)

2 Institute of Materials Research of SAS, Slovak Academy of Sciences, Watsonova 47, 040 01 Košice, Slovakia; molcanova@saske.sk (Z.M.); rdzunda@saske.sk (R.D.); bballokova@saske.sk (B.B.); dcsik@saske.sk (D.C.)

* Correspondence: iryna.cuperova@tuke.sk; Tel.: +421-55-602-2646

Abstract: Zinc (Zn)-based alloys are considered promising bioresorbable materials for intracorporeal implants due to their good biocompatibility and suitable degradation rate in physiological environments. However, their broader application is hindered by insufficient mechanical properties, which are essential for fulfilling the therapeutic function of bioresorbable implants. This study investigates the effect of severe plastic deformation on the microstructure and mechanical properties of as-cast Zn–0.1Mg (wt.%) alloy. The as-cast alloy, characterised by a coarse-grained microstructure with intermetallic phases at grain boundaries and low strength and ductility, was subjected to two passes of Equal Channel Angular Pressing (ECAP). The intense plastic deformation transformed the coarse-grained structure into an ultrafine-grained solid solution matrix. This substantial microstructural refinement led to a significant enhancement in mechanical performance. The yield strength (YS) and ultimate tensile strength (UTS) more than doubled, reaching 198 MPa and 215 MPa, respectively. Remarkably, the elongation increased from 2.2% to 187% in tensile testing. These findings confirm the beneficial effect of grain refinement and dynamic recrystallisation on the mechanical behaviour of bioresorbable Zn–0.1Mg alloy and highlight the high potential of ECAP processing for optimising the mechanical properties of Zn-based biodegradable materials.

Keywords: bioresorbable Zn–Mg alloys; severe plastic deformation; ECAP processing; microstructure; strength; plasticity; biomedical applications

1. Introduction

Nowadays, research in the field of biomaterials has increasingly focused on biodegradable materials intended for intracorporeal implants [1]. Many studies have recently been published on biodegradable alloys based on magnesium (Mg), iron (Fe), and zinc (Zn). The main motivation of the biodegradable implants' development is their ability to gradually degrade in physiological environments. The corrosion products of these materials are expected to elicit an appropriate host response, enabling their metabolism or assimilation by cells and tissues. Once the implant has fulfilled its therapeutic role in tissue regeneration, it is fully resorbed without leaving any residual material in the body [2,3].

In contrast to conventional corrosion-resistant alloys—such as stainless steels and Ni-, Ti-, Co-, and Cr-based alloys—which are commonly used in permanent implants for osteosynthesis, osteotomy, or the treatment of coronary artery stenosis [4–6], biodegradable alloys offer excellent biocompatibility without symptoms of local or systemic toxicity. Furthermore, they possess mechanical properties close to those of bone tissue and naturally resorb after fulfilling their therapeutic function, thus eliminating the need for secondary surgical explantation [1,3,7–9].

Zn-based alloys are increasingly being recognised as promising bioactive materials for the development of bioresorbable implants. Zinc, an essential trace element, plays a key role in various biological processes in the human body. Approximately 85% of total body Zn is found in bones and muscles, 11% in the skin and liver, and the remaining fraction in vital organs such as the brain, kidneys, and pancreas [10–12]. Its biodegradable character and excellent biocompatibility have been confirmed by several in vivo studies [9,13,14].

In addition to favourable biological properties, Zn-based alloys show potential to overcome several limitations associated with the more intensively studied Mg- and Fe-based biodegradable materials [15]. Although Mg alloys are attractive due to the fact that their Young's modulus (41–45 GPa) is close to that of natural bone [3], the biocompatibility of Mg^{2+} ions [16], and their ability to stimulate bone regeneration [17], their primary drawback is an excessively high degradation rate. This leads to intense hydrogen evolution and local alkalisation, potentially triggering inflammatory responses, implant destabilisation, and premature failure [1,18]. On the other hand, Fe and its alloys are popular among biomaterials researchers due to their high mechanical strength and the absence of hydrogen evolution during degradation. However, their extremely slow degradation rate and the accumulation of corrosion products that are difficult to eliminate limit their clinical applicability [3,19,20]. Additionally, their magnetic properties may interfere with magnetic resonance imaging, thereby reducing diagnostic quality post-implantation [21].

In this context, Zn alloys represent a balanced alternative. Their corrosion potential (-0.76 V) lies between that of Fe (-0.44 V) and Mg (-2.37 V) [1], ensuring an optimal degradation rate. Hydrogen is not released during Zn degradation, and the Zn^{2+} ions are naturally integrated into metabolic pathways without toxic effects [22,23]. An in vivo experiment by Yang et al. [14] demonstrated that a zinc stent implanted into a rabbit aorta retained mechanical integrity for the first 6 months and degraded by approximately 40% of its volume after 12 months, without signs of inflammation or thrombosis, indicating excellent biological tolerance.

On the other hand, the mechanical properties of pure as-cast Zn (ultimate tensile strength (UTS) $\approx$ 30 MPa, yield strength (YS) $\approx$ 27 MPa, elongation (A) < 1%) [24] fall significantly short of the minimum requirements for orthopaedic implant materials, where mechanical thresholds are typically set at YS > 230 MPa, UTS > 300 MPa, and elongation (A) > 15% [25]. One effective strategy to enhance the mechanical performance of pure Zn is alloying with biocompatible and non-toxic elements such as Mg, Cu, Ca, Sr, Li, Ag, or Mn. Among these, Mg and Li provide the most pronounced strengthening effect [24–26]. Even microalloying Zn with just 0.1 wt.% Mg increases its strength to ~80 MPa; however, elongation remains limited to approximately 1% [24]. Zn–Mg alloys are among the most extensively investigated biodegradable systems, with their biological safety repeatedly confirmed by cytotoxicity and genotoxicity studies [27,28]. The presence of Mg in the alloy contributes to regulating Zn^{2+} ion release and maintaining its concentration within a physiologically safe range. Therefore, Zn–Mg alloys exhibit lower cytotoxicity and reduced inflammatory responses compared to pure Zn.

To significantly improve the mechanical properties of Zn–Mg alloys, thermomechanical processing techniques based on plastic deformation have proven effective. For example, cold rolling of Zn–0.6Mg (wt.%) alloy with a 75% thickness reduction increased its tensile strength to ~380 MPa while maintaining sufficient ductility (~26%) [29]. Particular research attention has been directed toward the application of Equal Channel Angular Pressing (ECAP), which enables intense grain refinement through severe plastic deformation without altering the macroscopic geometry of the sample [30]. Several studies have demonstrated the high efficiency of ECAP in enhancing the strength while retaining good ductility in microalloyed Zn–Mg alloys [31–33]. Despite these promising outcomes, the synergistic effect of ECAP processing on the microstructural and mechanical evolution of Zn alloys—in contrast to the extensively studied Mg systems—remains insufficiently explored. Therefore, comprehensive research into the deformation processing of Zn–Mg alloys is necessary to deepen our understanding of their deformation mechanisms and to optimise processing parameters for biomedical applications [32].

The objective of the present study is to analyse the effect of ECAP processing of cold-deformed, as-cast microalloyed Zn–0.1Mg alloy on its microstructure and mechanical properties, and to assess the potential of this technique for optimising material characteristics of bioresorbable metallic implants.

2. Results

2.1. Microstructural Analysis of Zn–0.1Mg Alloy in As-Cast and Equal Channel Angular Pressing (ECAP)-Processed States

The microstructure of the as-cast Zn–0.1Mg alloy (Figure 1a) exhibits a polyhedric morphology consisting of coarse equiaxed grains of solid solution. Intermetallic secondary phase particles are located at the grain boundaries. Additionally, spherically shaped pores of varying sizes, uniformly distributed throughout the entire volume of the casting ingots, are present within this structure. Such pore morphology and distribution are characteristic of gravity-cast alloys. The average grain size of the solid solution, determined according to the STN EN ISO 643 standard [34], is 123 ± 18.8 μm.

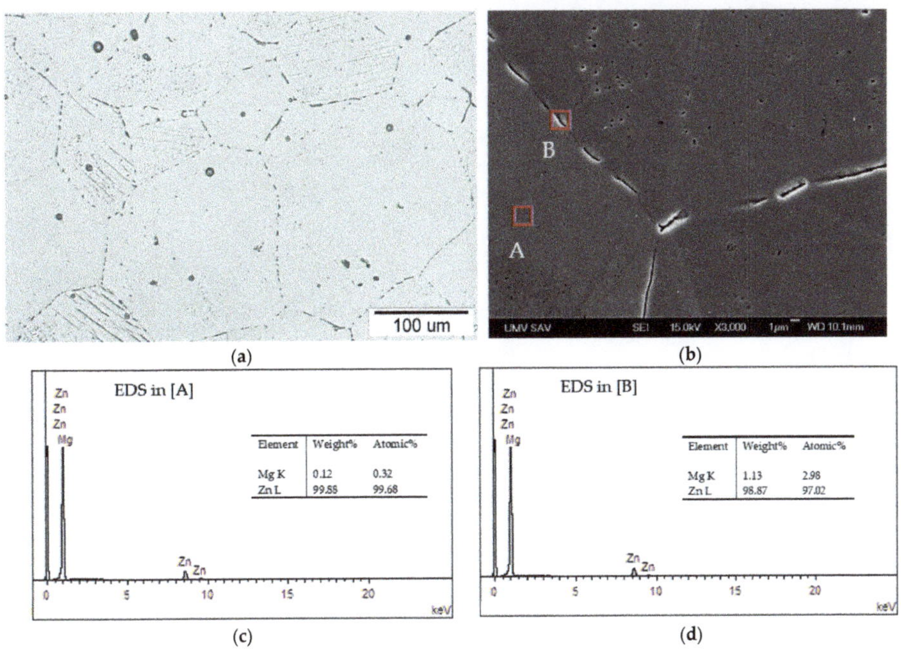

Figure 1. Microstructure of the as-cast Zn–0.1Mg alloy: (**a**) light microscopy (LM); (**b**) SEM microstructural analysis (region A representing the primary phase, and region B the secondary phase); (**c**) EDS analysis of the primary phase; (**d**) EDS analysis of the secondary phase.

Scanning Electron Microscope (SEM) analysis provided detailed insight into the morphology and distribution of the secondary phase (Figure 1b). This secondary phase predominantly precipitated at the grain boundaries in the form of elongated segmented structures. Semi-quantitative Energy-Dispersive X-ray Spectroscopy (EDS) analysis (Figure 1c) confirmed that solid solution grains near the grain boundaries contained slightly higher Mg content (0.12 wt.%) compared to the average Mg content of the alloy. The secondary particles at the grain boundaries contained significantly higher Mg content (1.13 wt.%) (Figure 1d). This observation aligns with the Zn–Mg binary phase diagram, indicating the low solubility of Mg in Zn at room temperature and the formation of the intermetallic Mg_2Zn_{11} phase upon cooling at grain boundaries [13,35].

Severe deformation of the as-cast Zn–0.1Mg alloy through two passes of Equal Channel Angular Pressing (ECAP) altered the microstructural morphology significantly. The longitudinal cross-section of the deformed sample showed elongation of polyhedric grains along the longitudinal axis of the sample (Figure 2a,b). Parallel deformation bands, etched and oriented in the slip direction induced by ECAP processing, were observed within the grains. Intermetallic phase particles were primarily located at the original grain boundaries (Figure 2c—point C), though occasionally found inside the grains at interfaces between deformation bands (Figure 2c—point B). The chemical composition of these deformation bands, analysed by EDS, corresponded to a slightly supersaturated solid solution of Mg in Zn (0.25 wt.% Mg—Figure 2d). Globular particles with elevated Mg content (6.26 wt.% Mg—Figure 2e) were sporadically observed. According to EDS analysis, elongated narrow intermetallic phase structures at the former grain boundaries also had higher Mg content (0.62 wt.% Mg—Figure 2f).

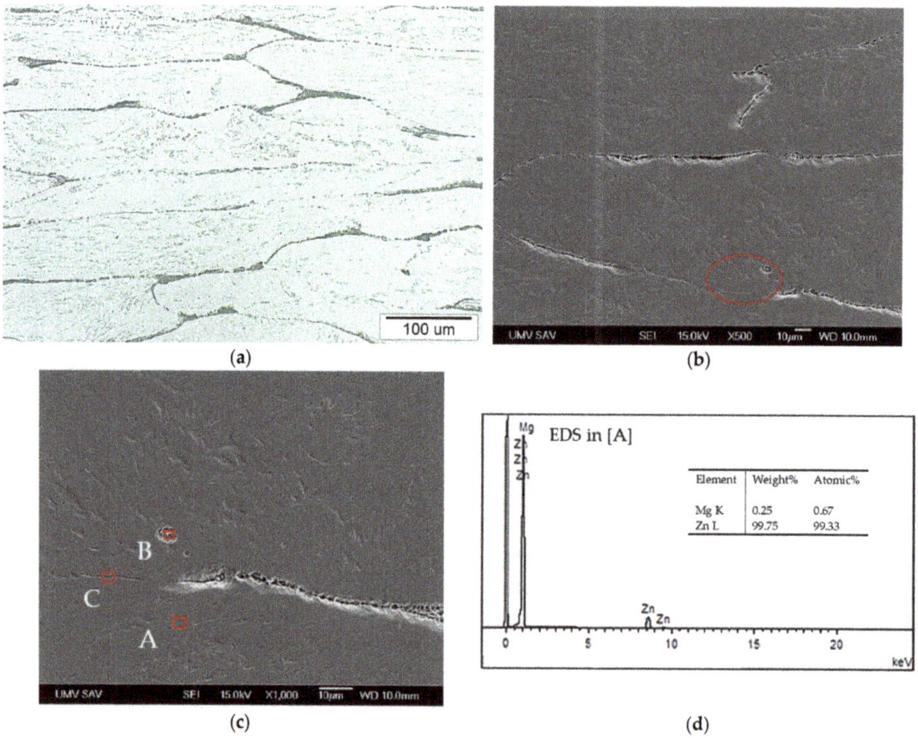

Figure 2. *Cont.*

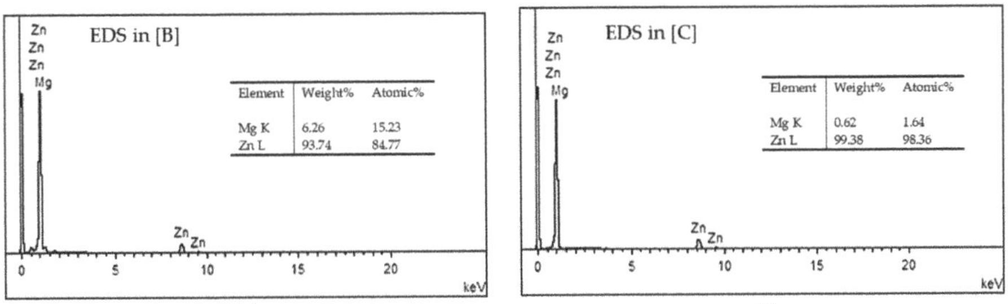

(e) (f)

Figure 2. Microstructure of the Zn–0.1Mg alloy after ECAP processing: (**a**) LM micrograph; (**b**) SEM micrograph (the area marked with a red circle indicates the region selected for detailed SEM observation); (**c**) detailed SEM view; (**d–f**) EDS analysis in regions [A], [B], and [C], respectively.

Detailed SEM analysis of the Zn–0.1Mg alloy microstructure after ECAP processing (Figure 3) revealed etched grain boundaries indicative of dynamically recrystallised ultrafine grains of the α-Zn solid solution within the deformation bands, exhibiting the highest degree of plastic deformation. These grains predominantly displayed equiaxed morphology, with slight elongation in the deformation direction and an average grain size around 1.5 μm.

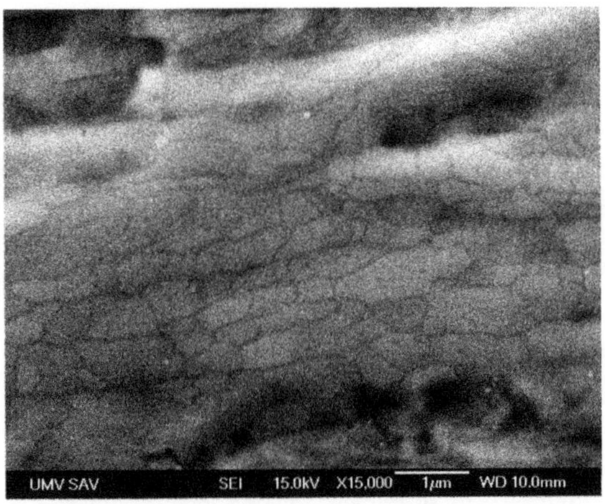

Figure 3. SEM micrograph of Zn–0.1Mg alloy after ECAP processing.

Additional phase and structural analysis of the Zn–0.1Mg alloy states was performed using X-ray diffraction (XRD). Diffraction patterns of both as-cast and ECAP-processed Zn–0.1Mg alloys (represented by blue symbols in Figure 4) revealed only Bragg peaks corresponding to the hexagonal hcp-Zn phase with the space group $P6_3/mmc$ (194). No diffraction peaks corresponding to Mg-rich intermetallic phases were detected, reflecting their low volume fraction, likely below the detection limit of XRD analysis. Results of microstructural refinement of the Zn phase using Rietveld analysis are graphically presented in Figure 4a,b by the solid green line, and refined microstructural parameters are summarised in Table 1.

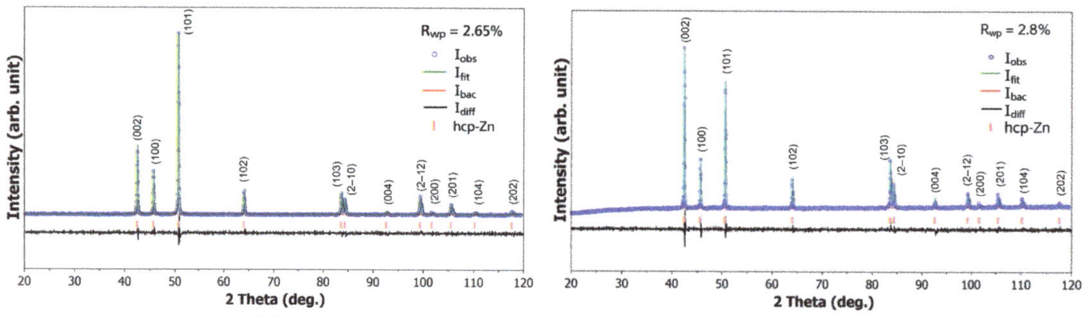

(a) (b)

Figure 4. The XRD patterns and Rietveld fit of: (**a**) the as-cast Zn–0.1Mg alloy sample; (**b**) the ECAPed Zn–0.1Mg alloy state.

Table 1. Microstructural parameters of the hcp-Zn phase refined by Rietveld analysis of the diffraction profile fitting. Lattice parameters (a, c), isotropic displacement parameter (U_{iso}), microstrain (ε), and texture index.

Sample	Phase	a [Å]	c [Å]	U_{iso} [Å^2]	ε [%]	Texture Index
As-cast Zn–0.1Mg alloy	hcp-Zn	2.66(4)	4.94(6)	0.0166	0.0020	1.018
ECAPed Zn–0.1Mg alloy	hcp-Zn	2.66(4)	4.94(6)	0.0167	0.0014	1.939

The lattice parameters of the hcp-Zn phase are identical for both samples and closely match the reference values for pure Zn (PDF card no. 00-004-0831, a = 2.665 Å, c = 4.947 Å). The most significant difference was observed in the texture index calculated using the spherical harmonics model. The texture index numerically characterizes the degree of preferred crystallite orientation in the material based on the Orientation Distribution Function (ODF). For an ideally random (isotropic) texture, the texture index equals 1, while higher values indicate stronger preferred crystallite orientation.

2.2. Mechanical Properties of the Zn–0.1Mg Alloys

The mechanical properties of the Zn–0.1Mg alloy in both as-cast and ECAP-processed states were evaluated through tensile testing at room temperature, resulting in the stress–strain curve depicted in Figure 5. The basic mechanical characteristics, including strength and ductility, are summarised in Table 2. The as-cast alloy demonstrated relatively low strength, with yield strength (YS) of 83 MPa and ultimate tensile strength (UTS) of 91 MPa, accompanied by extremely low tensile ductility (elongation measured on a gauge length L_0 = 2d, where d is the original specimen diameter) (A_2 = 2.2%) and reduction in area (Z = 0.2%).

Severe plastic deformation of the as-cast Zn–0.1Mg alloy using ECAP significantly enhanced its mechanical properties. The most pronounced effect of intense plastic deformation via ECAP processing was reflected in an extreme increase in elongation (A_2), reaching 187%, and a substantial improvement in reduction in area (Z) to 91%. Concurrently, the yield strength (YS) increased significantly to 198 MPa, and the ultimate tensile strength (UTS) reached 215 MPa. Furthermore, the conversion of tensile test results from the engineering stress–strain curve into the true stress–strain curve showed that the true ultimate tensile strength (true UTS) after ECAP processing reached up to 340 MPa.

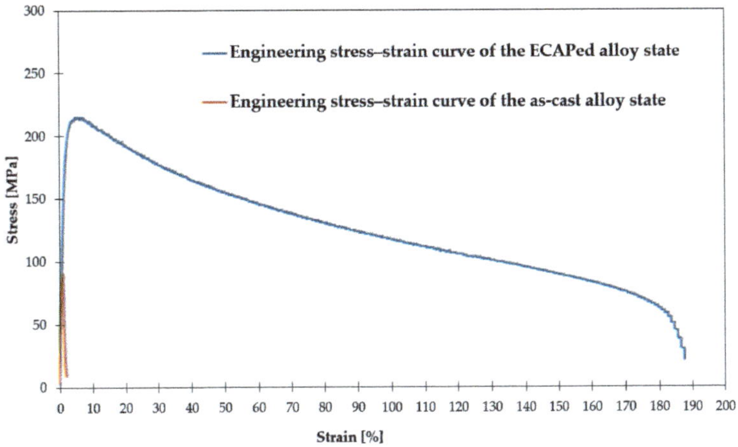

Figure 5. Stress–strain curves of the analysed Zn–0.1Mg alloy state.

Table 2. Mechanical properties of the analysed Zn–0.1Mg alloy state.

Sample	YS [MPa]	UTS [MPa]	A_2 [%]	Z [%]
As-cast Zn–0.1Mg alloy	83	91	2.2	0.2
ECAPed Zn–0.1Mg alloy	198	215	187	91

2.3. Fractographic Analysis of the Zn–0.1Mg Alloys

Fractographic analysis of the fracture surfaces of the as-cast Zn–0.1Mg alloy sample tested under tensile loading reflects the low ductility characteristics of this material state. The fracture surface, resulting from brittle fracture propagation across the grains of the solid solution, predominantly consists of transcrystalline cleavage facets oriented in various directions (Figure 6a). Intergranular areas formed by brittle separation along solid solution grain boundaries were also observed, exhibiting distinct intermetallic phase particles (Figure 6b). EDS analysis revealed elevated magnesium content (5.78 wt.% Mg—Figure 6c) in these particles, suggesting they are the same intermetallic phases identified at grain boundaries within the alloy microstructure (Figure 1).

The fracture surface morphology of the ECAP-processed Zn–0.1Mg alloy sample post-tensile testing exhibits exclusively ductile fracture characteristics (Figure 7a). The fracture surface is notably irregular, featuring large, elongated dimples aligned with larger particles along the original grain boundaries and very fine dimples resulting from transcrystalline ductile fracture through highly deformed solid solution grains. This fracture morphology corresponds with the significantly enhanced ductility and reduction in area observed during tensile testing. Coarser intermetallic phase particles arranged in rows were noted within the larger elongated dimples (Figure 7b). EDS analysis confirmed a high Mg content in these particles (10.54 wt.% Mg—Figure 7d), validating that they correspond to particles observed in the ECAP-processed microstructure along elongated grain boundaries. Elevated Mg concentrations (2.91 wt.% Mg—Figure 7e) were also detected at the ridges of larger dimples. Fine intermetallic phase particles were identified within the smaller dimples resulting from transcrystalline fracture (Figure 7c). These fine particles, formed within originally cast grains, also showed elevated Mg content (2.91 wt.% Mg—Figure 7f), likely corresponding to Mg_2Zn_{11} intermetallic phases.

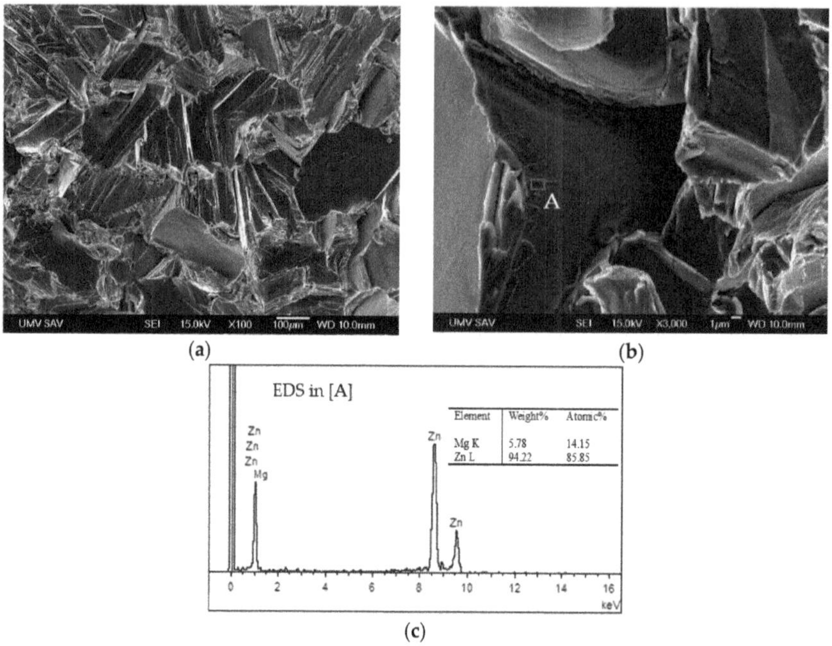

Figure 6. SEM micrographs of fracture surfaces of the as-cast Zn–0.1Mg alloy: (**a**) overview of the fracture surface; (**b**) detailed view of the selected area; (**c**) EDS analysis in area [A].

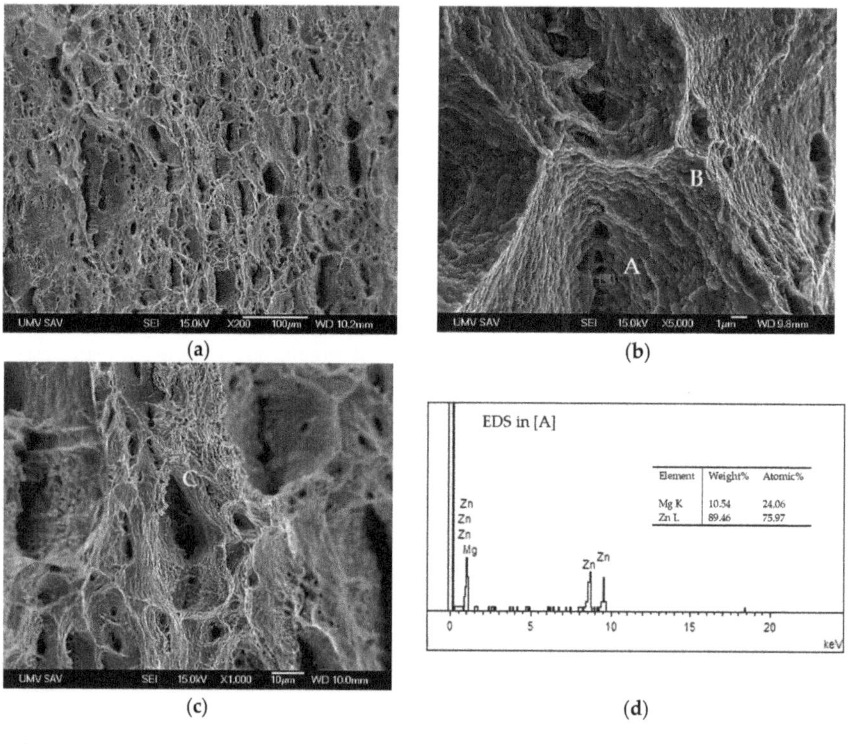

Figure 7. *Cont.*

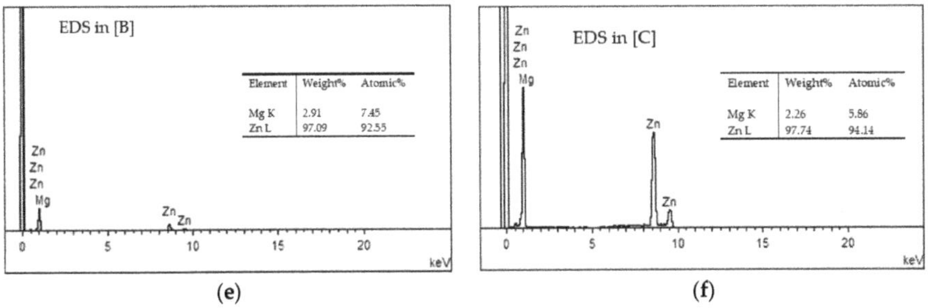

Figure 7. SEM micrographs of fracture surfaces of the Zn–0.1Mg alloy after ECAP processing: (**a**) overview of fracture surface; (**b**,**c**) detailed views of selected areas; (**d**–**f**) EDS analysis in areas [A], [B], and [C], respectively.

3. Discussion

The initial state of the analysed bioresorbable Zn–0.1Mg alloy consists of crystallised castings exhibiting a typical casting microstructure characterised by coarse-grained polyhedric morphology (Figure 1). The primary phase in the microstructure consists of polyhedric equiaxed α-Zn solid solution grains with an average size of approximately 123 μm, formed during alloy solidification and cooling. Upon cooling from the crystallisation temperature, a secondary intermetallic Mg-rich phase precipitated at the grain boundaries in segmented forms due to the decreasing solubility of Mg in Zn as temperature decreased, as described by the Zn–Mg phase diagram. These intermetallic particles represent a hard yet brittle component of the alloy's microstructure [13]. The EDS analysis confirmed increased Mg content within these particles (Figure 1d). However, their precise stoichiometric composition was not conclusively determined. X-ray diffraction (XRD) analysis (Figure 4) identified only the α-Zn phase with an HCP lattice structure in the as-cast alloy. The volume fraction of the intermetallic phase is presumably below the detection limit of diffraction analysis. Nevertheless, based on the Zn–Mg phase diagram [13,35] and literature data [36–39], the observed secondary particles can be presumed to be the Mg_2Zn_{11} intermetallic phase, frequently reported as dominant in Zn–Mg alloys with magnesium content at or above 0.1 wt.%. In alloys with higher Mg content (>0.1 wt.%), metastable $MgZn_2$ phase particles can also form due to localised Mg enrichment during solidification, consistent with the Zn–Mg phase diagram [13,39]. Porosity resulting from gas entrapment during gravitational casting is also a characteristic feature of the as-cast microstructure.

Indications of deformation twinning were observed in some primary α-Zn solid solution grains (Figure 2a), resulting from internal stresses generated during alloy solidification and cooling. Twinning is a dominant plastic deformation mechanism in metals with hexagonal close-packed (HCP) lattices, such as Zn and Mg [40,41]. This phenomenon arises from the limited number of active slip systems (basal and prismatic), insufficient to satisfy von Mises' criterion for plastic compatibility, requiring activation of at least five independent slip systems. Consequently, energetically more demanding pyramidal slip systems of type <c + a> must be activated [36,41]. For Zn and its alloys, the high lattice parameter ratio c/a (~1.856), exceeding the ideal value of 1.633, significantly increases the critical shear stress necessary for activating non-basal slip planes, thus favouring twinning, especially in coarse-grained microstructures and under cold deformation conditions [40,41].

The as-cast Zn–0.1Mg alloy structure was not macroscopically significantly altered by two ECAP passes (route A) at room temperature (Figure 2a,b). Polyhedric cast grains were elongated in the direction of extrusion axis during ECAP, with intermetallic phase particles remaining at grain boundaries. Intense plastic deformation resulted in the formation of deformation bands in the α-Zn grains oriented in the direction of shear within the

ECAP die, and eliminated porosity (Figure 1a). Redistribution of Mg-rich intermetallic phase particles occurred, observed both at original grain boundaries and within the grain interiors (Figure 2c), attributed to intense shear deformation within the ECAP die. XRD analysis confirmed identical phase compositions of Zn–0.1Mg alloy in both as-cast and ECAP-processed states. No intermetallic phases were detected through diffraction analysis. However, significant textural development with a texture index of 1.939 was confirmed by XRD, indicating preferred crystallographic orientation of grains/subgrains after ECAP, consistent with literature findings [29,31,42].

Detailed SEM analysis revealed ultrafine dynamically recrystallised grains/subgrains of solid solution smaller than 1.5 μm in deformation bands (Figure 3). This microstructural refinement is attributed to dynamic recrystallisation induced by severe plastic deformation. Room temperature recrystallisation was facilitated by high deformation intensity and extremely low recrystallisation temperatures (~−12 °C) of Zn–Mg alloys [40]. Mg-rich intermetallic particles, such as Mg_2Zn_{11}, significantly aided grain nucleation during dynamic recrystallisation by acting as effective nucleation sites due to localised stress concentrations [24]. In contrast to the as-cast condition, ECAP processing eliminated signs of twinning (Figure 2a), due to grain refinement, shifting the dominant deformation mechanism to the dislocation slip [40].

Mg-rich intermetallic particles, presumably the Mg_2Zn_{11} phase, play a significant role in the nucleation of new grains due to local stress concentration fields, acting as effective nucleation centres for dynamically recrystallised grains. Thus, they substantially enhance dynamic recrystallisation and accelerate microstructural refinement [24]. After ECAP processing, no signs of twinning were observed (Figure 2a), in contrast to the as-cast state, where twinning was present. This phenomenon is related to significant grain refinement, which suppresses twinning and transitions the deformation mechanism to the dominant dislocation slip [40]. The critical stress required for twinning activation increases significantly faster with decreasing grain size than the stress required for slip activation. According to the available literature [40,41,43–45], refining the microstructure of hexagonal close-packed (HCP) materials below a critical grain size (typically in the range of 1–10 μm) entirely suppresses twinning as a deformation mechanism. Consequently, dislocation slip mechanisms dominate, fundamentally affecting the plastic deformation behaviour. Li et al. [43], for example, demonstrated that reducing the grain size below 2.7 μm during ECAP processing of pure Mg not only suppresses twinning, but also results in reverse Hall–Petch behaviour characterised by reduced strength properties. Similar results were observed in Zn–Mn alloys, where grain size reduction from 4 μm to 2 μm led to a decrease in the volume fraction of twins from 14.2% to 2.3% [45]. Additionally, intermetallic particles present in the microstructure contribute to twinning suppression by acting as barriers to twin propagation [24]. When dislocation migration across grain boundaries and transcrystalline plastic deformation is significantly limited due to the absence of equivalent slip planes in adjacent grains, macroscopic deformation must proceed through alternative mechanisms. Under these conditions, grain boundary sliding (GBS) becomes the dominant mechanism ensuring continued plastic deformation [40].

The observed significant submicroscopic structural differences markedly influenced the mechanical behaviour of the Zn–0.1Mg alloy in the as-cast and ECAP-processed states. The coarse-grained microstructure of the as-cast state with twinning predisposed it to brittle fracture. In brittle HCP materials with only one active twinning mechanism, interaction between twins generates local stress concentrations, acting as initiators of microcracks and promoting further twinning, cyclically reducing material toughness [24,43]. This assumption was confirmed by tensile testing of the as-cast Zn–0.1Mg alloy, revealing relatively low yield strength (YS) of 83 MPa, low ultimate tensile strength (UTS) of 91 MPa, and extremely low elongation (A_2) of 2.2% (Table 2, Figure 5). The reduction in area (Z) during tensile testing reached only 0.2%, indicating minimal plastic deformation before fracture. Fractographic analysis of fracture surfaces in the as-cast state (Figure 6) demonstrated dominant transcrystalline cleavage facets and intercrystalline separation, fundamental signs of brittle fracture. Sharp and uneven edges of cleavage facets indicate local slip and grain rotation before fracture, consistent with minimal elongation and confirming macroscopic brittleness. Similar low mechanical properties of as-cast Zn–Mg alloys were reported by other studies [13,16,24,46], where UTS typically did not exceed 150 MPa, YS was below 110 MPa, and elongation was less than 2% for alloys containing Mg $\leq$ 1 wt.%.

Microstructural refinement induced by intense cold plastic deformation of the as-cast alloy significantly improved mechanical properties. The ECAP-processed Zn–0.1Mg alloy achieved YS of 198 MPa, UTS of 215 MPa, elongation (A_2) of 187%, and reduction in area (Z) of 91% (Table 2, Figure 5). Additionally, true ultimate tensile strength (true UTS) reached 340 MPa, significantly exceeding requirements for orthopaedic implant materials (YS > 230 MPa, UTS > 300 MPa, elongation (A) > 15%) [25]. Sufficient ductility of the ECAP-processed state allows for further improvement in mechanical properties through additional cold deformation methods, such as rolling. The obtained results notably surpass previously published data for Zn–Mg alloys processed through various technologies, including ECAP. Exceptional mechanical properties, particularly the remarkably high elongation (187%), significantly exceed literature-reported values for similarly processed Zn–Mg alloys (Table 3) [31,37,42,47,48].

Table 3. Comparison of mechanical properties of the Zn–Mg alloys.

Composition (wt.%)	Processing Technique	YS [MPa]	UTS [MPa]	A [%]	References
Zn–0.1Mg	RT-ECAP-2P	198	215	187	current study
Zn–0.1Mg	RT-ECAP-8P	329	383	45.6	[31]
Zn–0.1Mg	150T-ECAP-8P	274	291	7.8	[31]
Zn–0.15Mg	HE (250 °C)	114	250	22	[37]
Zn–0.08Mg	HE (150 °C) + WD	250	266	29.8	[38]
Zn–0.033Mg	RT-ECAP-4P	250	263	25.37	[42]
Zn–0.2Mg	HE (200 °C)	179	234	8	[47]
Zn–0.02Mg	HE (200 °C) + WD	388	455	5.4	[48]

HE: hot extrusion; WD: wire drawing; ECAP: equal channel angular pressing; RT: room temperature.

The fracture surface morphology of tensile-tested samples significantly changed following ECAP processing (Figure 7), correlating with improved mechanical properties of the analysed microalloyed Zn–Mg alloy. The presence of transcrystalline ductile fracture with dimples of various sizes and significant contraction clearly indicates considerable deformation energy absorption prior to fracture.

The significant enhancement in strength and ductility properties of the Zn–0.1Mg alloy following ECAP processing can be attributed to the synergistic effect of several microstructural factors. Key roles are played by grain refinement, elimination of porosity, and increased grain boundary sliding activity. The formation of a fine-grained microstructure primarily results in material strengthening, significantly influencing mechanical properties. According to the Hall–Petch relationship, yield strength increases with decreasing grain size due to the increased grain boundary area, effectively acting as barriers for dislocation movement. To verify the applicability of the Hall–Petch relationship in our case, we calculated the 0.2% strain offset yield strength ($YS_{0.2}$) for both the as-cast and ECAP-processed alloy states. The calculation was based on the empirical relation stated for Zn material with an average grain size (d_Z) in the range from 0.6 to 400 μm [49]:

$$YS_{0.2} = 40.8 + 104.8\ d_z{}^{-0.5} \quad (1)$$

The results demonstrated that the ECAP-processed alloy state exhibits a significantly higher calculated yield strength of 126.36 MPa, while the as-cast alloy state achieved a value of 50.25 MPa. The higher yield strength calculated for the ECAP-processed alloy, characterised by a grain-refined microstructure, supports the Hall–Petch relationship, suggesting that grain refinement enhances dislocation interactions with grain boundaries and thus improves resistance to plastic deformation. This strengthening mechanism markedly improves strength and is essential in increasing mechanical resilience of metals subjected to severe plastic deformation [38,40]. Additionally, the resultant fine-grained microstructure effectively suppresses twinning activation, often associated with microcracking during tensile loading. For instance, reduced twinning fraction from 14.2% to 2.3% in Zn–Mn alloys correlated with significant elongation improvement from 48% to 71% [45]. Moreover, grain boundaries enriched with finely dispersed secondary intermetallic phases significantly redistribute local stresses, reducing brittle fracture likelihood. These phases actively participate in coordinated deformation via grain boundary sliding, substantially increasing overall material ductility [29]. Increased activation of slip systems due to stress concentrations at grain boundaries leads to uniform plastic deformation, and grain boundary rotation facilitates continuous deformation without localised stress concentrations [31]. Additionally, dislocation movement along grain boundaries enhances toughness and, in some cases, enables superplastic behaviour, exemplified by extruded Zn–0.5Mn alloy achieving extraordinary tensile elongation of 236% with grain sizes around 0.35 μm [50].

4. Materials and Methods

The Zn–0.1Mg (wt.%) alloy was prepared by melting high-purity metals (Zn with a purity of 99.95% and Mg with a purity of 99.99%) in a Schmelzofen Goldbrunn 1000 electric resistance furnace (Goldbrunn Therm, Berlin, Germany) at a temperature of 730 °C. The molten alloy was mixed thoroughly for 10 min in a protective argon (Ar) atmosphere before being gravity-cast into graphite crucibles that had been preheated to 300 °C. After casting, the ingots were air-cooled naturally while still in the graphite crucibles. The resulting ingots had a cylindrical shape with a diameter of 20 mm and a length of 80 mm.

The chemical composition of the prepared as-cast alloy was analysed using spark optical emission spectrometry (OES) in accordance with the STN EN 12019 standard [51]. The analysis revealed that the Mg content in the alloy was 0.083 ± 0.017 wt.%. Additionally, trace concentrations of impurities were detected, with a total content not exceeding 0.013 wt.%, consisting of 0.006 wt.% Sn and 0.007 wt.% Cu.

The as-cast Zn–0.1Mg alloy cylindrical samples were then machined to dimensions of ø 8 mm × 60 mm for processing by severe plastic deformation using the ECAP technique. These as-cast alloy samples were pressed twice through an ECAP die (die parameters: $\Phi = 90°$, $\Psi = 37°$) at room temperature without rotation between passes, i.e., following route A (Figure 8), according to the standard classification outlined in the literature [52].

The microstructure of the as-cast and ECAP-processed samples of the Zn–0.1Mg alloy was analysed in the longitudinal direction, i.e., along the axis of the ingots and the ECAP-processed cylindrical specimens (Figure 8) by an OLYMPUS Vanox-T light microscope (Olympus, Tokyo, Japan) and a JEOL JSM 7000F scanning electron microscope (SEM) (JEOL, Tokyo, Japan). Samples for microstructural analysis were prepared using standard cold-mounting techniques, followed by grinding with waterproof abrasive papers and polishing with diamond paste of 0.7 μm grit size. Final polishing was completed using a colloidal silica suspension. The microstructure was subsequently revealed by chemical etching in a 5% Nital solution. The microstructural analysis focused on the morphology and size of the solid solution features as well as the distribution of secondary phases. For a more detailed microstructural examination, energy-dispersive X-ray spectroscopy (EDS) was used to perform a semi-quantitative chemical composition analysis of the phases present. Phase analysis of the Zn–0.1Mg alloy in both the as-cast and ECAP-processed states was conducted using X-ray diffraction (XRD). X-ray diffraction measurements were performed using the Philips X'Pert Pro powder diffractometer (Malvern Panalytical, Almelo, The Netherlands) equipped with a Co X-ray tube within the 2θ range of 20° to 120°, with a step size of 0.03° and a dwell time of 10 s per step. Rietveld analysis of the measured diffractograms was performed using the software package GSAS-II (version 5.4.2) [53].

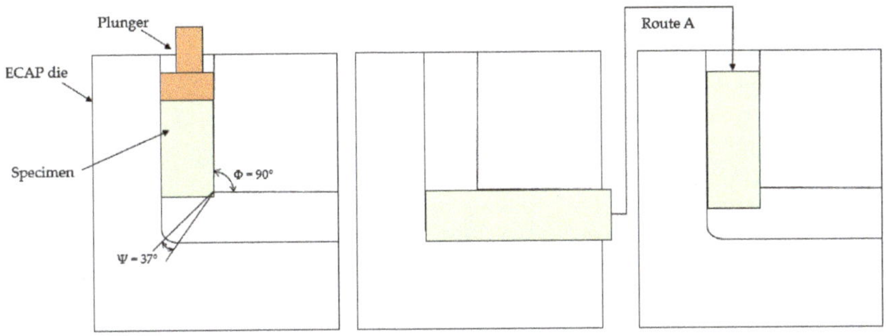

Figure 8. Two-dimensional scheme of the alloy sample ECAP processing following route A.

The mechanical properties of the studied alloy, in both the as-cast and ECAP-processed states, were evaluated by uniaxial tensile testing realised at room temperature using a Tinius Olsen H300K-V universal testing machine (Tinius Olsen, Horsham, PA, USA). The tensile tests were conducted according to Method A of the STN EN ISO 6892-1 standard [54], at a strain rate of 2.5×10^{-4} s^{-1}. The tensile tests were performed on cylindrical samples with threaded heads, featuring a gauge diameter of 5 mm and a gauge length of 10 mm. The evaluated mechanical properties included yield strength (YS), ultimate tensile strength (UTS), and ductility parameters: elongation (A_2) and reduction in area (Z). The tensile strength of ECAP-processed alloy state was assessed not only by the conventional engineering approach based on the initial cross-sectional area of the specimen, but also as true value reflecting the actual deformation state and the real cross-sectional reduction during tensile testing. This comprehensive method enables a more accurate evaluation of the material's deformation behaviour under tensile loading. Following tensile testing, the morphology of the fracture surfaces was examined by fractography analysis using a JEOL JSM 7000F SEM scanning electron microscope, with EDS analysis performed to characterise the phases present on the fracture surfaces.

5. Conclusions

Zn–Mg alloys have high potential for use in biodegradable implants. Their chemical composition and processing methods significantly influence properties such as biocompatibility, degradation rate, and particularly mechanical characteristics. The desired strength characteristics and sufficient ductility of these alloys can be achieved through various forming techniques, including severe plastic deformation methods. This study investigated the effect of Equal Channel Angular Pressing (ECAP), a severe plastic deformation technique, on the microstructure and mechanical properties of the as-cast Zn–0.1Mg alloy. The main conclusions from this study are summarised as follows:

1. The as-cast Zn alloy with 0.1 wt.% Mg exhibited a polyhedric casting microstructure consisting of coarse (average grain size d_Z = 123 μm) equiaxed grains of Mg solid solution in Zn with an H12 lattice and segmented secondary intermetallic phase formations, such as Mg_2Zn_{11}, precipitated at the grain boundaries of the solid solution.
2. Severe cold plastic deformation of the as-cast alloy, achieved by two passes through the ECAP die using route A, led to significant changes in the microstructure of the solid solution grains. Deformation by shear in the ECAP die channel formed deformation bands in the original cast grains, within which ultrafine dynamically recrystallised substructures of solid solution with an average grain size not exceeding 1.5 μm were observed. ECAP processing also resulted in the formation of texture with preferred crystallographic orientation of grains/subgrains.
3. Significant refinement of the alloy's solid solution substructure induced by severe plastic deformation substantially improved its mechanical properties determined by tensile testing. The ECAP-processed Zn–0.1Mg alloy achieved a yield strength of 198 MPa and an ultimate tensile strength of 215 MPa, more than twice the values observed in the as-cast alloy. However, the most dramatic improvement was observed in ductility parameters. The ECAP-processed alloy reached elongation up to 187% and reduction in area up to 91%. Additionally, true ultimate tensile strength reached as high as 340 MPa. Sufficient ductility of the deformed alloy state allows for further strength enhancement through additional cold forming techniques, such as rolling, after ECAP processing.
4. Significant improvement in ductility induced by ECAP also altered the fracture mechanism observed during tensile testing. Brittle fracture in the as-cast Zn–0.1Mg alloy, characterised by transcrystalline cleavage and intercrystalline separation, transformed into ductile transcrystalline fracture in the ECAP-processed alloy, evidenced by dimples of varying sizes on the fracture surface and high reduction in area. This indicates that the severely deformed alloy absorbed considerable deformation energy before fracture.
5. Microstructural refinement in materials with a hexagonal close-packed (HCP) lattice below a critical grain size (typically in the range of 1–10 μm) completely suppresses twinning as a deformation mechanism. Consequently, alternative deformation mechanisms begin to dominate, fundamentally changing plastic deformation behaviour. Formation of an ultrafine-grained microstructure primarily strengthens the material, and significantly enhances ductility, as the refined microstructure provides an increased number of active slip systems and enables coordinated plastic deformation through grain boundary sliding of the solid solution grains.

Author Contributions: Conceptualization, I.C., M.F. and K.S.; methodology, I.C., M.F., Z.M., M.M., R.D., B.B., D.C., K.G. and K.S.; software, K.S., R.K. and P.P.; validation, I.C., K.S., R.K. and P.P.; formal analysis, I.C., M.F., R.K., P.P., Z.M., M.M., R.D., B.B., D.C., K.G. and K.S.; investigation, I.C., M.F., R.K., P.P., Z.M., M.M., R.D., B.B., D.C. and K.S.; resources, Z.M. and B.B.; data curation, I.C.; writing—original draft preparation, I.C.; writing—review and editing, I.C., M.F. and K.S.; visualization, I.C., M.F., K.G. and K.S.; supervision, M.F. and K.S.; project administration, K.G.; funding acquisition, K.S. All authors have read and agreed to the published version of the manuscript.

Funding: This work was supported by the Slovak Research and Development Agency under Contract No. APVV-23-0030, APVV-20-0205, VEGA project No. 1/0122/25, VEGA project No. 2/0039/23 and KEGA project No. 011TUKE-4/2025. This research was funded in part by the international project M-ERA.NET 3/2022/235/H2MobilHydride.

Institutional Review Board Statement: Not applicable.

Informed Consent Statement: Not applicable.

Data Availability Statement: The data presented in this study are available in the article.

Conflicts of Interest: The authors declare no conflicts of interest.

Abbreviations

The following abbreviations are used in this manuscript:

ECAP	Equal Channel Angular Pressing
YS	Yield Strength
UTS	Ultimate Tensile Strength
True UTS	True Ultimate Tensile Strength
A	Elongation
A_2	Elongation measured on a gauge length L_0 = 2d (d—specimen diameter)
Z	Reduction in Area
SEM	Scanning Electron Microscopy
EDS	Energy-Dispersive X-ray Spectroscopy
XRD	X-ray Diffraction
LM	Light Microscopy
ODF	Orientation Distribution Function
HCP	Hexagonal Close-Packed
GBS	Grain Boundary Sliding
OES	Optical Emission Spectrometry

References

1. Zheng, Y.F.; Gu, X.N.; Witte, F. Biodegradable metals. *Mater. Sci. Eng. R Rep.* **2014**, *77*, 1–34. [CrossRef]
2. Liu, Y.; Zheng, Y.; Chen, X.H.; Yang, J.A.; Pan, H.; Chen, D.; Wang, L.; Zhang, J.; Zhu, D.; Wu, S.; et al. Fundamental theory of biodegradable metals—Definition, criteria, and design. *Adv. Funct. Mater.* **2019**, *29*, 1805402. [CrossRef]
3. Chen, Y.; Xu, Z.; Smith, C.; Sankar, J. Recent advances on the development of magnesium alloys for biodegradable implants. *Acta Biomater.* **2014**, *10*, 4561–4573. [CrossRef]
4. Moravej, M.; Mantovani, D. Biodegradable metals for cardiovascular stent application: Interests and new opportunities. *Int. J. Mol. Sci.* **2011**, *12*, 4250–4270. [CrossRef]
5. Xu, L.; Wei, C.; Deng, L.; Wang, P.; Zhong, W.; Huang, W. A review of non-biodegradable alloys implantation induced inflammatory and immune cell responses. *J. Alloys Compd.* **2024**, *977*, 173086. [CrossRef]
6. Ormiston, J.A.; Serruys, P.W.S. Bioabsorbable coronary stents. *Circ. Cardiovasc. Interv.* **2009**, *2*, 255–260. [CrossRef]
7. Chen, Q.; Thouas, G.A. Metallic implant biomaterials. *Mater. Sci. Eng. R Rep.* **2015**, *87*, 1–57. [CrossRef]
8. Witte, F. The history of biodegradable magnesium implants: A review. *Acta Biomater.* **2010**, *6*, 1680–1692. [CrossRef]
9. Bowen, P.K.; Drelich, J.; Goldman, J. Zinc exhibits ideal physiological corrosion behavior for bioabsorbable stents. *Adv. Mater.* **2013**, *25*, 2577–2582. [CrossRef]
10. Li, H.; Yang, H.; Zheng, Y.; Zhou, F.; Qiu, K.; Wang, X. Design and characterizations of novel biodegradable ternary Zn-based alloys with IIA nutrient alloying elements Mg, Ca and Sr. *Mater. Des.* **2015**, *83*, 95–102. [CrossRef]
11. Plum, L.M.; Rink, L.; Haase, H. The essential toxin: Impact of zinc on human health. *Int. J. Environ. Res. Public Health* **2010**, *7*, 1342–1365. [CrossRef]
12. Kambe, T.; Tsuji, T.; Hashimoto, A.; Itsumura, N. The physiological, biochemical, and molecular roles of zinc transporters in zinc homeostasis and metabolism. *Physiol. Rev.* **2015**, *95*, 749–784. [CrossRef]
13. Vojtech, D.; Kubásek, J.; Šerák, J.; Novák, P. Mechanical and corrosion properties of newly developed biodegradable Zn-based alloys for bone fixation. *Acta Biomater.* **2011**, *7*, 3515–3522. [CrossRef]
14. Yang, H.T.; Wang, C.; Liu, C.Q.; Chen, H.W.; Wu, Y.F.; Han, J.T.; Jia, Z.C.; Lin, W.J.; Zhang, D.Y.; Li, W.T.; et al. Evolution of the degradation mechanism of pure zinc stent in the one-year study of rabbit abdominal aorta model. *Biomaterials* **2017**, *145*, 92–105. [CrossRef]
15. Li, G.; Yang, H.; Zheng, Y.; Chen, X.H.; Yang, J.A.; Zhu, D.; Ruan, L.; Takashima, K. Challenges in the use of zinc and its alloys as biodegradable metals: Perspective from biomechanical compatibility. *Acta Biomater.* **2019**, *97*, 23–45. [CrossRef]
16. Shen, C.; Liu, X.; Fan, B.; Lan, P.; Zhou, F.; Li, X.; Wang, H.; Xiao, X.; Li, L.; Zhao, S.; et al. Mechanical properties, in vitro degradation behavior, hemocompatibility and cytotoxicity evaluation of Zn–1.2Mg alloy for biodegradable implants. *RSC Adv.* **2016**, *6*, 86410–86419. [CrossRef]

17. Seitz, J.-M.; Lucas, A.; Kirschner, M. Magnesium-based compression screws: A novelty in the clinical use of implants. *JOM* **2016**, *68*, 1177–1182. [CrossRef]
18. Tong, X.; Dong, Y.; Han, Y.; Zhou, R.; Zhu, L.; Zhang, D.; Dai, Y.; Shen, X.; Li, Y.; Wen, C.; et al. A biodegradable Zn–5Gd alloy with biomechanical compatibility, cytocompatibility, antibacterial ability, and in vitro and in vivo osteogenesis for orthopedic applications. *Acta Biomater.* **2024**, *177*, 538–559. [CrossRef]
19. Guo, H.; Cao, R.H.; Zheng, Y.F.; Bai, J.; Xue, F.; Chu, C.L. Diameter-dependent in vitro performance of biodegradable pure zinc wires for suture application. *J. Mater. Sci. Technol.* **2019**, *35*, 1662–1670. [CrossRef]
20. Li, S.; Ren, J.; Wang, X.; Ding, Y.; Li, P.; Hu, Y.; Yang, Y. Dilemmas and countermeasures of Fe-based biomaterials for next-generation bone implants. *J. Mater. Res. Technol.* **2022**, *20*, 2034–2050. [CrossRef]
21. Schinhammer, M.; Hänzi, A.C.; Löffler, J.F.; Uggowitzer, P.J. Design strategy for biodegradable Fe-based alloys for medical applications. *Acta Biomater.* **2010**, *6*, 1705–1713. [CrossRef]
22. Bowen, P.K.; Shearier, E.R.; Zhao, S.; Guillory, R.J.; Zhao, F.; Goldman, J.; Drelich, J.W. Biodegradable metals for cardiovascular stents: From clinical concerns to recent Zn-Alloys. *Adv. Healthc. Mater.* **2016**, *5*, 1121–1140. [CrossRef]
23. Liu, X.; Sun, J.; Qiu, K.; Yang, Y.; Pu, Z.; Li, L.; Zheng, Y. Effects of alloying elements (Ca and Sr) on microstructure, mechanical property and in vitro corrosion behavior of biodegradable Zn–1.5Mg alloy. *J. Alloys Compd.* **2016**, *664*, 444–452. [CrossRef]
24. Liu, S.; Kent, D.; Doan, N.; Dargusch, M.; Wang, G. Effects of deformation twinning on the mechanical properties of biodegradable Zn-Mg alloys. *Bioact. Mater.* **2019**, *4*, 8–16. [CrossRef]
25. Venezuela, J.; Dargusch, M.S. The influence of alloying and fabrication techniques on the mechanical properties, biodegradability and biocompatibility of zinc: A comprehensive review. *Acta Biomater.* **2019**, *87*, 1–40. [CrossRef]
26. Li, Z.; Shi, Z.-Z.; Hao, Y.; Li, H.-F.; Zhang, H.-J.; Liu, X.-F.; Wang, L.-N. Insight into role and mechanism of Li on the key aspects of biodegradable ZnLi alloys: Microstructure evolution, mechanical properties, corrosion behavior and cytotoxicity. *Mater. Sci. Eng. C* **2020**, *114*, 111049. [CrossRef]
27. Bao, G.; Fan, Q.; Ge, D.; Wang, K.; Sun, M.; Zhang, Z.; Guo, H.; Yang, H.; He, B.; Zheng, Y. In vitro and in vivo studies to evaluate the feasibility of Zn-0.1Li and Zn-0.8Mg application in the uterine cavity microenvironment compared to pure zinc. *Acta Biomater.* **2021**, *123*, 393–406. [CrossRef]
28. Murni, N.S.; Dambatta, M.S.; Yeap, S.K.; Froemming, G.R.A.; Hermawan, H. Cytotoxicity evaluation of biodegradable Zn–3Mg alloy toward normal human osteoblast cells. *Mater. Sci. Eng. C* **2015**, *49*, 560–566. [CrossRef]
29. Ji, C.; Ma, A.; Jiang, J.; Wu, H.; Liu, H.; Guo, S.; Yuan, Y. Regulating mechanical properties and degradation behavior of biodegradable Zn–0.6Mg alloy via ECAP plus cold rolling. *J. Alloys Compd.* **2023**, *937*, 168487. [CrossRef]
30. Wang, C.; Ma, A.; Sun, J.; Liu, H.; Huang, H.; Yang, Z.; Jiang, J. Effect of ECAP process on as-cast and as-homogenized Mg–Al–Ca–Mn alloys with different Mg_2Ca morphologies. *J. Alloys Compd.* **2019**, *793*, 259–270. [CrossRef]
31. Ye, L.; Liu, H.; Sun, C.; Zhuo, X.; Ju, J.; Xue, F.; Bai, J.; Jiang, J.; Xin, Y. Achieving high strength, excellent ductility, and suitable biodegradability in a Zn–0.1Mg alloy using room temperature ECAP. *J. Alloys Compd.* **2022**, *926*, 166906. [CrossRef]
32. Liu, H.; Huang, H.; Zhang, Y.; Xu, Y.; Wang, C.; Sun, J.; Jiang, J.; Ma, A.; Xue, F.; Bai, J. Evolution of Mg–Zn second phases during ECAP at different processing temperatures and its impact on mechanical properties of Zn–1.6Mg (wt.%) alloys. *J. Alloys Compd.* **2019**, *811*, 151987. [CrossRef]
33. Yang, Z.; Liu, H.; Ren, K.; Ye, L.; Zhuo, X.; Ju, J.; Xue, F.; Bai, J.; Jiang, J.; Xin, Y. A high strength and ductility Zn–Cu–Mg alloy achieved by bandlike distribution of ultra-fine $CuZn_5$ and Mg_2Zn_{11} particles. *Mater. Sci. Eng. A* **2022**, *850*, 143584. [CrossRef]
34. *STN EN ISO 643*; Steels. Micrographic Determination of the Apparent Grain Size. Slovak Office of Standards, Metrology and Testing: Bratislava, Slovakia, 2024.
35. Kubásek, J.; Vojtěch, D.; Pospíšilová, I.; Michalcová, A.; Maixner, J. Microstructure and mechanical properties of the micrograined hypoeutectic Zn–Mg alloy. *Int. J. Miner. Metall. Mater.* **2016**, *23*, 1167–1176. [CrossRef]
36. Li, R.; Ding, Y.; Zhang, H. Toughness and strength coordination in a low-alloy Zn–0.5 Mg alloy via extrusion and post-deformation annealing. *Met. Mater. Int.* **2023**, *29*, 2807–2825. [CrossRef]
37. Mostaed, E.; Sikora-Jasinska, M.; Mostaed, A.; Loffredo, S.; Demir, A.; Previtali, B.; Mantovani, D.; Beanland, R.; Vedani, M. Novel Zn-based alloys for biodegradable stent applications: Design, development and in vitro degradation. *J. Mech. Behav. Biomed. Mater.* **2016**, *60*, 581–602. [CrossRef]
38. Jin, H.; Zhao, S.; Guillory, R.; Bowen, P.K.; Yin, Z.; Griebel, A.; Schaffer, J.; Earley, E.J.; Goldman, J.; Drelich, J.W. Novel high-strength, low-alloy Zn–Mg (<0.1 wt% Mg) and their arterial biodegradation. *Mater. Sci. Eng. C* **2018**, *84*, 67–79.
39. Ye, L.; Huang, H.; Sun, C.; Zhuo, X.; Dong, Q.; Liu, H.; Ju, J.; Xue, F.; Bai, J.; Jiang, J. Effect of grain size and volume fraction of eutectic structure on mechanical properties and corrosion behavior of as-cast Zn–Mg binary alloys. *J. Mater. Res. Technol.* **2022**, *16*, 1673–1685. [CrossRef]
40. Bednarczyk, W.; Wątroba, M.; Kawałko, J.; Bała, P. Can zinc alloys be strengthened by grain refinement? A critical evaluation of the processing of low-alloyed binary zinc alloys using ECAP. *Mater. Sci. Eng. A* **2019**, *748*, 357–366. [CrossRef]

41. Cepeda-Jiménez, C.M.; Molina-Aldareguia, J.M.; Pérez-Prado, M.T. Origin of the twinning to slip transition with grain size refinement, with decreasing strain rate and with increasing temperature in magnesium. *Acta Mater.* **2015**, *88*, 232–244. [CrossRef]
42. Wang, X.; Ma, Y.; Meng, B.; Wan, M. Effect of equal-channel angular pressing on microstructural evolution, mechanical property and biodegradability of an ultrafine-grained zinc alloy. *Mater. Sci. Eng. A* **2021**, *824*, 141857. [CrossRef]
43. Li, J.; Xu, W.; Wu, X.; Ding, H.; Xia, K. Effects of grain size on compressive behaviour in ultrafine grained pure Mg processed by equal channel angular pressing at room temperature. *Mater. Sci. Eng. A* **2011**, *528*, 5993–5998. [CrossRef]
44. Lapovok, R.; Thomson, P.F.; Cottam, R.; Estrin, Y. The effect of grain refinement by warm equal channel angular extrusion on room temperature twinning in magnesium alloy ZK60. *J. Mater. Sci.* **2005**, *40*, 1699–1708. [CrossRef]
45. Sun, S.; Ren, Y.; Liqing, W.; Yang, B.; Li, H.; Qin, G. Abnormal effect of Mn addition on the mechanical properties of as-extruded Zn alloys. *Mater. Sci. Eng. A* **2017**, *701*, 129–133. [CrossRef]
46. Li, H.F.; Xie, X.H.; Zheng, Y.F.; Cong, Y.; Zhou, F.Y.; Qiu, K.J.; Wang, X.; Chen, S.H.; Huang, L.; Tian, L.; et al. Development of biodegradable Zn–1X binary alloys with nutrient alloying elements Mg, Ca and Sr. *Sci. Rep.* **2015**, *5*, 10719.
47. Wang, L.Q.; Ren, Y.P.; Sun, S.N.; Zhao, H.; Li, S.; Qin, G.-W. Microstructure, mechanical properties and fracture behavior of as-extruded Zn–Mg binary alloys. *Acta Metall. Sin. (Engl. Lett.)* **2017**, *30*, 931–940. [CrossRef]
48. Wang, L.Q.; He, Y.F.; Zhao, H.; Xie, H.B.; Li, S.; Ren, Y.P.; Qin, G.W. Effect of cumulative strain on the microstructural and mechanical properties of Zn-0.02 wt%Mg alloy wires during room-temperature drawing process. *J. Alloys Compd.* **2018**, *740*, 949–957. [CrossRef]
49. Balog, M.; Krížik, P.; Školáková, A.; Švec, P.; Kubásek, J.; Pinc, J.; de Castro, M.M.; Figueiredo, R. Hall-Petch strengthening in ultrafine-grained Zn with stabilized boundaries. *J. Mater. Res. Technol.* **2024**, *33*, 7458–7468. [CrossRef]
50. Guo, P.; Zhu, X.; Yang, L.; Deng, L.; Zhang, Q.; Li, B.Q.; Cho, K.; Sun, W.; Ren, T.; Song, Z. Ultrafine- and uniform-grained biodegradable Zn–0.5Mn alloy: Grain refinement mechanism, corrosion behavior, and biocompatibility in vivo. *Mater. Sci. Eng. C* **2021**, *118*, 111391. [CrossRef]
51. *STN EN 12019*; Zinc and Zinc Alloys. Analysis by Optical Emission Spectrometry. Slovak Office of Standards, Metrology and Testing: Bratislava, Slovakia, 2000.
52. Sun, Y.; Li, W.; Shi, X.; Tian, L. A uniformly fine-grained Mg–Y–Nd–Zr magnesium alloy with simultaneously optimized strength and ductility processed by forging and ECAP. *Mater. Res. Express* **2020**, *7*, 116520. [CrossRef]
53. Toby, B.H.; Von Dreele, R.B. GSAS-II: The genesis of a modern open-source all-purpose crystallography software package. *J. Appl. Crystallogr.* **2013**, *46*, 544–549. [CrossRef]
54. *STN EN ISO 6892-1*; Metallic Materials. Tensile Testing. Part 1: Method of Test at Room Temperature. Slovak Office of Standards, Metrology and Testing: Bratislava, Slovakia, 2022.

MDPI AG
Grosspeteranlage 5
4052 Basel
Switzerland
Tel.: +41 61 683 77 34

Inorganics Editorial Office
E-mail: inorganics@mdpi.com
www.mdpi.com/journal/inorganics

www.ingramcontent.com/pod-product-compliance
Lightning Source LLC
Chambersburg PA
CBHW061514210926
51850CB00008B/1918